AF602033

The Authors

Prof. T. Pullaiah obtained his M. Sc. (1973) and *Ph.D* (1976) degrees in Botany from Andhra University. He was a Post-doctoral Fellow at Moscow State University, Russia, during 1976-1978. He travelled widely in Europe and visited Universities and Botanic Gardens in about 17 countries. He joined Sri Krishnadevaraya University as Lecturer (1979) and became Professor (1993). He held several positions in the University as Dean, Faculty of Life Sciences; Head, Department of Botany; Chairman, Board of Studies in Botany; Head, Department of Sericulture; Co-ordinator and Chairman, Board of Studies in Biotechnology; Principal, Sri Krishnadevaraya University College. He retired from active Service in 2011 and was selected by UGC as UGC-BSR Faculty Fellow in the Department of Botany, Sri Krishnadevaraya University. He has published 75 books, 335 research papers and 35 popular articles. His books include Flora of Andhra Pradesh (5 volumes), Biodiversity in India (8 volumes), Flora of Eastern Ghats (4 volumes), Flora of Telangana (3 volumes), Encyclopaedia of World Medicinal Plants (5 volumes), Ethnobotany of India (5 volumes), Global Biodiversity (4 volumes), Orchids of Eastern Ghats and Monograph on *Brachystelma* and *Ceropegia* in India. He was the Principal-Investigator of 20 major Research Projects totalling more than a Crore of Rupees funded by DBT, DST, CSIR, UGC, BSI, WWF, GCC etc. Under his guidance, 54 students obtained their Ph. D. degrees and 34 students their M. Phil. Degrees. He is the recipient of Best Teacher Award from the Government of Andhra Pradesh, Prof. P. Maheshwari Gold Medal and Dr. G.Panigrahi Memorial Award of Indian Botanical Society and Prof. Y. D. Tiagi Gold Medal of the Indian Association for Angiosperm Taxonomy. He was President of Indian Association for Angiosperm Taxonomy (2013) and President of the Indian Botanical Society (2014). He was a member of Species Survival Commission of International Union for Conservation of Nature and Natural Resources (IUCN).

Dr. S. Karuppusamy is Assistant Professor in the Department of Botany, The Madura College, Madurai, Tamil Nadu. He obtained his M.Sc. degree from Madurai Kamaraj University, Madurai and Ph.D. degree from Gandhigram Rural Institute, Deemed Univerrsity, Tamil Nadu. Later on he worked as Post Doctoral Fellow in the Department of Biotechnology, and Research Associate in the Department of Botany, Sri Krishnadevaraya University, Anantapur. He has published more than 120 research papers in reputed journals and co-authored five books – Flora of Eastern Ghats volume 3 and 4, Flora of Andhra Pradesh volume 5, Revised Flora of Andhra Pradesh Volume 3 and *Caralluma (Sensu lato*), Orchids of Eastern Ghats and Monograph on *Brachystelma* and *Ceropegia* in India.. He is the recipient of Prof. Y.S.Murthy Gold Medal from Indian Botanical Society and Rolla S. Rao award from Indian Association for Angiosperm Taxonomy. He is a Fellow of Indian Association for Angiosperm Taxonomy and Indian Botanical Society. He is the Principal-Investigator of 3 major Research Projects funded by UGC, SERB and NIF. Under his guidance, 4 students obtained their *Ph.D* degrees.

Flora of Eastern Ghats

Volume 6

Hydrocharitaceae – Cyperaceae

Flora of Eastern Ghats

Volume 6

Hydrocharitaceae – Cyperaceae

T. Pullaiah
and
S. Karuppusamy

2020

Regency Publications

A Division of

Astral International Pvt. Ltd.

New Delhi – 110 002

ISBN: 9789390371181 (Int. Edition)

Publisher's Note:

Every possible effort has been made to ensure that the information contained in this book is accurate at the time of going to press, and the publisher and author cannot accept responsibility for any errors or omissions, however caused. No responsibility for loss or damage occasioned to any person acting, or refraining from action, as a result of the material in this publication can be accepted by the editor, the publisher or the author. The Publisher is not associated with any product or vendor mentioned in the book. The contents of this work are intended to further general scientific research, understanding and discussion only. Readers should consult with a specialist where appropriate.

Every effort has been made to trace the owners of copyright material used in this book, if any. The author and the publisher will be grateful for any omission brought to their notice for acknowledgment in the future editions of the book.

Published by : **Regency Publications®**
A Division of
Astral International Pvt. Ltd.
– ISO 9001:2015 Certified Company –
4736/23, Ansari Road, Darya Ganj
New Delhi-110 002
Ph. 011-43549197, 23278134
E-mail: info@astralint.com
Website: www.astralint.com

Digitally Printed at : **Replika Press Pvt. Ltd.**

Abbreviations Used in General Text and Enumeration

BLAT: Blatter Herbarium, St. Xavier College, Mumbai

BSID: Botanical Survey of India, Deccan Circle, Hyderabad

CAL: Central National Herbarium, Howrah, Calcutta (now called as Kolkata)

cm: centimeter

DD: Herbarium, BSI, Dehradun

et al.: *et alia* (and others, of people)

FBI: Hooker's Flora of British India

Fig.: Figure

Fl.: Flowering season

Fr.: Fruiting season

Fischer: Gamble & Fischer, Flora of Presidency of Madras

m: meter

MH: Madras Herbarium, BSI, Southern Circle, TNAU campus, Coimbatore mm: millimeter(s)

Ori.: Oriya

p.p.: *pro-parte*

sine. coll.: Without collector's name

s.n.: Without number

SGH: Sri Ganesan Herbarium, The Madura College, Madurai

SKU: Sri Krishnadevaraya University, Anantapur

Sq. km: Square kilo-meter(s)

SVU: Sri Venkateswara University, Tirupati

Tam.: Tamil

Tel.: Telugu

var.: varities (variety)

Vern.: Vernacular name (s)

Abbreviations – Districts

ATP: Anantapur (now called Anantapuramu)

BGR: Bolangir

CDP: Cuddapah (Now called as Kadapa)

CPT: Chengalpat

CTK: Cuttack

CTR: Chittoor

DGL: Dindigul

DGR: Deogarh

DKL: Dhenkanal

DMP: Dharmapuri

EG: East Godavari

GJM: Ganjam

GJT: Gajapathi

GNT: Guntur

KGR: Krishnagiri

KHD: Kalahandi

KMM: Khammam

KNL: Kurnool

KSN: Krishna

KRT: Koraput

MBJ: Mayurbhanj

MDS: Madras

MDU: Madurai

MBNR: Mahaboobnagar

NA: North Arcot

NLR: Nellore

NMK: Namakkal

PLN: Pulbhani

PRK: Prakasam

RGD: Rayagadh

SA: South Arcot

SBP: Sambalpur

SKLM: Srikakulam

SLM: Salem

TRP: Tiruchirapalli

TRV: Tiruvannamalai

VJN: Vijayanagaram

VLR: Vellore

VSKP: Visakhapatnam

WG: West Godavari

Abbreviations – Plant Explorers

AAA: A.A.Ansari

AL: A. Laxmaiah

AM: A.Mohan

AMK: A.Mukherji

AMR: A.Madhusudana Reddy

ANH: A.N.Henry

ANS: A. Narayana Swamy

APD & SP: A.P.Das & Sauris Panda

AVNR: A.V.N. Rao

AWL: A.W.Lushington
BR: B.Ravi Prasad Rao
BSS: B.Sadasivaiah
BSN: B.Suryanarayana
CAB: C.A.Barber
CECF: C.E.C.Fischer
CS: C.Subbarayudu
DAM: D.Ali Moulali
DBD: D.B.Deb
DCSR: D.C.S.Raju
DN: D.Narasimhan
DRC: D.Rangachari
EC: E.Chennaiah
EKK: E.K.Krishnan
EV: E.Vajravelu
GP: G.Panigrahi
GVS: G.V.SubbaRao
HFM: H.F.Mooney
HHH: H.H.Haines
HOS: H.O.Saxena
HS: H.Santapau
JHFB: J.H.Franklin Benjamin
JLE: J.L.Ellis
JS: J.Joseph
JSG: J.S.Gamble
KCJ: K.C.Jacob
KCM: K.C.Mohan
KH: K.Hanumanthappa
KMM: K.M.Matthew
KMS: K.M.Sebastine
KNS: K.N.Subramanian
KP: K.Prasad
KRM: K.Ramamurthy

KS: K.Subramanyam
KSK: K.Sampath Kumar
KSM: K.Sri Rama Murthy
KT: K.Thothathri
MB: M.Brahmam
MBV: M.B.Viswanatham
MCB: M.Chandrabose
MCK: M.Chennakesavulu
MM: M.Mohanan
MRR: M.Rama Rao
MSR: M.S.Ramaswami
NPBK: N.P.Balakrishnan
NRR: N.Rama Rao
NV: N.Venkatappa
NVG: N.Venugopal
NY: N.Yesoda
PMR: P.M.Rajeswaramma
PP: P.Perumal
PVK: P.Venkanna
PVP: P.V.Prasanna
PVS: P.V.Sreekumar
RCS: R.Chandrasekaran
RHB: R.H.Beddome
RKM: R.Krishnamohan
RSR: Rolla S.Rao
RVK: R.VijayaKumar
RVR: R.R.Venkata Raju
R.V. Reddy: R.VenkateswaraReddy
SKK: S.Karthikeyan
SKS: S.Karuppusamy
SKW: S.K.Wagh
SP & AP: Sauris Panda & A.P.Das
SPG: Sai Prasad Goud

SRSR: S.R.S.Reddy
SS: S.Sunitha
SSR: S.Sandhya Rani
SX: Saxena
TDCK: T.Dharma Chandra Kumar
TP: T.Pullaiah
TRS: T. Ravisankar
VBH: V.B.Hosagoudar
VNS: V.Narayanaswami
VRK: V.Ramakrishnaiah
VS: V. Srinivasulu

MONOCOTYLEDONS

HYDROCHARITACEAE

1. Acaulescent herbs; spathes peduncled:
   2. Spathes winged; stamens 6-9 .... **Ottelia**
   2. Spathes unwinged; stamens 1-3:
      3. Perianth lobes 6; peduncle straight .... **Blyxa**
      3. Perianth lobes 3; peduncle coiled .... **Vallisneria**
1. Caulescent herbs; spathes sessile:
   4. Leaves whorled .... **Hydrilla**
   4. Leaves not whorled .... **Nechamandra**

BLYXA Noronha ex Thouars

1. Leaves broad at base, gradually attenuate; flowers bisexual; spathes 1-flowered; stamens 3:
   2. Seeds tailed at ends .... **B. echinosperma**
   2. Seeds not tailed at ends .... **B. aubertii**
1. Leaves narrow at ends; flowers unisexual; spathes several–flowered; stamens 8-12 .... **B. octandra**

Blyxa aubertii Rich., Mem. Cl. Sci. Math. Inst. Natl. France 12: 77. t.4. 1812; Hartog in Steenis, Fl. Males. I. 5: 390. t.5. 1957; Matthew, Fl. Tamilnadu Carnatic 3: 1542. 1983. *B. ceylanica* Hook.f., Fl. Brit. India 5: 661. 1888. *B. griffithii* Planch. ex Hook.f., Fl. Brit. India 5: 660. 1888. *B. oryzetorum* (Decne) Hook.f., Fl. Brit. India 5:661. 1888. *Diplosiphon oryzetorum* Decne., Voy. Inde 4: 166. 1844.

Acaulescent submerged fresh water herb. Leaves linear, membranous, 20-30 x 0.5-1 cm, 5-7-nerved, glabrous, base broad, margin hyaline, apex gradually tapering, setaceous. Spathes to 6.5 cm long, 1-flowered, bisexual; peduncle to 12 cm long, shorter than leaves; sepals 3, oblong, 1.2 x 0.15 cm; petals linear, to 2 cm long, white; stamens 3; anthers linear; ovary as long as spathe; rostrum to 8 cm long, subterete, rigid. Fruit linear; seeds not tailed, obscurely tubercled or smooth.

In stagnant waters in Eastern Ghats of Tamil Nadu up to 900 m. Fl. & Fr.: November – May.

Kolli hills, on the way to Tampcol (NMK), *SKS* 1322 (SGH).

INDIA: All the states.

WORLD: Madagascar, Malaysia, New Guinea, Japan, Korea, N. Australia.

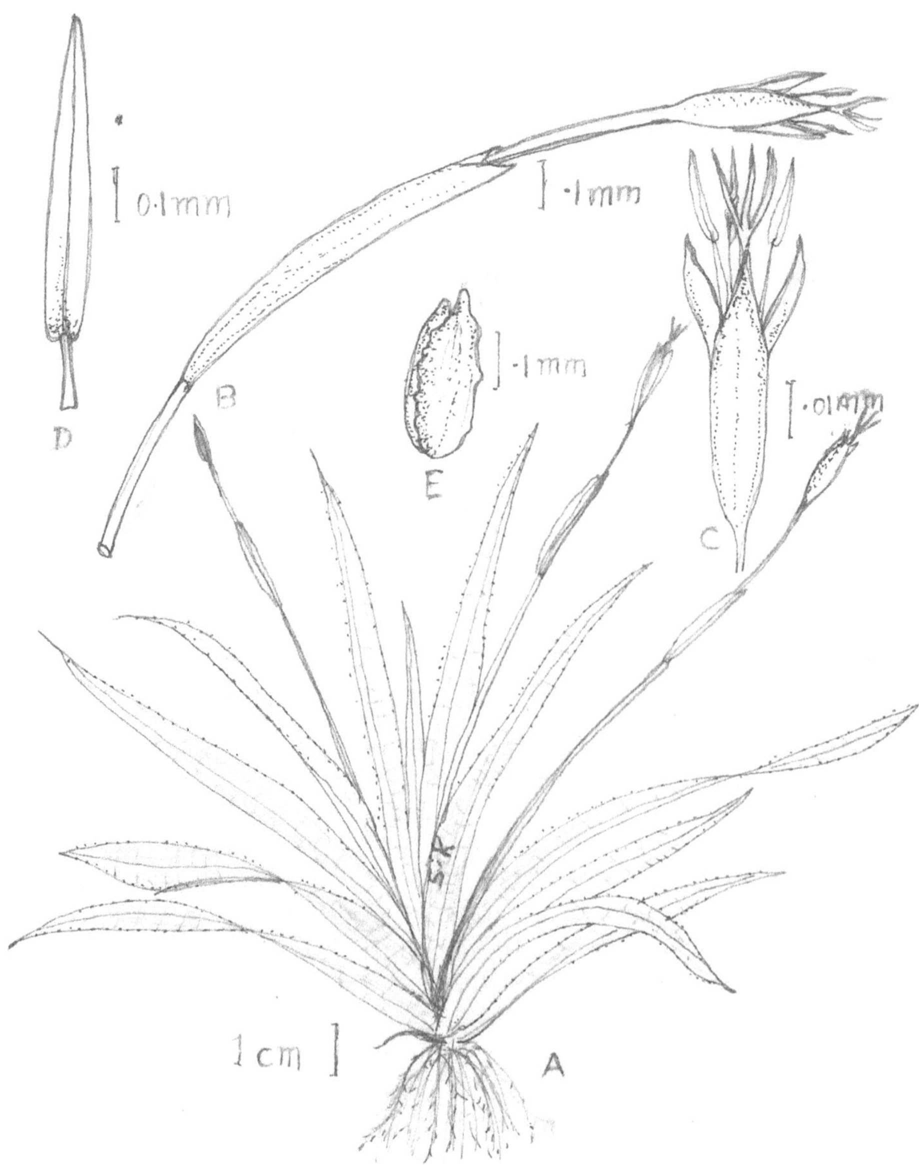

Figure 1. **Blyxa aubertii** Rich.

A. Habit; B. Inflorescence; C. Flower-entire; D. Stamen; E. Seed.

Blyxa echinosperma (C.B. Clarke) Hook. f., Fl., Brit. India 5: 661. 1888; Fischer 3: 1397. 1928; Hartog in Steenis, Fl. Males 1, 5: 391. 1957; Matthew, Fl. Tamilnadu Carnatic 3: 1542. 1983. *Hydrotrophus echinospermus* C.B. Clarke, J. Linn. Soc. Bot. 14: 8. t. l. 1873. *Blyxa talbotii* Hook. f., Fl. Brit. India 5: 661. 1888; Fischer. 3: 1397. 1928. *B. aubertii* Rich. var. *echinosperma* (C.B.Clarke) Cook & Lound, Aq.Bot. 15(1): 14. 1983. *B. lancifolia* Hook.f., Fl. Brit. India 5: 661. 1888. *B. ceratosperma* Maxim., Nat. Pflanzenfam. 2(1): 253 1889.

An acaulescent aquatic herb. Leaves linear, 10-20 x 0.3-1.2 cm. membranous, glabrous, base broad, margin entire or obscurely serrulate, apex acuminate; petioles sheathing at base. Flowers bisexual, solitary; spathes 2-4 cm long; sepals 3, linear; petals 3, linear; stamens 3; ovary terete, styles 3-fid. Fruit linear, torulose, ribbed, to 40-seeded; seeds ellipsoid, spinescent, tailed at the ends, spines in longitudinal rows.

In stagnant waters. Fl. & Fr.: June – September.

Tirumala (CTR), *MHR* 13367, 13937; Chinnamantanala (KNL), *JLE* 42288 (MH); Araku valley (VSKP), *DDSR* 21372 (MH).

INDIA: Peninsular, E. & N.E.India.

WORLD: Sri Lanka, Bangladesh, Myanmar, Nepal, China, Indonesia. Japan, Korea, Malaysia, Philippines, Thailand, Vietnam, Australia.

Blyxa octandra (Roxb.) Planch. ex Thwaites, Enum. Pl. Zeyl. 332. 1864; Fischer 3: 1397. 1928; Hartog in Steenis, Fl. Males. 1.5: 392. 1957; Matthew, Fl. Tamilnadu Carnatic 3: 1543. 1983. *Vallisneria octandra* Roxb., Pl. Coromandel t. 165. 1802 & Fl. Ind. 3: 752. 1832. *Blyxa roxburghii* Rich., Mem. Inst. Paris 12(2): 77. t. 5. 1812; FBI 5: 660. 1888.

An acaulescent aquatic herb. Leaves linear, 18-45 x 0.2-1.2 cm, membranous, glabrous, base broad, margin entire, apex acuminate; petioles sheathing at base. Flowers unisexual, on 5-7 cm long spathes; sepals 3; petals 3. Male flowers: 6-9 per spathe, spathes to 7 x 0.3 cm, stamens 8-12, filaments unequal, pistillodes 3, basally connate, Female flowers: solitary in spathes, sessile. Fruits linear, to 6 x 1.5 cm, torulose, glabrescent; seeds many, very shortly tailed, distinctly tubercled, in 6-8 rows.

Occasional in stagnant waters. Fl. & Fr.: June – October.

Near Satyavedu (CTR), *MCB* 45243 (MH); Kolleru lake (WG), *KS* 5072 (MH).

INDIA: Peninsular and E. India.

WORLD: Sri Lanka, Bangladesh, China, Myanmar,Vietnam, Australia.

HYDRILLA L.C. Richard

Hydrilla verticillata (L.f.) Royle, Ill. Bot. Himal. t. 376. 1839; FBI 5: 659. 1888; Fischer 3: 1396. 1928; Hartog in Steenis, Fl. Males 1, 5. 385. 1957; Matthew, Fl. Tamilnadu Carnatic 3: 1545. 1983. *Serpicula verticillata* L. f., Suppl. Pl. 416. 1781.

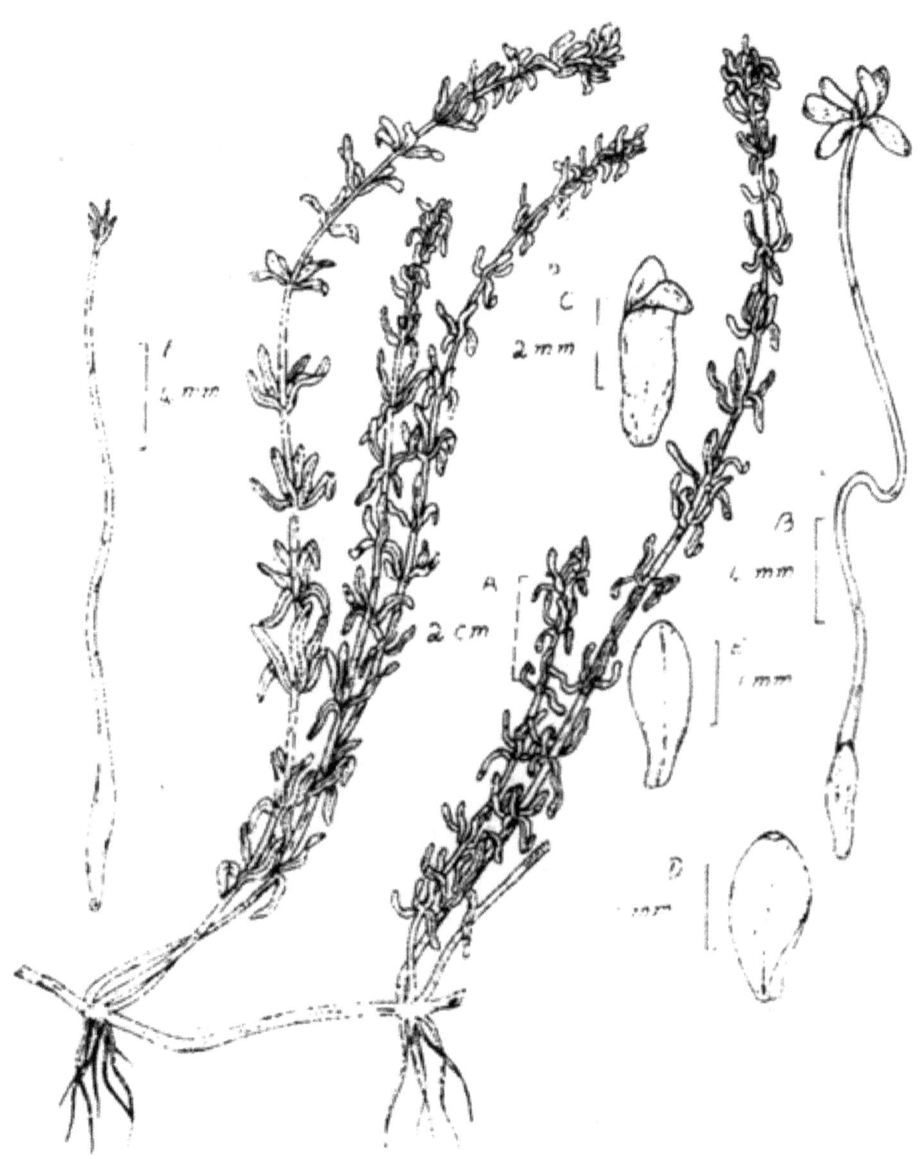

Figure 2: **Hydrilla verticillata** (L.f.) Royle.

A. Twig, B. Female flower, C. Spathe, D. Sepal, E. Petal, F. Pistil with rostrum.

Acaulescent submerged herb; branchlets 15-75 cm long; basal nodes distant. Leaves linear-lanceolate, whorled, 3-6 per node, basal ones linear-oblong, upper ones ovate, 0.6-1.5 x 0.2-0.4 cm, membranous, glabrous, base rounded, margin entire-sharply serrulate, scaberulous on indentations, apex apiculate. Flowers unisexual, monocious or dioecious; sepal 3; petals 3. Male flowers: Spathe solitary and sessile; ovary oblong, beaked, to 3 cm, style 3-fid. Fruit subulate, subterete, to 5-seeded; seeds oblong.

Very common in stagnant waters, freshwater ponds, puddles, tanks, lakes and rivers throughout Eastern Ghats. Fl. & Fr.: September-January.

Narpala (ATP), *KNR & DAM* 10628; Gani *RF* (KNL), *RVR* 1549; Palakonda Hills (KDP), *CS* 8273; Burrakayalakota (CTR), *KNR & DAM* 9834; Chintaparthi river (CTR), *MHR* 13375; Udayagiri(NL), *BS* 4244 (BSI); Achampeta (GNT), *VRK* 5873; Tatipudi reservoir (VZN), *MCK* 18866; Gudur (NLR), *KCJ* 18520 (MH); Dummagudem forest (KMM), *RCS* 99061 (BSID); Poosalipatty (DMP), *P.Venu* 111750 (MH); Mukundapur (GJM), *VNS* 4685 (MH).

INDIA: Throughout India.

WORLD: SE Europe, Africa, Asia, Australia.

NECHAMANDRA Planchon

Nechamandra alternifolia (Roxb.) Thwaites, Enum. Pl. Zeyl. 332. 1864; Subramanyam & N.P. Balakr., Bull. Bot. Surv. India 3: 23. 1962; Matthew, Fl. Tamilnadu Carnatic 3: 1546. 1983. *Vallisneria alternifolia* Roxb., Fl. Ind 3: 750. 1832. *Lagarosiphon alternifolia* (Roxb.) Druce, Rep. Bot. Exch. Club. Brit. Isles 1916. 630. 1917; Fischer 3: 1396. 1928. *Nechamandra roxburghii* Planch., Ann. Sci. Nat. Bot. Ser. 3. 11: 78. 1849. *Lagarosiphon roxburghii* (Planch.) Benth., Gen. Pl. 3: 451. 1880; FBI 5: 659. 1888.

A fresh water submerged herb; branchlets 20-50 cm, glabrous. Leaves alternate, basally decussate, linear, 1-9 x 0.3-1 cm, membranous, glabrous, 3-5-nerved, base amplexicaul, margin entire, apex acuminate. Flowers dioecious, in axillary, solitary, sessile spathes; sepals 3; petals 3. Male flowers: stamens 2, divergent, anthers ovate. Female flowers: Spathe subulate, turgid, to 8 mm; ovary obovoid, beaked, styles 3, notched; stigma papillose. Fruit ovoid, included in spathe, many-seeded; seeds ellipsoid, closely pitted in longitudinal rows.

Common in stagnant waters, ponds, tanks, lakes, rivers throughout Eastern Ghats. Fl. & Fr.: September-February.

Guvvalacheruvu RF (KDP), *RVR* 8147; Macherla (GNT), *VR* 5894; Repalle (GNT), *GVN* 16811 (MH); Krishna, *PV* 5811 (MH); Tiruporur to Tiruvennainallur (SA), *KR* 51170 (MH); Veeranum (SA), *KR* 58108 (MH).

INDIA: Throughout India.

WORLD: Sri Lanka, Bangladesh, Myanmar, Nepal, China, Vietnam.

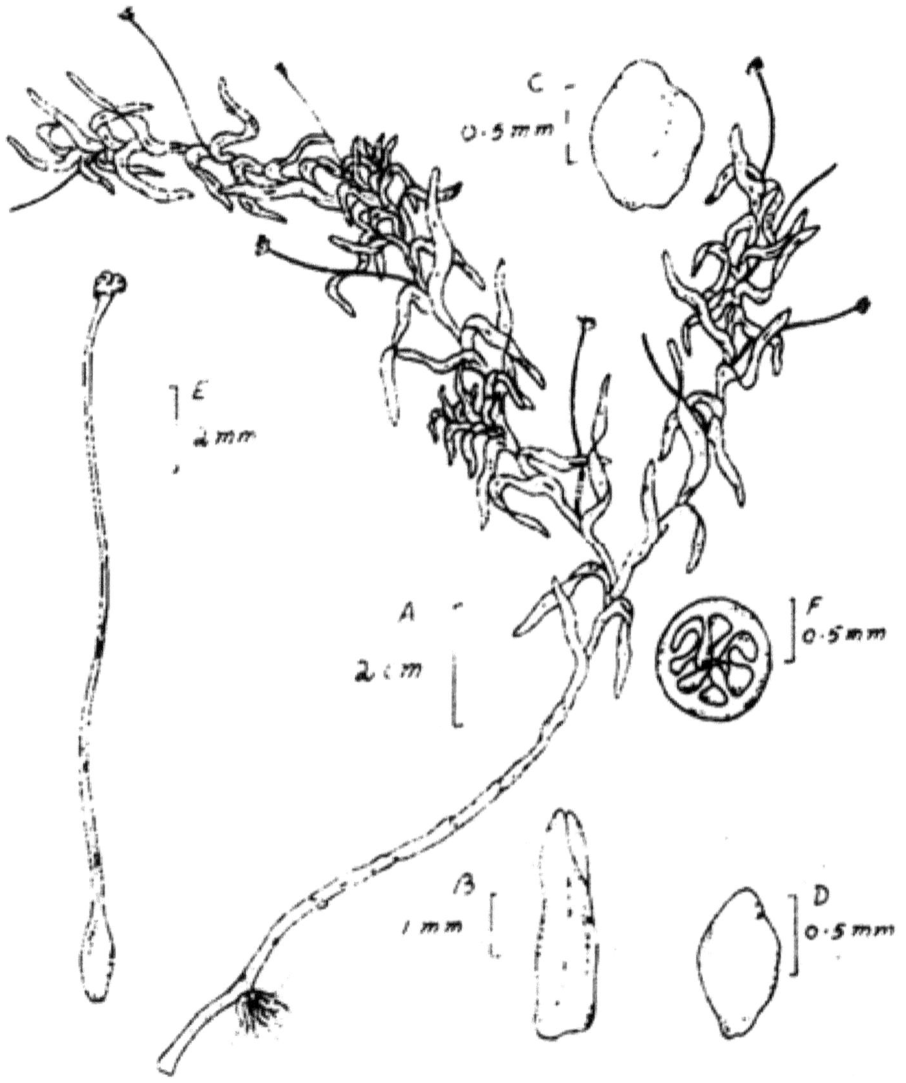

Figure 3. **Nechamandra alternifolia** (Roxb.) Thw.
[Syn.: *Lagarosiphon alternifolia* (Roxb.) Druce]
A. Twig, B. Spathe, C. Sepal, D. Petal, E. Pistil with rostrum, F. Ovary C.s.

OTTELIA Persoon

Ottelia alismoides (L.) Pers., Syn. Pl. 1: 400. 1805; FBI 5: 662. 1888; Fischer 3: 1398. 1928; Hartog in Steenis, Fl. Males 1, 5: 398. 1957; Matthew, Fl. Tamilnadu Carnatic 3: 1546. 1983. *Stratiotes alismoides* L., Sp. Pl. 535. 1753.

An acaulescent submerged herb. Leaves submerged and floating; submerged leaves oblong, 4-9 x 3-18 cm, chartaceous, base cuneate, truncate, margin entire, apex rounded, obtuse. Flowers unisexual or/and bisexual. Spathe elliptic-ovate, greenish, 6-winged, apex 2-lobed; peduncle 10-50 cm, 3-angled. Flowers solitary, sessile; sepals 3, linear, persistent; petals 3, white with yellow spots, oblong, stamens 9-15, ovary oblong, beaked, 6-9-celled, ovules arranged on parietal placentation, styles 10-15, bifid. Fruit oblong, crowned by sepals.

Common in tanks and slugglish streams. Fl. & Fr.: May – November.

Jambugumbala RF (ATP), *NY* 575; Aluru Kona (ATP), *TP* 921; Gulamallebad RF (KNL), *RVR* 3085; Podalakur (NLR), *BR* & *PMR* 23018; Kambakkam (CTR), *MVS* & *BR* 31198; Bodi (PKM), *BSS* & *SKB*. 34408; Sompalli reservoir (VSKP), *MHR* & *MCK* 14940; Siepanaidupeta (SKLM), *MCK* 18882; Krishnanandi (KNL), *VBH* 86697 (BSID); Guvvalacheruvu (KDP), *KS* 7801 (MH & CAL); Balapalle (KDP), *JSG* 11060 (CAL); Kolleru lake (WG), *KS* 5074 (MH); Kovvur (EG), *JLE* 37199 (MH); Majhipada section, Satkosia Tiger reserve, *KCM* 6164 (BSID).

INDIA: Throughout India.

WORLD: Africa, Sri Lanka, Myanmar, Nepal, China, Indonesia, Japan, Korea, Philippines, Thailand, Malaysia, Vietnam, Australia.

VALLISNERIA Linnaeus

Vallisneria natans (Lour.) H. Hara, J. Jap. Bot. 49: 136. 1974; Matthew, Fl. Tamilnadu Carnatic 3: 1747. 1983. *Physkium natans* Lour., Fl. Cochinch. 663. 1970. *Vallisneria spiralis* auct. non. L. 1753; FBI 5: 660. 1888. *p.p.*; Fischer 3: 1369. 1928.

Acaulescent, stoloniferous freshwater herbs; stolons to 5.5 cm, terete. Leaves radical, very long and linear, ribbon like, 8-36 x 0.4-1.2 cm, narrow at base, margin denticulate, apex obtuse. Flowers unisexual (dioecious). Male flowers: minute, many together in a shortly peduncled ovoid, 3-lobed spathe; spathes oblong, 6 x 4 mm. Female flowers: solitary, in a tubular, 3-toothed spathe at the end of a very long, filiform, spirally coiled scape, 50-60 cm long, sepals 3; petals absent; ovary narrow, ovules arranged on marginal placentation, styles 3, 2-lobed, stigmas 3, broad, notched. Fruit linear, to 2 cm, included in the spathe, many-seeded.

A common weed rooting at the bottom of pools and tanks. Fl. & Fr.: September-March.

Narpala (ATP), *KNR & DAM* 10629; Sunkesula (KNL), *RVR* 3149; Guvvalacheruvu RF (KDP), *R.V.Reddy* 8018; Macherla RF (GNT), *VRK* 3885, 5896; near Rayachoti (KDP), *KS* 7814 (MH & CAL) & *EK* 79267 (DD); Gunjana river, Kodur (KDP), *JLE* 15793 (MH); On way to Somasila (NLR), *PV* 110192 (BSID); Kolleru lake

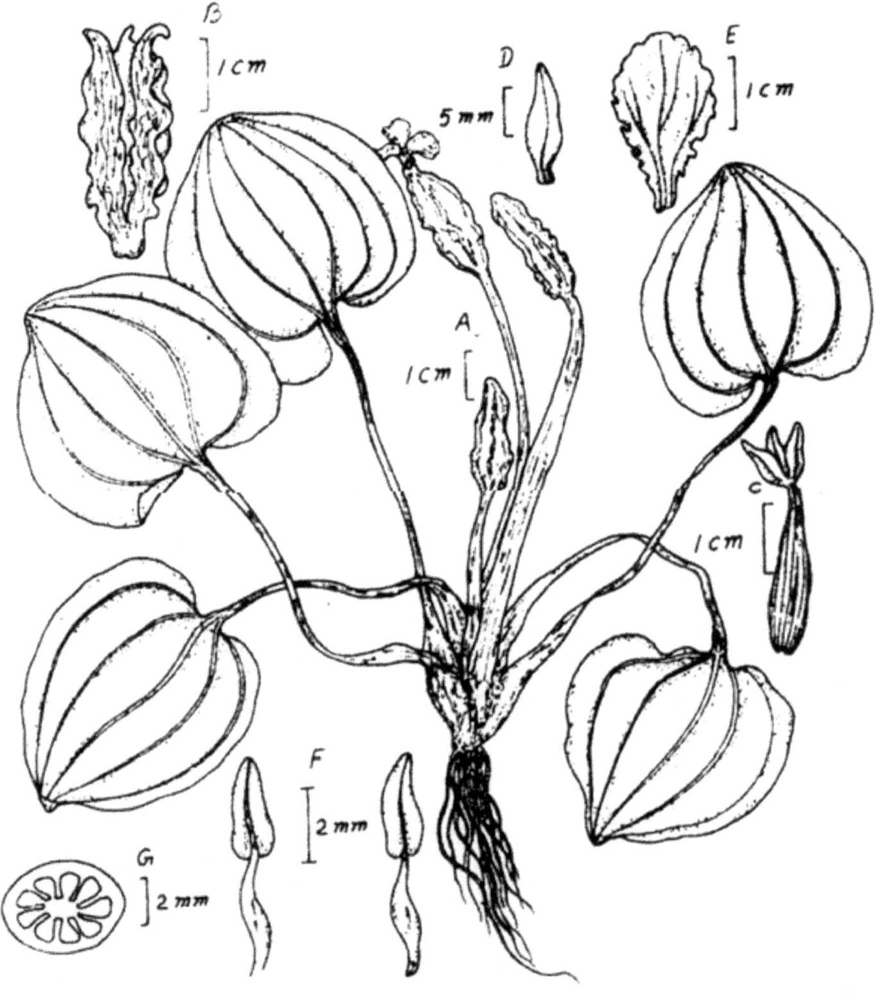

Figure 4. **Ottelia alismoides** (L.) Pers.

A. Twig, B. Spathe, C. Calyx with pistil, D. Sepal, E. Petal, F. Stamens, G. Ovary T.s.

(WG), *KS* 5063 (MH); Araku (VSKP), *NPBK* 10871 (MH); Gingee (SA), *KRM* 60324 (MH); Tikarpada section, Satkosia Tiger Reserve, *KCM* 6190 (BSID).

INDIA: Throughout India.

WORLD: SW Asia, Nepal, China, Korea, Malaysia, Vietnam, Russia, Australia.

Note: Lowden (*loc. cit.*) revised the genus *Vallisneria* L. He followed Mikono, who erected var. *dense-serrulata* for *Vallisneria spiralis.* Accordingly var. *spiralis* is distinguished from the var. *dense-serrulata* in having the adnation of staminodia of the pistillate flowers to the shallowly cleft apex of the fused stigmatic lobes. In the latter case the staminodia are adnate to deeply cleft base of the stigmatic lobes. In the present study Hara (*loc. cit.*) concept is followed who treated *Vallisneria natans* as the correct and for the taxon, which is considered as synonym to *Vallisneria spiralis* var. *dense-serrulata* by Lowden.

BURMANNICEAE

BURMANNIA Linnaeus

Burmannia coelestis D. Don, Prodr. Fl. Nepal. 44. 1825; FBI 5: 665. 1888; Fischer 3: 1399. 1928; Matthew, Fl. Tamilnadu Carnatic 3: 1549. 1983. *B. candida* Griff. ex Hook. f., FBI 5: 565. 1888. *Gonyanthes pusillus* Miers, Trans. Linn. Soc. 18. 537, t. 38, f.3. 1841. *Burmannia pusilla* (Miers) Thw., Enum. Pl. Zeyl. 325. 1864; FBI 5: 565. 1888. *B. bifurca* Ham. ex Hook.f., Fl. Brit. India 5: 565. 1888.

A slender erect herb, to 30 cm high; stem simple, slightly margined, glabrous. Leaves simple, linear, 3-4 x 0.3-6 mm. Bracts lanceolate, perianth purplish blue with yellow lobes; outer perianth lobes tubular, winged, and fused with ovary, inner perianth lobes ovate, perianth tube 5 mm long; stamens 3, connectives dorsally crested, spurred at base; ovary oblong-ovoid, 3-celled, ovules arranged on axile placentation; style stout, 3-fid; funnel-shaped. Capsules 3-winged; seeds minute, oblong.

In moist places of Araku valley in Visakhapatnam district. Fl. & Fr.: December-April.

Araku valley (VSKP), *NPBK* 10771 (MH & CAL); Araku valley-Anjoda (VSKP), *GVS* 32888 (MH).

INDIA: Peninsular, N.E.India, Tropical Himalaya.

WORLD: Bangladesh, Myanmar, Nepal, China, Cambodia, Laos, Indonesia, Thailand, Vietnam, Australia.

Note: Fischer (1928) reported one more species of *Burmannia, viz. B. disticha* L. in Presidency of Madras with Visakhapatnam district (At Ventala at 1500 m) distribution. However, no specimens of this taxon with Andhra Pradesh distribution is seen in any herbaria. Hajra (1989) while revising the Burmanniaceae of India reported this taxon occurrence only in Karnataka in Peninsular India. Hence *B. disticha* is excluded from the description.

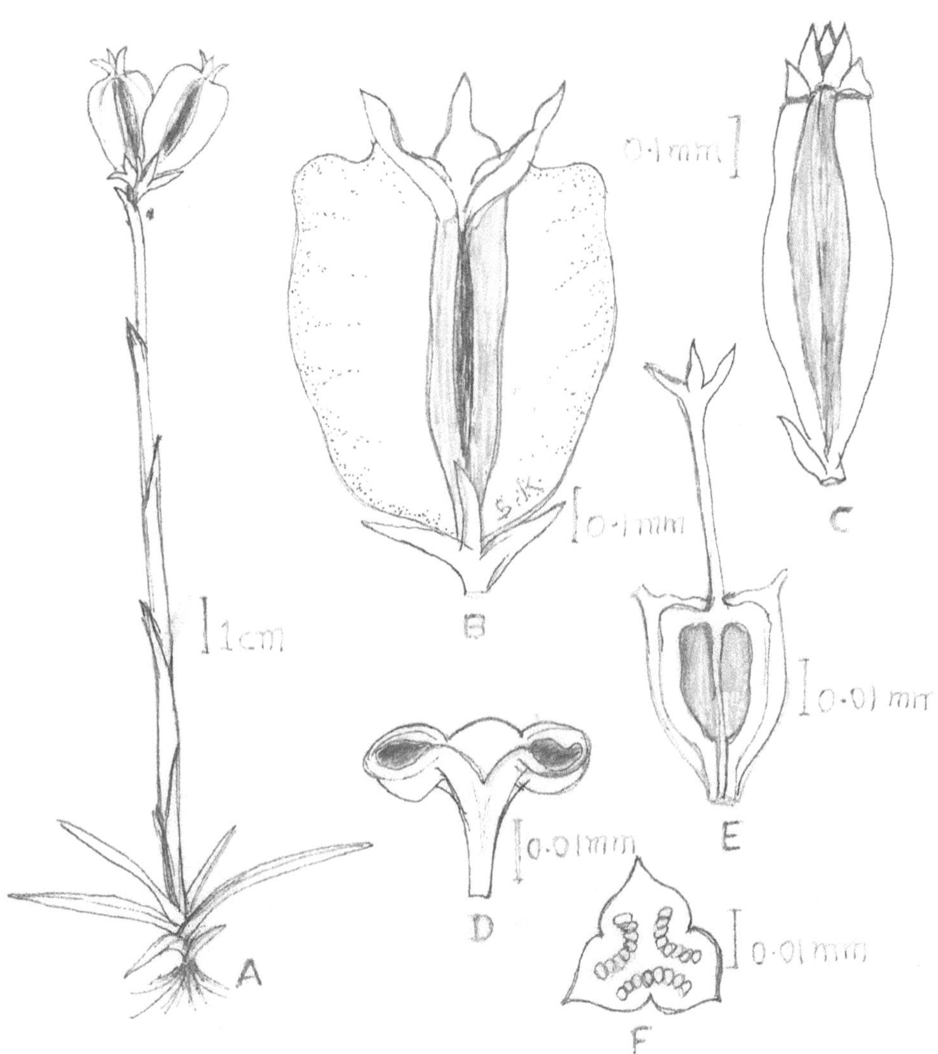

Figure 5. **Burmannia coelestis** D. Don

A. Habit; B. Female flower; C. Male flower; D. Stamen; E. Pistil; F. Ovary-c.s.

ORCHIDACEAE

We brought out a separate book on Orchids of Eastern Ghats.

Prasad, K., S. Karuppusamy and T. Pullaiah. 2019. Orchids of Eastern Ghats. Scientific Publishers, Jodhpur.

Readers are requested to refer this book for orchids.

ZINGIBERACEAE

1. Leaves spiral; staminodes absent .... **Cheilocostus**
1. Leaves distichous; staminodes petaloid (except *Alpinia*):
    2. Inflorescence terminal on a leafy shoot:
        3. Anthers winged .... **Globba**
        3. Anthers not winged:
            4. Flowers exceeding 8 cm in length .... **Hedychium**
            4. Flowers not exceeding 4 cm in length .... **Alpinia**
    2. Inflorescence on a leafless scape:
        5. Lateral staminodes large, petaloid; inflorescence not radical..... **Curcuma**
        5. Lateral staminodes minute; inflorescence radical .... **Zingiber**

ALPINIA Roxb. *nom. cons.*

Alpinia malaccensis (Burm. f.) Roscoe, Trans. Linn. Soc. London 8: 345. 1807; FBI 6: 255. 1892; Fischer 3: 1493. 1928. *Maranta malaccensis* Burm. f., Fl. Ind. 2. 1768. *Alpinia malaccensis* var. *sericea* Baker, Fl. Brit. India 6: 256. 1892.

A perennial herb, stem 2-3 m long; root stock perennial. Leaves shortly petioled, narrowly oblong or oblong-lanceolate, to 90 x 20 cm, base cuneate, margin undulate and fringed with tawny-tomentum, apex acuminate, glabrous above, pubescent beneath. Flowers in simple, to 30 cm long racemes, rachis fulvous-hairy; pedicels up to 1 cm long, villous; bracteoles white, coriaceous, enveloping the bud, pubescent at apex without, caducous; calyx coriaceous, pubescent without towards apex, gibbous, campanulate, split on one side, 3-lobed at apex, lobes deltoid; corolla white, tube up to 0.6 cm long, lobes 3, unequal, dorsal lobe lanceolate, ciliate, lateral lobes to dorsal lobe, smaller; lip red with yellow margins; stamen 1, staminodes 2, lateral; ovary densely pubescent; stigma funnel-shaped, villous. Capsules 2.5 cm across, yellow.

Rare in damp forests close to streams. Fl. & Fr.: April – July.

Sunkarimetta (VSKP), *NPBK* 745 (CAL); Vankachinta (VSKP), *GVS* 30047 (MH); Gurtedu (EG), *SS* 3617 (AU); Chahala, Similipahar (MBJ), *HOS* & *MB* 5410 (RRL-B).

INDIA: N.E.India, W.Himalaya, Andhra Pradesh.

WORLD: Bangladesh, Myanmar, Bhutan, China, Indonesia, Malaysia, Thailand.

Alpinia calcarata Roscoe is rarely cultivated. V.P.South (GNT), *VRK* 3947. Similarly **Alpinia zerumbet** (Pers.) B.L.Burt. & R.M.Sm. is also cultivated in gardens. Salem *A.Ravi Kiran* & *J.V.Sudhakar s.n.* (MH).

BOESENBERGIA Kuntze

Boesenbergia longiflora (Wall.) Kuntze, Revis, Gen. Pl. 2: 685. 1891. *Gastrochilus longiflorus* Walll., Pl. Asiat. Rar. 1: 22. 1829; FBI 6: 217. 1890. *Curcumorpha longiflora* (Wall.) A.S.Rao & Verma, Bull. Bot. Surv. India 13: 339. 1974. & 14: 123. 1975; Saxena & Brahmam, Fl. Orissa 3: 1900. 1995.

Perennial rhizomatous herb, rhizome short, root fibres tufted. Leaves oblong or ovate, 20-40 x 7-15 cm, apex acute, acuminate or caudate-acuminate, base rounded or slightly cordate, glabrous or puberulous beneath, purplish when young; petiole up to 30 cm long, deeply channelleed or almost winged above; sheaths purplish or green. Spikes radical, one to several, successively developing, narrow, cylindrical, 3-10 cm long, peduncle 0.5-4 cm long. Flowers white, variegated red; bracts 2-5, lanceolate, 6-8.5 x 1-1.8 cm, purplish or green; bracteoles elliptic, *c.* 4 cm long; calyx 2-2.8 cm long; corolla tube white or red, 7.5-12.5 cm long, corolla lobes white or with pinkish tips, oblong-or oblong-lanceolate, 2.5-3.5 cm long, lateral staminodes spathulate, obtuse, as long as the corolla lobes; labellum very large, suborbicular, 5 cm long; stamen large, truncate, filament 2 mm long; ovary cylindric.

Occasional in wet places along streams in Similipahar in Odisha. Fl. & Fr.: June – Septemebt.

Similipahar (MBJ), *MB* & *Prabhakar Rao* 4587 (RRL-B); Purunakote section, Satkosia Tiger reserve, *KCM* 6171 (BSID).

INDIA: Sikkim, Meghalaya, Assam, Bihar, Orissa.

WORLD: Myanmar, Malacca.

CHEILOCOSTUS C. Specht

Cheilocostus speciosus (J.König) C. Specht, Taxon 55: 159. 2006. *Banksia speciosa* J. König, Observ. Bot. 3: 75. 1783. *Costus speciosus* (Koen.) Smith, Trans. Linn. Soc. London 1: 249. 1800; FBI 6: 249. 1892; Fischer 3: 1490. 1928. *C. speciosus* var. *nipalensis* (Rosc.) Baker in Hook. f., Fl. Brit. India 6: 249. 1892.

An erect herb, up to 2.5 m high; root stock tuberous, insipid; stem subwoody at the base, spirally twisted so that the leaves appear spirally arranged. Leaves subsessile, oblong or lanceolate, 10-30 x 3-8 cm, base rounded, margin entire, apex acute or acuminate, often cuspidate, glabrous above, silky-pubescent beneath, ligule short, to 2 cm; sheath tubular. Flowers white, numerous in terminal spikes; bracts oblong-ovate, reddish, acuminate, often pungently mucronate; bracteole solitary below the calyx; calyx tubular, glabrous, 3-lobed; lobes triangular, corolla white, lobes elliptic-oblong, labellum white, with a yellow centre, broadly obovate with incurved margins, base densely hairy; staminodes medium on an oblong, petaloid process; ovary globose, ovules arranged on axile placentae. Capsules globose-ovoid, tardily-dehiscing; seeds black with white aril.

More or less common throughtout Eastern Ghats in undergrowth in deciduous and moist evergreen forests. Fl. & Fr.: July – October. Vern.: Tel.: *Besika*

Peccheruvu RF (KNL), *RVR & PVP* 2191; Papanasanam (CTR), *BR* & *BSS* 30214; Ananthagiri – Bispur (VSKP), *KSK* 23430; Near Devarapalli, Rampa hills (EG), *KSK* 23462; Punyagiri (VZN), *BR* & *KNR* 9896; Rollapenta (KNL), *JLE* 42245 (MH); Gundlakamma (KNL), *KNS* 3042 (DD); Near Rollapenta (PKM), *VBH* 83965 (BSID); Polavaram (WG), *CAB* 4784 (MH); Tadepalli (EG), *NRR* & *DN* 84306 (BSID); Rampa hills (EG), *MSR* 1568 (CAL); Muzumamidivalasa, Maredumilli (EG), *MM* 105018 (MH); Barmakonda RF (VSKP), *KCJ* 17146 (MH); Forest near Anantagiri (VSKP), *NPBK* 10996 (MH); Madgole to Paderu (VSKP), *GVS* 28097 (MH); Srungavarapukota (VZN), *GVS* 21815 (MH); Kurupam (VZN), *MV* 2009 (AU); Salur forest (SKLM), *NPBK* 1101 (CAL); Bandirevu forest (KMM), *RCS* 98928 (BSID); Edrallapalli RF (KMM), *RCS* 102402 (BSID); Purankote, Satkosia Wildlife Sanctuary, *D.Hazra* & *D.Das s.n.* (BSID).

INDIA: Throughout India.

WORLD: Sri Lanka, Myanmar, Bhutan, Nepal, China, Laos, Malaysia, Indonesia, Philippines, Thailand, Vietnam, Australia.

CURCUMA Linnaeus

1 Plants small; pseudostem 5-25 cm tall:

  2. Flowers never exceeding bracts; lip purple towards base .... **C. decipiens**

  2. Flowers longer than bracts; lip yellow .... **C. neilgherrensis**

1. Plants large; pseudostem more than 30 cm tall:

  3. Rhizome with the smell of green mango .... **C. amada**

  3. Rhizome without smell of green mango:

    4. Inflorescence central to the leaf-tuft and appearing with the leaves:

      5. Rhizome poorly developed; all bracts deep orange yellow .... **C. reclinata**

      5. Rhizome well developed; bracts pink, white or pale green:

        6. Rhizome deep orange yellow within .... **C. longa**

        6. Rhizome light yellow .... **C. montana**

    4. Inflorescence lateral to the leaf-tuft, often appearing before the leaves:

      7. Leaves with persistent purple flush on either side of the midrib .... **C. zedoaria**

      7. Leaves uniformly green:

        8. Corolla lobes pink .... **C. aromatica**

        8. Corolla lobes white or yellowish

          9. Rhizome without sessile tubers .... **C. pseudomontana**

          9. Rhizome with sessile tubers .... **C. angustifolia**

Curcuma amada Roxb., Asiat. Res. 11: 341. 1810; FBI 6: 213. 1890; Fischer 3: 1483. 1928.

A stemless herb, 60-75 cm high, with tuberous, pale yellow root stock. A tuft of leaves arising from the base of root stock, long petioled, oblong- lanceolate or elliptic, to 25-60 x 10-17 cm, apex acute. Inflorescence central to the leaf-tuft and appearing with the leaves. Bracts white or pale green, 2.5 cm long, those of the coma very few, white or tinged with pink. Flowers whitish with pale yellow in spikes; calyx very thin, split above, lobes 3, rounded shallow, anteriorly puberulous; corolla with tubular portion 1.8 cm long, then ventricose, labellum slightly exceeding the corolla, 1.2 cm broad, 3-lobed, midlobe emarginated.

In the undergrowth of deciduous forests, especially cool, shady regions in northern Eastern Ghats. Fl. & Fr.: June – October. Vern.: Tel.: *Mamidi allam.*Ori.: *Amada;* Tam.: *Mangainji.*

Araku valley (VSKP), *NPBK* 627 (CAL); Valamuru (EG), *MM* 102540 (BSID); Rukikonda (GJM), *VNS* 6087 (MH); Pachidya (GJM), *VNS* 6047 (MH); Sabakota (GJM), *VNS* 5563 (MH).

INDIA: Peninsular & E. India.

WORLD: Malaysia.

Curcuma angustifolia Roxb., Asiat. 11: 338. 1810; FBI 6: 210. 1890; Brahmam & Saxena, Fl. Orissa 3: 1896. 1995.

Perennial herb, rhizome pale yellow inside, root stock *c.* 3 cm long, whitish inside. Leaves appearing after the flowers or very young at the time of flowering, arising laterally from the base of the rootstock. Leaves few, narrowly lanceolate to elliptic-lanceolate, lamina 15-40 x 5-11 cm, base cuneate, uniformly green, glabrous or young pubescent beneath; petiole with sheath 15-30 cm. Plant in flower 30 cm high, peduncle clothed with puberulous, often coloured leafsheaths. Spike 7.5-12.5 cm, lower bracts 3.2 cm, the free part about two-thirds the whole length, tip rounded, coma small with only 5-9, oblong, usually deep magenta, glabrous or minutely pubescent bracts. Flowers exserted, 3.7 cm long; calyx 1.5 cm, puberulous, colourless with 3 rounded lobes, deeply split dorsally; corolla lobes white or yellowish, oblong, dorsal a little longer, somewhat pubescent; labellum longer than the corolla lobes, yellow, rounded, deeply bifid or emarginated; lateral staminodes yellow, free portion 1.2 cm long, oblong, slightly exceeding the petals; ovary villous.

Frequent in forest of Eastern Ghats of Odisha (Brahmam and Saxena, 1995). Fl.: May. Vern.: Tel.: *Ararut gaddalu,* Tamil.: *Ararut kilangu, Ararut kizhangu;* Ori.: *Palma, Paalagunda.* Engl.: East Indian arrow root.

INDIA:West Bengal, Bihar, Odisha, Uttar Pradesh, Maharashtra, Himalayas from Kumaon to NE India.

WORLD: Nepal.

Curcuma aromatica Salisb., Parad. Lond. t. 96. 1807; FBI 6: 210. 1890; Fischer 3: 1483. 1928.

A plant with large root-stock, of palmately branched, sessile, annulate, biennial tubers, yellow and aromatic inside. Leaves elliptic or lanceolate-oblong, 30-60 cm long, caudate-acuminate, green, often variegated above, pubescent beneath, base deltoid; petioles as long as or longer than the blade. Racemose spike 15-30 cm long; flowers pink, fragrant, shorter than the bracts, bracts ovate, recurved, cymbiform, obtuse, fertile, pale green, coma more or less tinged with red or pink; calyx irregularly 3-lobed; corolla tube 1.5 cm long, the upper half funnel-shapped, lateral segments oblong, upper longer, ovate, concave; lip yellow, obovate, deflexed, subentire or obscurely 3-lobed; staminodes obtuse, as long as the corolla-segments; ovary oblong; style filiform; stigma obscurely 2-lobed. Capsules globose, seeds ovoid.

Rare in Rampa hills. Fl. & Fr.: June – September. Vern.: Tel.: *Kasturi pasupu;* Ori.: *Bano haldi;* Tam.: *Kastui manjal.*

Chodavaram (EG), *MSR* 1475 (CAL).

INDIA: Throughout India.

WORLD: Sri Lanka, Myanmar, Bhutan, Nepal, China.

Curcuma decipiens Dalz., in Hooker's J.Bot. Kew Gard. Misc. 2: 144. 1850; FBI 6: 215. 1890; Fischer 3: 1483. 1928.

A stemless herb with tuberous root stock and often bearing numerous almond like tubers. Leaves broadly elliptic, 15-25 x 7-12 cm, shortly deltoid, base rounded or subcordate; margin entire, acuminate, glabrous, petiole as long as the leaf. Racemose spikes 5-12 cm long, those which first emerge from the soil lateral, the later spikes central; flowering bracts ovate, obtuse, saccate, purple; bracts of the coma numerous; oblong, purple, calyx obtusely toothed, puberulous below the teeth and near the base; corolla purple, tubular, 3-lobed, oblong-lanceolate and tube twice as long as the calyx, lip obovate, 2-fid, with crisped margin, stamen 1 perfect; staminodes oblong, petaloid; ovary 3-celled; ovules many on axile placentation; style filiform and stigma 2-lipped, ciliate. Fruit a tardily dehiscent globose 2-valved capsule; seeds ovoid or oblong, usually arillate.

Occasional in forest undergrowth in Northern Eastern Ghats. Fl. & Fr.: June – September.

Ravigudem-Polavaram Agency (WG), *DCSR* 447 (CAL).

INDIA: Peninsular India. Endemic.

Curcuma longa L., Sp. Pl. 2. 1753; FBI 6: 214. 1890; Fischer 3: 1183. 1928. *C. domestica* Valenton, Bull. Jard. Bot. Buitenzorg (ser. 2) 27: 31. 1918; Matthew, Fl. Tamilnadu Carnatic 3: 1615. 1983.

Perennial herb, rhizome branched, deep orange yellow with in, aromatic. Leaf tuft 1.2-1.5 m high. Leaves elliptic-oblong or oblong, 40-60 x 13-15 cm, apex caudate-acuminate, base attenuate, glabrous on both sides, petiole 7.5-25 cm long, sheaths longate. Inflorescence central to the leaf-tuft and appearing with the leaves, spike

8-15 cm long, coma-bracts pinkish, fertile bracts pale green, white or sometimes with purple tip and margins, oblong, to 4 x 1.5 cm; flowers few to a bract, yellow, not exserted, calyx-tube to 1.5 cm, scarcely 3-lobed, split about one-fourth way down, labellum obovate, subentire, apex retuse, pale yellow, ovary hairy at the top.

Cultivated occasionally. Fl.: September – October. Vern.: Tel.: *Pasupu*, Ori.: *Haldi*; Tam.: *Manjal*.

Aalangavai, Thandikudi (DGL), *K.Ravikumar* 92362 (MH, BSID).

INDIA: Cultivated throughout India.

WORLD: Widely cultivated throughout the tropics.

Curcuma longa L. var. **vanaharidra** Vela. & *al.*, J. Econ. Taxon. Bot. 33: 173. 2009.

Plants perennating by underground rhizome, *c.* 1 m high, mother rhizome oval/oblong with light orange yellow cortex and darker core, 7-10 x 3.5-4.5 cm in size, palmately branched, giving forth one or two secondary rhizomes, 6-8 x 2-3 cm in size and oblong, primary fingers 8-10.5 x 1.7-2.2 cm in size, pointed downwards, secondary branches present on primary fingers, 3 – 6.7 x 1.3-1.7 cm in size, inside colour of fingers light mustard-yellow or light orange yellow, flesh mildly scented and slightly acrid in taste; stipitate tubers few in number. Leaves semierect, *c.* 60.5 x 20.8 cm, lamina thin, shining green, more fragile than that of turmeric, very broadly lanceolate, leaf tip acuminate, base gradually tapering with faint purple tint along the midrib which fades completely at maturity and the region acquires a very dark green colour all along the midrib from top to base, margin slightly wavy, glabrous on dorsal side, minutely hairy along the veins on ventral side, veins prominent and distantly placed, leafsheath green, margin green with brown tint, 28.5 cm or slighly more; petiole green, glabrous, 19.5 cm long; ligule truncate, less than 0.5 cm. Peduncle *c.* 13.5 cm long; spike central surrounded by a tuft of 4-6 leaves, 13.5 x 9.5 cm, sterile bracts present, one in number, ovate, whitish-green, hairy below and glabrous above, tip obtuse, *c.* 5.3 x 4.2 cm; floral bracts 16-17 in number, ovate, whitish-green, tip round, margin light green, minutely hairy above and below, *c.* 4.5 x 4 cm; coma bracts up to 5 in number, lanceolate, spreading, whitish with very light pink towards the top, tip obtuse or acute, minutely hairy below and above; inner bracts transparent white, three in number, ovate, bifid, emarginate, glabrous, 2.3 x 1.3 cm, Flowers up to three per bract, either shorter than bract or as equal, slightly laterally appressed, 4.5 cm long; ovary white, minutely hairy, hairs minutely three-toothed, one side deeply cleft; corolla tube 3 cm long, whitish to cream; corolla-lobes three, the main upper lobe white, prominently beaked, white with very light pink tint, beak minutely hairy, 1.1 x 1.3 cm in size; side lobes white, oblong, conical with acute tip, 1 x 1 cm, staminodes whitish, side lobes oblong, 1.2 x 1 cm, lip three-lobed, cream in colour, 1.4 x 1.2 cm, shallowly emarginate, lip whitish-cream with a yellow band; anther horizontally placed, glabrous, white, 0.3 cm long, spur long, converging downwards; stigma white, glabrous, highly protruding, no fruit setting.

Rare as forest undergrowth in Araku valley in Visakhapatnam district, wild.

Araku (VSKP), *SJ* 4055 (NBPGR).

INDIA: Andhra Pradesh, Endemic.

Curcuma montana Roxb., Pl. Coromandel 2: 28. t. 151. 1802; FBI 6: 214. 1890 *p.p.*; Fischer 3: 1483. 1928; A.S.Rao & Verma, Bull. Bot. Surv. India 14: 122. 1972.

Herbs, up to 70 cm high; rhizome and sessile tubers yellowish inside and aromatic; root fibres with elliptic tubers, 4 cm long, white inside and not aromatic. Leaf tuft 50-70 cm high; leaves oblong, elliptic-oblong or oblong-lanceolate, up to 50 x 20 cm, acute, midrib purplish, base tapering. Inflorescence rising in the centre of leaf tuft, spike 10 x 4.5 cm, coma bracts pink or purplish. Flowers pale yellow, longer than bracts; labellum suborbicular, entire, margins undulate.

Rare in shady, moist localities of forests in northern Eastern Ghats. Fl. & Fr.: June – September. Vern.: Tel.: *Maccha Pasupu*, Ori.: *Sakuta*; Tam.: *Kaatumanjal.*

Tadepalli (EG), *NRR* & *DN* 84354 (MH); Althi (SKLM), *PVS* 76883 (MH); Kasipur (KPT), *Patnaik* & *Party* 8508 (RRL-B); Similipahar (MBJ), *HOS* & *MB* 4196, 4589 (RRL-B); Pampasar section, Satkosia Tiger reserve, *KCM* 5131 (MH).

INDIA: Assam, Andhra Pradesh, Madhya Pradesh, Odisha, Uttar Pradesh, Kerala.

WORLD: Throughout Tropics.

Curcuma neilgherensis Wight, Icon. Pl. Ind. Orient. t. 2006. 1853; FBI 6: 210. 1890; Fischer 3: 1482. 1928; Matthew Fl. Tamilnadu Carnatic 3: 1615. 1983.

A tuberous herb. Rootstock small, white inside; root-fibres numerous, slender, terminating in small tubers. Leaves lanceolate or oblong-lanceolate, shortly petioled, 15-25 x 2-5 cm, acuminate, glabrous, narrowed at the base. Inflorescence racemose, vernal spike of variable length, flowers bright yellow; fertile bracts oblong-lanceolate, acute, pale yellowish-green; coma dense, pink; calyx-tube 1.5 cm long, 3-toothed; corolla yellow; lobes lanceolate, labellum broadly ovate, apex 2-lobed; staminodes oblong; ovary oblong, pubescent, style filiform. Capsules globose, glabrous; seeds ovoid or oblong.

Occasional in Tirumala hills. Fl. & Fr.: July – August. Vern.: Tam.: *Kaatumanjal.*

Eswaramala hills (ATP), *B. Tirupal Reddy* 325 (BSID); Gundlabrahmeswaram (KNL), *BR* & *BSS* 32346; Tirumalla hills (CTR), *MVS* & *VS Rao* 31008, *DRC* 1530 (SVU); Yercaud (SLM), *SKK* 26999 (MH); Kolli hills, Kulivalavu (KGR), *SKS* 816 (SGH).

INDIA: Peninsular India. Endemic.

Curcuma pseudomontana Graham, Cat. Pl. Bombay 210. 1839; Fischer 3: 1483. 1928. *C. montana* sensu Baker in Hook. f., FBI 6: 214. 1890 *p.p.*,non Roscoe 1807.

An erect herb, 75-80 cm long with small rootstock bearing small almond-like or subglobose tubers at the ends of the fibres, tubers pure white inside and edible.

Leaves uniformly green, reaching 30-60 cm or more long (including the petiole), lanceolate-oblong, margin entire, apex acuminate, tapering to the base. Flowers bright yellow, in racemose, oblong spikes on 7-10 cm long peduncle, flowering bracts obovate-lanceolate, the lowest with purple edges only; the upper more or less uniformly mauve-purple, stamen 1 perfect, filament short; anthers not crested, with contiguous, spurred at the base; lateral staminodes oblong, petaloid, lip orbicular, with a deflexed tip, ovary 3-celled, ovules many on axile placentation, style filiform, stigma 2-lipped. Fruit a dehiscent, globose, 3-valved capsule; seeds ovoid or oblong, usually arillate.

Common in shady localities. Fl. & Fr.: May – September.

Rollamadugu (KDP), *BSS* & *SKB* 30306; Tyada RF (VSKP), *BR & KNR* 9895; Mamandur –Tirumala border (CTR), *GVM* & *JHFB* 117040 (MH); Maredumilli (EG), *GVS* 68530 (MH); Addatigala (EG), *SS* 449 (AU), *GVS & GRK* 68530 (MH & CAL); Near Puttagondhilanka, Tedepalli (EG), *NRR* 84354 (MH); Madgole to Paderu (VSKP), *GVS* 28084 (MH); Busikonda (VSKP), *GVS* 42699 (MH); On the way to Althi (SKLM), *PVS* & *NRR* 76883 (BSID); Mannanur RF (MBNR), *VBH* 86622 (BSID); Bhadrachalam forest (KMM), *RCS* 104311 (BSID); Edrallapalli RF (KMM), *RCS* 102410 (BSID).

INDIA: Andhra Pradesh, Telangana, Karnataka, Kerala. Endemic.

Curcuma reclinata Roxb., Asiat. Res. 11: 342. 1810; FBI 6: 214. 1890; Saxena & Brahmam, Fl. Orissa 3: 1899. 1995.

Perennial herb, up to 30 cm high; rhizome poorly developed, root fibres with subglobose or ellipsoid tubers, 0.6-2.5 cm diam. Leaves elliptic, to 25 x 10 cm, base unequal, apex acuminate or cuspidate; petiole 10-17 cm long. Inflorescence central to leaf-tuft; spike 7.5-10 cm long; bracts 3.7-5 cm, with recurved, rounded limb, yellowish, coma scarcely any and of the same colour. Flowers small, orange-yellow; calyx inflated, obscurely 1-toothed.

Bankura and Kanjpani in Odisha (Panigrahi *et al.*, 1964). Fl.: July.

INDIA: Central India, Bihar, Odisha.

WORLD: Myanmar and Malaysia.

Curcuma zedoaria (Christm.) Roscoe, Trans. Linn. Soc. London 8: 354. 1807; FBI 6: 210. 1890; Gamble 3: 1482. 1928; Saxena & Brahmam, Fl. Orissa 3: 1899. 1995. *Amomum zedoaria* Christm., Vollst. Pflanzensyst. 5: 12. 1779.

Perennial herb, 0.9-1.2 m high; rhizome well developed, pale yellow or straw-coloured inside, faintly aromatic, root-fibres with ovoid or ellipsoid tubers. Leavs oblong or oblong-lanceolate, 30-60 cm long, base narrowed, apex acuminate, glabrous, with a persistent purple flush on either side of the midrib, petiole long. Inflorescence lateral to the leaf-tuft, appearing before the leaves; flowering bracts ovate, recurved, cymbiform, 3.7 cm long, green tinged with red, bracts of coma many, spreading, crimson or purple. Flowers yellow, shorter than the bracts. Lateral staminodes obovate, labellum suborbcular, 1.2 cm broad, yellow, recurved, emarginated; ovary hairy. Capsule ovoid-trigonous, dehiscing irregularly; seed oblong, aril lanceolate, white, cut into slender unequal segments.

Occasional in open grassy places, often cultivated. Fl.: April. Vern.: Tel.: *Kuchur, Kichili gaddalu*; Tam.: *Kichilikizhangu*.

Near Rollapenta (PKM), *VBH* 83961 (BSID).

INDIA: Eastern Himalaya; cultivated and naturalized through out India.

WORLD: SE Asia.

Elettaria cardamomum Maton is occasionally cultivated in Kolli hills (NMK), A.A.Ansari & P.Dwarakan 56325 (MH)

GLOBBA Linnaeus

1. Leaves glabrous beneath .... **G. racemosa**
1. Leaves more or less pubescent beneath:
    2. Inflorescence without bulbils .... **G. orixensis**
    2. Inflorescence with bulbils:
        3. Anther with bifid wings; bracts persistent .... **G. marantina**
        3. Anther without wings; bracts deciduous .... **G. sessiliflora**

Globba marantina L., Mant. Pl. 170. 1771; FBI 6: 206. 1890. *G. bulbifera* Roxb., Asiat. Res 11: 358. 1810; FBI 6: 206. 1890; Fischer 3: 1481. 1928; Matthew, Fl. Tamilnadu Carnatic 3: 1617. 1983.

A slender herb, 40-50 cm long, with creeping rhizome and erect stem; roots fleshy. Leaves distichous, oblong or oblong-lanceolate, 9-15 x 3-5 cm, base cuneate, margin entire, apex caudate; ligule broad, ciliate; sheath ciliate along the apical margin; petiole reduced or short. Spike terminal, 10-15 cm long; flowers orange-yellow, fragrant and buds sometimes replaced by bulbils, peduncle 6-8 cm long, exerted from leaf-sheaths; lower sterile bracts oblong, upper one persistent, calyx funnel-shaped, 3-lobed, ciliate; corolla yellow, tubular, slender, longer than the calyx, lobes ovate, subequal, spreading or reflexed; labellum small, 2-fid, base auricled, narrow; lateral staminodes petaloid; stamen with a long 2-appendaged filament; anther oblong, with spreading 2-fid wings, ovary globose, 1-celled; ovules arranged on 3 parietal placentae, style filiform, stigma turbinate. Capsules globose, tardily dehiscent, to 4 mm; seeds ovoid, often tomentose.

In shady localities of forests. Fl. & Fr.: August – October. Vern.: Ori.: *Chota rasna*.

Upper Ahobilam (KNL), *TP* & *RVR* 2598; Lankamala RF (KDP), *SRSR* 14635; Thalakona RF (CTR), *KNR* & *PSPB* 9845 & *GVS* 46953 (MH); Maredumilli RF (EG), *BR* & *KNR* 10607; Gundlabrahmeswram (KNL), *JLE* 16958 (MH); Ahobilam (KNL), *JLE* 25574 (MH); Talakona RF (CTR), *GVS* 46006 (MH); Papikonda RF, Polavaram Agency (WG), *DCSR* 502 (CAL); Bodalanka (EG), *DN* 85564 (BSID); Chinthiru, Maredumilli (EG), *MM* 102511 (BSID); Sunkarimetta (VSKP), *NPBK* 712 (CAL), *GVS* 21637 (MH); Araku valley (VSKP), *NPBK* 10754 (MH); Galikonda (VSKP), *GVS*

21637 (MH); Salur – Jeypore road side forest (SKLM), *NPBK* 1095 (CAL); Rama hills (MDS), *VNS* 76 (CAL); Purunakote RF, Satkosia Tiger Reserve, *KCM* 5217 (BSID); Satkosia Wildlif Sanctuary, *D.Hazra* & *D.Das s.n.* (BSID).

INDIA: S., E. & N.E.India, Andaman & Nicobar Islands.

WORLD: Indo-Malesia.

Globba orixensis Roxb., Asiat. Res. 11: 358. 1810; FBI 6: 201. 1890; Fischer 3: 1480. 1928.

1. Leaves glabrous on both surface .... var. **orixensis**
1. Leaves pubescens and glaucous on ventral surface .... var. **pubescens**

Globba orixensis Roxb. var. **orixensis**

Erect perennial herbs; rhizome creeping with fleshy roots. Leafy shoot 30-35 cm high, green, swollen at base; sheaths 10-15 cm broad at base, puberulous outside and glabrous inside. Leaves simple, alternate, broadly lanceolate, 5-7 x 1.5-2 cm, subsessile, cuneate at base, caudate at apex, entire at margin, glabrous on both surfaces; midrib prominent with 8-15 parallel nerves; ligules 1-3 mm long, green with ciliate margin. Panicles terminal, 6.5 cm long; bracts lanceolate, 5-7 x 2-3 mm, acute at apex, deciduous. Flowers 4-5 cm long, orange-yellow; calyx infundibuliform, 0.5-1 x 0.2-0.3 cm, glabrous, yellow; teeth minute, tridendate; corolla tube *c.* 1.5 cm long, slender, 3-lobed; lobes subequal, 6-8 x 3 mm, orange-yellow; staminodes 2, subequal, 5 x 2 mm, petaloid, orange-yellow with deflexed lip; labellum narrow, shallowly bifid, glabrous, orange-yellow with reddish brown spots at throat; filament of the fertile stamen 2.4 cm long, yellow, glabrous, arched; anther 2-celled, oblong, 2 mm long, nearly acute at apex, pale-yellow, dorsifixed; ovary unilocular; ovules many on parietal placenta; style linear, 2-2.5 cm long, glabrous, white; stigma cupular with ciliate mouth; nectar glands linear, 4-5 mm long. Infructescence 6-6.5 cm long, dark maroon at base with persistent calyx. Capsule globose, 1 cm in diameter, warted; seeds numerous, brownish red, arillate, faintly ciliate at margin; aril black.

Common in northern Eastern Ghats in shade. Fl. & Fr.: August – October.

Anantagiri (VSKP), *NPBK* 790 (CAL); Near Surhiguda, Lakshmipuram (VSKP), *NRR* & *DN* 84390 (BSID); Seethi (SKLM), *PVS* & *NRR* 79411 A (BSID); Bodalanka (EG), *DN* 85557 (BSID); Purunakote RF, Satkosia Tiger Reserve, *KCM* 5216 (BSID).

INDIA: Andhra Pradesh, Madhya Pradesh, Odisha, Jharkhand, West Bengal, Sikkim, Assam, Meghalaya, Mizoram and Tripura.

WORLD: Myanmar, Malaysia and Thailand.

Globba orixensis Roxb., Asiat. Res. 11: 358. 1810 var. **pubescens** Subba Rao & Kumari, Fl. Visakhapatnam dist. 259. 2008.

Erect herbs, up to 70 cm high. Leaves few, oblong-lanceolate, up to 24 x 7.8 cm, base cuneate, margin entire, apex acuminate or caudate, glabrous above, pubescent and glaucous beneath; petiole up to 0.8 cm long, puberulous, ligule membranous, ciliate, sheaths up to 15.5 cm long, puberulous. Flowers yellow, in panicles; bracts ovate-lanceolatee, membranous, each subtending a flower, caducous; calyx funnel-shaped, tridentate, teeth minute; corolla-tube up to 1.8 cm long, slender, 3-lobed, lobes subequal, ovate; staminodes 2, petaloid, subequal, ovate, lip up to 0.9 cm long, deeply bifid, narrow, deflexed, spotted red; filaments of the fertile stamen up to 2.5 cm long, arched, anther up to 0.2 cm long; ovary ovate, warted, glandular, style up to 2.6 cm long, stigma clavate; nectary glands linear.

Common in northern Eastern Ghats in shade. Fl. & Fr.: June.

Maredumilli (EG), *GVS* 24263 (holotype CAL, isotype, MH); Chintapalli (VSKP), *GVS* 30081 (MH, paratype).

INDIA: Eastern Ghats. Endemic.

Note: *Globba orixensis* is closely allied to *G. racemosa* but differs in deep orange, smaller flowers, shorter corolla tube, lip with a reddish-brown spot on the throat and smaller capsules. Both the varieties of *G. orixensis* differs by their long caudate nature of leaves with pubescent and glaucous nature of ventral surface in var. *pubescens.*

Globba racemosa Smith., Exot. Bot. 2: 115. t. 117. 1804; FBI 6: 201. 1890. *G. clarkei* Baker in Hook. f., Fl. Brit. India 6: 201. 1890. *G. hookeri* C.B.Clarke ex Baker in Hook. f., Fl. Brit. India 6: 201. 1890.

An erect herb, 30-90 cm long with creeping rhizome, roots fleshy. Leaves oblong or elliptic-oblong, subcaudate, glabrous, 20-25 x 5-10 cm, slightly hairy or quite glabrous beneath. Racemose panicles narrow, lowest branches 3-flowered, bulbils never present; flowers orange-yellow; bracts small, deciduous; calyx yellowish, shortly lobed; corolla tube about twice the length of the calyx, petals broadly ovate, longer than the staminodes, lip ovate, as long as the petals, shallowly 2-lobed; stamen with a long, 2-appendaged filament, anther oblong, staminodes ovate, equal in length; ovary globose, 1-celled, ovules arranged on parietal placentation, style filiform, stigma turbinate. Capsules globose, dehiscent; seeds ovoid and tomentose.

Occasional in forests of northern Eastern Ghats. Fl. & Fr.: July – October. Vern.: Ori.: *Chan silndi.*

Seethi (SKLM), *PVS* & *NRR* 79411 A (BSID); Anantagiri (VSKP), *NPBK* 790 (CAL); Laxmipuram (VSKP), *NRR* & *DN* 84390 (BSID); Bodalanka (EG), *DN* 85557 (BSID).

INDIA: C. to E. Himalaya, N.E.India.

WORLD: Bangladesh, Myanmar, Bhutan, Nepal, China, Thailand, Vietnam.

Note: Karthikeyan & *al.* (1989) cited the taxon distribution in Central to East Himalayas and North East India.

Globba sessiliflora Sims, Bot. Mag. 35: t. 1428. 1811. *G. ophioglossa* Wight, Icon. Pl. Ind. Orient. 6: 16. 1853; FBI 6: 202. 1890; Fischer 3: 1480. 1928. *G. canarensis* Baker in Hook.f., Fl. Brit. India 6: 206. 1890.

An erect herb, 30-50 cm long with creeping rhizome, roots fleshy. Leaves linear to elliptic or oblong-lanceolate, 10-15 x 4-6 cm, usually conspicuously and finely caudate and pubescent below. Racemose panicles narrow, lowest branches 4-6-flowered, sometimes few bulbils present in the lower axils; flowers orange yellow, about 2 cm across, bracts small, deciduous; calyx broadly funnel-shaped, pale green, teeth broad ovate, corolla orange-yellow, corolla tube thrice the length of the calyx; petals ovate, shorter than the lanceolate staminodes; lip twice the length of the corolla, very deeply bifid, segments narrow; stamen with a long 2-appendaged filament; anther oblong, connective produced beyond the cells, simple, winged; staminodes petaloid; ovary globose, 1-celled, ovules arranged on pariental placentae; style filiform, 1-2 cm, stigma turbinate. Capsules globose, dehiscent, smooth; seeds ovoid, white and tomentose.

In damp localities in northern Eastern Ghats. Fl. & Fr.: July – October.

Rampa hills (EG), *BR & KNR* 10606; Galikonda hills (VSKP), *AJR* 20655; Sesharayi (EG), *SS* 677 (AU); Mudurlanka, Maredumilli (EG), *MM* 102562 (BSID); Kamsupaka (VZN), *MV* 7171 (AU); Pachidya (GJM), *VNS* 5850 (MH).

INDIA: North India, Andhra Pradesh.

WORLD: Myanmar, Thailand.

HEDYCHIUM J. Koenig

1. Lip orbicular – obcordate:
    2. Stamens shorter than the lip .... **H. coronarium**
    2. Stamens twice as long as the labellum .... **H. coccineum**
1. Lip obcordate; stamens longer than the lip .... **H. flavescens**

Hedychium coccineum Buch.-Ham. ex Sm. in Rees Cycl. 17: 5. 1811; FBI 6: 231. 1892; Saxena & Brahmam, Fl. Orissa 3: 1994. *H. angustifolium* var. *coccineum* Baker in Hook.f., Fl. Brit. India 6: 231. 1892. *H. coccineum* var. *roscoei* Baker in Hook.f., Fl. Brit. India 6: 231. 1892.

Perennial herb, 1-2 m tall. Leaves linear-oblong, 30-50 x 3-3.7 cm, long-acuminate, pubescent on midrib below and at margin or tip, otherwise glabrous. Flowers red, brick-red or flesh-coloured in 30 cm long, lax spikes. Bracts oblong or lanceolate, 3 cm long, each bract subtending 2-4 flowers; corolla lobes twisted, narrowly linear; staminodes lanceolate, labellum suborbicular, split into 2, ovate-oblong or irregularly subovate, clawed; stamens twice as long as the labellum. Capsule elliptic.

Occasional in Eastern Ghats of Odisha. Fl.: August – November; Fr.: December – April.

Similipahar (MB), *SX* 481 (RRL-B), *SX & MB* 4233, 4749 (RRL-B), *GPG* 12420 (ASSAM).

Hedychium coronarium Koen. in Retz., Obs. Bot. 3: 73, 1793; FBI 6: 225. 1892; Fischer 3: 1485. 1928; Matthew, Fl. Tamilnadu Carnatic 3: 1617. 1983. *H. chrysoleucum* Hook., Bot. Mag. 76: t. 4516. 1850. *H. coronarium* var. *chrysoleucum* (Hook.) Baker in Hook.f., Fl. Brit. India 6: 226. 1892.

A tuberous herb; root stock horizontal, 1.5-5 cm in diameter, fleshy, jointed; stem 1-1.5 m high, erect. Leaves distichous, lanceolate or oblong-elliptic, 20-40 x 4-9 cm, glabrous, base cuneate-sub-acute, margin entire, apex finely acuminate; ligule 2 cm, densely pubescent; sheath elongate. Racemose spikes bearing pure white, fragrant flowers; bracts closely imbricate, 3-4-flowered, bracteoles 3, membranous. Flowers to 10 cm long; calyx tubular, apex unequally lobed, pubescent, green, shorter than the bracts; corolla-tube 6 cm, lobes linear equal, reflexed; labellum broadly obovate-suborbicular, 2-fid, clawed; staminodes petaloid, oblanceolate, apex retuse, ovary 3-celled, globose, pubescent; ovules many, arranged on axile placentae, stigma funnel-like, ciliate. Capsules oblong, glabrous; seeds many, with a crimson aril.

In moist localities in the hills. Fl. & Fr.: July – August. Vern: Ori.: *Bhod siludhi*

Sesharayi (EG), *SS* 672 (AU); Way to Guvvalaguda, Lakshmipuram (SKLM), *NRR* & *DN* 84386 (BSID); Bhimavaram (VZN), *MV* 19860 (AU); Similipahar (MBJ), *SX* & *MB* 4782, 5230 (RRL-B); Ganjam (GJM), *SX* 3305 (RRL-B); Marapalam, Nagalur road, Yercaud, *AVNR* 26935 (MH); Similipal reserve (MBJ), *A.R.K.Sastry* & *G.P.Singh* 12245 (BSID).

INDIA: Throughout India.

WORLD: Bhutan, China, Indonesia, Malaysia, Myanmar, Nepal, Sikkim, Sri Lanka, Thailand, Vietnam; Australia.

Hedychium flavescens Carey ex Roscoe, Monandra. Pl. Scitam. t. 50. 1825; Fischer 3: 1485. 1928. *H. coronarium* Koen. var. *flavescens* (Carey ex Roscoe) Baker in Hook. f., Fl. Brit. India 6: 226. 1892.

A robust plant, to 2.5 high. Leaves lanceolate, 20-25 x 6-8 cm, caudate, glabrous or pubescent along the midrib below and at the mouth of the sheath, ligule large. Spike dense-flowered; bracts imbricate, 3-4-flowered, large, oblong, obtuse, more or less villous at the apex and pubescent on the back; calyx as long as or slightly longer than the bract, apex puberulous; corolla tube up to 7.5 cm long, lip obcordate, narrowed to the base into a distinct claw, staminodes narrowly oblanceolate, often notched at the apex; stamen longer than the lip; ovary 3-celled, globose; ovules arranged on axile placentae. Capsules oblong, glabrous, seeds many, minute.

In moist localities in Kollimalais, Shevaroys and hills of East Godavari and Visakhapatnam districts. Fl. & Fl.: July – August.

Dummakonda RF (EG), *GVS* 68626 (MH); Nulakamaddi (EG), *VNS* 600 (CAL); Sapparlagedda (EG), *GVS* 42784 (MH); Cherukonda (VSKP), *GVS* 28192 (MH); Brooklyn-Yercaud (SLM), *KS* 6573(MH).

INDIA: S. & N.E.India.

WORLD: Nepal, China.

KAEMPFERIA Linnaeus

1. Root tubers rounded; leaves linear-lanceolate .... **K. angustifolia**
1. Root tubers oblong; leaves oblong-obovate .... **K. rotunda**

Kaempferia angustifolia Roscoe, Trans. Linn. Soc. London 8: 351 1807. *K. roxburghiana* Schult., Mant. 1: 33 1822. *K. undulata* Link, Enum. Hort. Berol. Alt. 1: 3 1821.

A small herb to 15 cm tall; rhizome thick with many small rounded tubers. Leaves elliptic-linear-lanceolate, 5-20 x 3-5 cm, glabrous on both surfaces, base obtuse, acuminate at apex. Scape terminal, single-flowered. Flowers white, labellum deep purplish blotched.

Purankote, Satkosia Tiger Reserve, *KCM* 6566 (BSID).

INDIA: Throughout the states of India.

WORLD: Myanmar, Malaysia, Sumatra and Java.

Kaempferia rotunda L., Sp. Pl. 3. 1753; FBI 6: 222. 1890; Fischer 3: 1484. 1928; Sunitha *et al.*, J. Econ. Taxon. Bot. 26: 156. 2002.

A stemless herb with succulent roots ending with oblong tubers. Leaves few (3-4), arising after 2-3 spathes, erect, 25-35 x 5-6 cm, distichous, 10-15 ribbed, oblong-obovate, base narrowed, margin undulate, apex acuminate, green above and slightly purple below, petiole channeled. Flowers fragrant, in 2-3-flowered spike, arise before leaves, floral bracts green, pubescent; calyx cylindric, splitting spathaceously, cream-coloured; petals cream-coloured; stamen 1, perfect, staminodes 3, broad, petaloid, spreading, tip slightly acuminate; ovary 3-celled, epigynous.

Rare as forest undergrowth in Gundlabrahmeswaram in Nallamalais. Fl. & Fr.: April – July.

Gundlabrahmeswaram (KNL), *SS* & *AMR* 24488 (SKU), *BR* & *BSS* 32343 (BSID).

INDIA: Throughout India.

WORLD: Sri Lanka, Myanmar, China, Indonesia, Malaysia, Thailand.

ZINGIBER Boehmer *nom. cons.*

1. Spikes terminating the leafy stem .... **Z. capitatum**
1. Spikes radical, produced directly from the root stock:
    2. Spikes on long peduncles:
        3. Leaves linear, or linear-lanceolate, 1-2.5 cm broad; labellum dark purple .... **Z. officinale**
        3. Leaves oblong-lanceolate, 3-8.7 cm broad; laabellum white or yellow:
            4. Leaves pubescent beneth; ligule 1-2 mm long; flowering bracts usually deep red during flowering, pubescent or hairy .... **Z. montanum**

4. Leaves glabrous; ligule 10-25 mm long; flowering bracts usually green during flowering, glabrous or glabrescent .... **Z. zerumbet**

2. Spikes sessile, short and dense:

5. Ligule of leaf membranous, exceeding 1 cm in length:

6. Corolla tube less than 3.5 cm, hairy; lateral petals free......... **Z. roseum**

6. Corolla tube more than 4.5 cm, glabrous; lateral petals fused... **Z. rubens**

5. Ligule of leaf coriaceous, not exceeding 1 cm in length ... **Z. wightianum**

Zingiber capitatum Roxb., Asiat. Res. 11: 348. 1810 & Fl. Ind. ed. 2, 1: 55. 1832; FBI 6: 248. 1892 var. **elatum** (Roxb.) Baker in Hook.f., Fl. Brit. India 6: 249. 1892; Sabu, Zingiberaceae South India 229. 2006.

A tuberous herb. Leaves linear, ascending, 20-40 x 3-5 cm, tapering gradually to the point, usually pubescent beneath. Racemose spike at the end of the leafy stem, densefid, erect, oblong-cylindrical, 7-15 x 3-5 cm; bracts closely imbricate, subcoriaceous, green with a narrow brown edge; corolla tube as long as the bract, pale yellow, unspotted, midlobes orbicular, emarginated, basal auricles large, oblong, obtuse, bright red; stamen pale yellow, nearly as long as the lip; ovary 3-celled, ovules many on axile placentae. Capsules bright red; seeds black, shining, aril large, lacerated white.

Rare in forest clearings and edges of forests and open grasslands in northern Eastern Ghats. Fl. & Fr.: July – August. Vern.: Ori.: *Machi kanda.*

Way to Gudem, Chintapalli (VSKP), *JLE* 37129 (MH); Similipahar (MBJ), *SX & MB* 4863, 5242 (RRL-B); Manaskonda, Koraput (KPT), *SX & MB* 6784 (RRL-B); Gandhmardan hilltop (BGR), *SX & MB* 6290 (RRL-B); Majhipada section, Satkosia Tiger reserve, *KCM* 6141 (BSID).

INDIA: Himalaya, E. & N.E.India.

WORLD: Nepal.

Zingiber montanum (J.König) Link ex Dietr., Sp. Pl. 1: 52. 1831: Sabu, Zingiberaceae South India 231. 2006. *Amomum montanum* J.König in Retz., Observ. Bot. 3: 51. 1783. *Zingiber purpureum* Roscoe, Trans. Linn. Soc. London 8: 348. 1807. *Z. cassumunar* Roxb., Asiat. Res. 11: 347. t. 5. 1810; FBI 6: 248. 1802; Fischer 3: 1490. 1925.

A large slender aromatic herb, root stock perennial, yellow inside, with an aromatic, warm, somewhat camphoraceous taste, without bitterness. Leaves distichous, numerous, oblong-lanceolate, 20-35 x 2-4 cm long, margin entire, apex acute, glabrous above, pubescent beneath; sheaths pubescent. Flowers in racemose spikes; bracts broadly ovate, subacute, bright red, pubescent, margin narrowly membranous; calyx *c.* 1.5 cm long, truncate, white, membranous, split, glabrous; corolla tube 2.3-2.5 cm long, pale yellow; lobes lanceolate, pale yellow, dorsal lobe 3.2 x 1.5 cm, cymbiform, lateral lobes 2.5 x 1 cm, linear-lanceolate, reflexed; labellum 3-lobed, *c.* 2.5 cm wide, yellowish white, suborbicular, lateral lobes *c.* 8 x 5 cm, obliquely obovate, erect. Capsules subglobose, dehiscent, to 1.8 x 2 cm, membranous; seeds many, minute, puple.

Occasional in moist localities on banks of streams. Fl. & Fr.: August – September. Vern.: Tel.: *Karu pasupu.*

Maredumilli (EG), *BR & KNR* 9890; Tyada RF (VSKP), *BR* & *KNR* 9890; Talakona RF (CTR), *GVM* & *JHFB* 112841, 112966 (MH); Kutravada (EG), *GVS & GRK* 68536 (MH); Rampa hills (EG), *VNS* 87 (CAL); Tikarpada RF, Satkosia Tiger Reserve, *KCM* 6000 (BSID).

INDIA: West Benga, Odisha, Andhra Pradesh, Tamil Nadu, Kerala

WORLD: Sri Lanka, Malay Peninsula, widely cultivated in Tropical Asia.

Zingiber officinale Roscoe, Trans. Linn. Soc. London 8: 348. 1807; FBI 6: 246. 1892; Fischer 3: 1489. 1928. *Amomum zingiber* L., Sp. Pl. 1. 1753.

Perennial herb, 30-120 cm high. Leaves distichous, subsessile on the sheath, linear or linear-lanceolate, 12-15 x 1-2.5 cm, glabrous, apex acuminate. Flowers greenish or greenish-yellow, in 4-7.5 cm long spikes, floral bracts closely imbricate, labellum small, dark purple, 3-lobed, midlobe oblong-obovate, lateral lobes short, obtuse; stamens dark purple, as long as the labellum, shorter than corolla.

Occasionally cultivated for the rhizome which is extensively used for spice. Fl.: November. Vern.: Tel.: *Allamu*; Ori.: *Ada*; Tam.: *Sukku, Inchi*. English: Ginger.

Maredumilli (EG), *GVS* 68528 (MH).

INDIA: Widely cultivated.

WORLD: Cultivated in Tropical countries.

Zingiber roseum (Roxb.) Roscoe, Trans. Linn. Soc. London 8: 348. 1807; FBI 6: 244. 1892; Fischer 3: 1489. 1928. *Amomum roseum* Roxb., Pl. Coromandel t. 126. 1800.

A tall herb. Leaves oblong or oblong-lanceolate, 30-40 x 6-10 cm, margin entire, apex acuminate, sessile, more or less pubescent below. Racemose spikes arising directly from tuberous root, bracts red, exterior ones broadly ovate, interior linear-lanceolate, more or less hairy; calyx cylindric, shortly 3-lobed; corolla red or scarlet, lip 3-lobed, midlobes oblong-cuneate, margins recurved, crisped, lateral lobes short, broad; stamens arching over the lip and equaling it in length, ovary 3-celled, ovules many on axile placentation, style filiform, stigma small. Capsules subglobose, dehiscent; seeds many, minute.

Common in shade on damp soil in northern Eastern Ghats at 1000 m. Fl. & Fr.: August – September. Vern.: Tel.: *Besika*

Maredumilli (EG), *BR & KNR* 10611 & *GVS* 67589 (MH & CAL); Chinthiru, Maredumilli (EG), *MM* 100711 (MH); Tyada RF (VSKP), *BR & KNR* 9889;

Madugukota (EG), *SS* 9925 (AU); Bodalanka (EG), *DN* 85555 (BSID); Barnakonda RF (VSKP), *KCJ* 17165 (MH); Forest near Sunkarimetta (VSKP), *NPBK* 10944 (MH); Cherukonda (VSKP), *GVS* 28162 (MH), *GVS* 42753 (MH); On the way to Rusaraguda (VZN), *PVS* & *NRR* 79413 (BSID).

INDIA: S. to E.India.

WORLD: China, Myanmar, Thailand.

Zingiber rubens Roxb., Asiat. Res. 11: 348. 1810; FBI 6: 244. 1892; Saxena & Brahmam, Fl. Orissa 3: 1909. 1995.

Rhizomatous herb, up to 1.2 m high. Leaves narrowly elliptic-oblong, erecto-patent, 25-41 x 4-11.5 cm, acuminate; ligule membranous, 1-1.2 cm. Spikes small, 3.7-6 x 3.7 -7 cm, sessile or with small peduncle up to 3 cm from the base of the stem; bracts not closely imbricate, red, linear-oblong, 3 cm long; calyx spathaceous, membranous; corolla tube 3-5 cm, lobes linear, acuminate, red, 2 cm long; stamen shorter, red, arching over the labellum, appendage or beak incurved. Capsule reddish, slightly compressed or angled, straight or slightly curved, hairy; seeds 3 in each cell, oblong, 5 mm long, red streaked, enclosed in a white aril, which is lobed at the apex.

Rare in semi-evergreen forests of northern Eastern Ghats. Fl. & Fr.: September – November.

Upper Sileru (VSKP), *CSR* & *VSR* 2365 (KU); Tulka RF, Satkosia Tiger Reserve, *KCM* 5932 (BSID).

INDIA: Andhra Pradesh, E. to N.E.India.

WORLD: Bangladesh, Myanmar, China, Thailand, Vietnam.

Zingiber wightianum Thw., Enum. Pl. Zeyl. 315. 1861; FBI 6: 244. 1892; Fischer 3: 1489. 1928.

A perennial herb; root stock horizontal, tuberous; leafy stem elongated, 1-1.5 m long. Leaves lanceolate or oblong-lanceolate, margin entire, apex acuminate, 10-35 x 3-8 cm long, pubescent beneath. Racemose spikes oblong or subglobose; peduncle very short; floral-bracts pubescent, green, outer ovate and inner lanceolate; corolla-segments pale, lip obovate, cuneate, pale yellow, veined and spotted with purple, as long as the corolla-segments, emarginate, basal auricles small, ovate; stamens shorter than the lip, arching over it; ovary 3-celled, ovules many on axile placentation, stigma small. Capsules ellipsoid, to 2.5 x 1.5 cm; seeds many, minute.

Rare in places with considerably heavy rainfall in East Godavari (cf. Fischer, *loc. cit.*) and Visakhapatnam districts. Fl. & Fr.: July – August.

Path to Bison hill (EG), *CAB* 5030 (MH); Towards Borra (VSKP), *GVS* 44265 (MH).

INDIA: South India.

WORLD: Sri Lanka.

Zingiber zerumbet (L.) Roscoe ex Smith, Exot. Bot. 2: 105.t. 112.1805; FBI 6: 267. 1892; Fischer 3:1490. 1928. *Amomum zerumbet* L., Sp. Pl. 1. 1753.

Rhizomatous, perennial herb, up to 1.2 m high; rhizome whitish outside, pale-yellow inside. Leaves narrowly elliptic-oblong or elliptic-lanceolate, 8-12 each side, erecto-patent, glabrous below, 25-37 x 4-9 cm, acute or acuminate, thinly silky when young. Flowers pale yellow, 4.5 cm long in elongate spikes arising from the rhizome. Spikes cylindrical or ellipsoid-oblong, 6.2-11 x 3.7-4.3 cm; bracts closely imbricate, green in flower, turning to red afterwards; calyx 1.8 cm long, sheathing; corolla tube slender, 1.8 cm; labellum with lateral lobes suborbicular, 1.5-1.8 cm diam., thin, wrinkled; stamen 1.5-1.8 cm, filament very short.

Infrequent, in moist hilly areas of northern Eastern Ghats. Fl. & Fr.: July – October. Vern.: Tel.: *Bomiki, Santa pasupu, Karrallamu;* Ori.: *Gada, Parsu Kedar.*

On the way to Vangara, Kedaripuram (VZN), *NRR* 83639 (BSID); Bhanjanagar (GJM), *Mishra* 1087 (RRL-B).

INDIA: Throughout India.

WORLD: Bangladesh, Myanmar, China, Cambodia, Laos, Malaysia, Indonesia, Philippines, Thailand, Vietnam.

MARANTACEAE

1. Ovary 1-celled .... **Maranta**
1. Ovary 3-celled:
    2. Leaves radical; inflorescence condensed globose spike .... **Stachyphrynium**
    2. Leaves cauline; inflorescence lax spreading panicle:
        3. Outer staminodes 2, fertile stamen with conspicuous a petaloid appendage........................ **Indianthus**
        3. Outer staminodes absent, fertile stamens without a petaloid appendage........................ **Schumannianthus**

INDIANTHUS Suksathan & Borchs.

Indianthus virgatus (Roxb.) Suksathan & Borchs., Bot. J. Linn. Soc. 159: 393. 2009. *Schumannianthus virgatus* (Roxb.) Rolfe, J. Bot. 45: 244 1907. *Phrynium virgatum* Roxb., Asiat. Res. 11: 324 1810. *Arundastrum virgatum* (Roxb.) Kuntze, Revis. Gen. Pl. 2: 684 1891. *Clinogyne virgata* (Roxb.) Benth., Gen. Pl. 3: 651 1883. *Donax virgata* (Roxb.) Schum., Pflanzenr. IV 48 (Heft 11): 33 33 1902. *Maranta virgata* (Roxb.) A.Dietr., Sp. Pl. 1: 21 1831. *Phyllodes virgata* (Roxb.) Kuntze, Revis. Gen. Pl. 2: 695 1891.

Perennial erect, diffuse shrubs, to 4 m high; stem thickened at nodes. Leaves distichus, to 40 x 18 cm, ovate-oblong, cuspidate at apex, rounded at base; petiole to 1 cm long; sheath to 20 cm long, open. Panicle to 80 cm long, diffuse, branches slender, dichotomous; bracts 3-5 x 0.6 cm, lanceolate. Flowers paired in each bract; sepals small; corolla white, tube short; lobes oblong; stamen single, outer staminodes

Figure 6. **Indianthus virgatus** (Roxb.) Suksathan & Borchs

A. Twig; B. Inflorescence; C. Corolla opened; D. Pistil; E. Stamen; F. Fruit.

petaloid, white, to 1.5 cm long; inner staminodes smaller; ovary densely hairy, 3-celled, solitary in each cell, basal. Berry obovoid, rugose.

Common in high altitude ravines of Sirumalai hills in Tamil Nadu. Fl. & Fr.: June – September.

Sirumalai hills, (DGL), *SKS* 811 (SGH).

INDIA: Peninsula.

WORLD: Sri Lanka.

Note: A revised generic classification of Asian Marantaceae by Suksathan *et al.* (2009) provided the description of a new genus *Indianthus*, to which *Schumannianthus virgatus* is transferred.

MARANTA Linnaeus

Maranta arundinacea L., Sp. Pl. 2. 1753; Fischer 3: 1495. 1928; Saxena & Brahmam, Fl. Orissa 3: 1913. 1995.

Branched herb, 0.9-1.8 m high, with creeping root stock and fleshy, cylindric-obovoid tubers, covered with pale scales, which leave scars after falling. Leaves ovate-obong, up to 25 X 11 cm at the base of the stem, upper leaves ovate-lanceolate to narrowly lanceolate, 10-15 cm long, base rounded or cuneate. Flowers white in spike, sepals green, 1.2 cm long.

Occasionally cultivated for the starch and often as a pot plant in gardens. Fl.: November. Vern.: Tel.: *Palaguntha*, Tam.: *Aruruttuk-kilangu*. English: Arrowroot.

INDIA: Cultivated.

WORLD: South America.

SCHUMANNIANTHUS Gagnepain

Schumannianthus dichotomus (Roxb.) Gagnep., Bull. Soc. Bot. France 51: 176. 1904; Saxena & Brahmam, Fl. Orissa 3: 1915. *Phrynium dichotomum* Roxb., Asiat. Res. 11: 324. 1810. *Clinogyne dichotoma* Salisb., Trans. Hort. Soc. 1: 276. 1812; FBI 6: 258. 1891.

Bamboo-like undershrub, 3-4 m high, stems slender, cylindric, polished green, up to 2.5 cm diam., root creeping, woody. Leaves ovate to elliptic-oblong or ovate-oblong, 10-15 cm long, cuspidate, finely nerved, sometimes pubescent, base rounded, sheath produced into a short ligule; petiole 5-6 mm long. Flowers white, in short, 2-, rarely 3-flowered lateral, bracteates, shortly peduncled racemes; bracts laceolate, dry, 3.7-5 cm long; bracteole at the base of the pedicel, subulate, hard, 2 mm long, angular when dry; sepls 7.5 mm long; petals 3 cm long; staminal tube elongate, labellum with a hard, saccate base; ovary tomentose, style fleshy. Fruit 2-3-lobed, subglobose with flattened top, 1-1.2 cm diam; seeds 2-3.

In marshy places under shade in Eastern Ghats of Odisha. Fl.: April – June; Fr.: July – Agust. Vern.: Ori.: *Khorson*.

Kurab, Athmallik (Anugul District), *HFM* 2866 (DD); Kadalidih, Puri hills, Bamra, *HFM* 3440 (DD).

STACHYPHRYNIUM K. Schumann

Stachyphrynium placentarium (Lour.) Clausager & Borschs., Philipp. J. Sci. 15: 230. 1919. *Phyllodes placentaria* Lour., Fl. Cochinch. 13. 1790. *Phrynium parviflorum* Roxb., Fl. Ind. 1: 7. 1820; FBI 6: 259. 1892; Fischer 3: 1495. 1928. *P. placentarium* (Lour.) Merrill, Philip. J. Sci. 15: 230. 1919 & Trans. Amer. Philos. Soc. N. S. 24: 120. 1935; Saxena & Brahmam, Fl. Orissa 3: 1914. 1995.

A bushy herb up to 1.7 m high, with creeping tuberous, perennial root stock. Leaves radical, usually solitary, rarely two, oblong or ovate-lanceolate, 30-45 x 15-20 cm, base rounded or truncate, margin entire, apex acuminate or cuspidate; petiole longer than the blade. Spike globose arising from the side of the petiole; bracts lanceolate, pale green, acute, 2-3-fid; sepals 3-4, lanceolate, pale-green, glabrous; corolla tube as long as the bract, segments linear-oblong; staminal lobes small, orbicular, white tipped with yellow; ovary 3-celled, ovule usually solitary in each cell. Fruit usually 1-seeded.

Rare, found growing in shade as forest undergrowth in the hills of Northern Eastern Ghats. Fl. & Fr.: September – December.

Maredumilli (EG), *BR & KNR* 10614; Bison hill (EG), CAB 5054 (MH); Sunkarimetta (VSKP), *NPBK* 734 (CAL), *GVS* 21698 & 32883 (MH); Puldumar (KHD), *HFM* 2445 (DD).

INDIA: Throughout India except N.W.India.

WORLD: Myanmar, Bhutan, China, Indonesia, Philippines, Thailand, Vietnam.

MUSACEAE

1. Pseudostem cylindric with offsets .... **Musa**
1. Pseudostem swollen at the base without offsets .... **Ensete**

ENSETE Horan

Ensete glaucum (Roxb.) Cheesman, Kew Bull. 1947: 101. 1947; Subba Rao & Kumari, Bull. Bot. Surv. India 14: 164. 1972. *Musa glauca* Roxb., Pl. Coromandel t. 300. 1820; FBI 6: 232. 1892. *M. nepalensis* Wall. in Roxb., Fl. Ind. 2: 492. 1824; FBI 6: 261. 1892.

A monocarpic unbranched non-stoloniferous plant, 3-3.5 m high; pseudostem swollen at base. Leaves oblong-lanceolate, 1-1.5 x 0.35-0.4 m, ascending, apex acute, base unequal, green above, glaucous beneath; petiole 30-35 cm, shallow and broad-channeled. Inflorescence pendulous; bracts broadly ovate, acute, persistent, basal flowers fertile, hermaphrodite, gradually transforming to staminate flowers towards apex through hermaphrodite and neuter flowers in the middle. Flowers white or transluscent, outer perianth striped, inner perianth membranous. Fruit bunch compact, 12-14 hands per bunch and up to 14 fingers in two rows per hand. Fruits geotropic, green, oblong, 5-9 cm long, pulp scanty; seeds smooth, black on drying.

Rare in isolated localities on well exposed rocky slopes of mountains with some undergrowth in East Godavari and Visakhapatnam districts. Fl. & Fr.: June – August.

Bodalanka (EG), *DN* 85560 (BSID); Adapavalasa (VSKP), *GVS* 19702 (MH); Errakonda (VSKP), *GVS* 24527.

INDIA: Eastern Ghats, Eastern Himalaya.

WORLD: Myanmar, Nepal, China, Indonesia, New Guinea, Philippines, Thailand.

MUSA Linnaeus

1. Inflorescence erect. .... **M. ornata**
1. Inflorescence drooping:
   2. Petiole open; bracts yellow streaked without and not splitting at apex; anthers almost twice the length of filaments; inner tepals shorter with brown stripes; fingers 9-13 per hand .... **M. shankarii**
   2. Petiole closed; bracts otherwise, inner tepals otherwise .... **M. balbisiana**

Musa balbisiana Colla, Mem. Gen. Musa 56. 1820; Hara & *al.*, Enum. Fl. Pl. Nepal 1: 63. 1978. *M. sapientum* auct. (non L., Syst. Nat. Ed. 10. 1303. 1759); FBI 6: 262. 1892 *p.p.*; Fischer 3: 1497. 1928.

Plant with cylindrical pseudostem with offsets, about 3 m high. Leaves large, erect or ascending; leaf blade adaxially green and slightly pruinose or not, ovate-oblong, 2.9 m x 90 cm, base auriculate, asymmetric. Inflorescence pendulous, up to 2.5 m; peduncle and rachis glabrous. Bracts of bisexual and male flowers adaxially purple-red, abaxially brownish purple to yellow-green and pruinose, ovate to lanceolate, persistent, apex obtuse, reflexed after flowering; bracts of female flowers deciduous. Male flowers up to 20 per bract, in 2 rows. Compound tepal adaxially pale purple, abaxially pale purple-white, 4-5 cm, striate, teeth yellow to orange; free tepal milky white, translucent, obovate, apex emarginate, shortly mucronate-apiculate. Infructescence pendulous, with 8 clusters each of 15 or 16 berries in 2 rows. Berries grey-green, obovoid, 13 x 4 cm, distinctly angled at maturity, base narrowed into a stalk, 2.5 cm long, apex contracted or not into a short column. Seeds numerous, brown, oblate, 5-10 mm in diam., minutely warty.

Cultivated for fruit throughout Eastern Ghats. Vern. Tel.: *Arati*; Tam.: *Vazhai pazham*.

Paderu agency (VSKP), *NRR* & *DN* 84223 (MH).

INDIA: Cultivated throughout India.

WORLD: Cultivated in Many Tropical countries.

Note: The application of binomial for the sweet starchy South Indian "Banana" fruit or 'Plantain,' *Musa sapientum* L. (*loc. cit.*) and the binomial for "Arati kaya" used as cooking vegetable. Both the names are considered to be closely allied triploid hybrids between *Musa acuminata* Colla and *M. balbisiana* Colla. The identification of a given clone depends on an analysis of the contribution made by each parent and ploidy.

Both the triploid hybrids, the sweet plantain and the vegetable 'Aratikaya', are cultivated by the Girijans in their village areas and their cultivation on commercial scale is quite successful together with other selected hybrids.

Musa ornata Roxb., Fl. Ind. ed. 1. 2: 488. 1824. *M. rosacea* sensu Baker in FBI 6: 263. 1892; Fischer 3: 1497. 1928.

A plant with slender, cylindric stem to 2.5 m high, stoloniferous; root perennial. Leaves linear-oblong, firm in texture, 1-3-1.8 x 30-45 cm, petiole 30-60 cm long. Flowers in drooping or erect spikes reaching 15 cm, long; flowers of the basal bracts pistillate; bracts ovate, lilac or reddish. Pistillate flowers 6-8 cm long; calyx yellowish-white, 5-toothed; petal as long as the calyx; ovary inferior and many-ovuled, staminodes 5. Staminate flowers: stamens 5 (rarely 6) perfect, 6th usually rudimentary or absent. Fruit linear-oblong, slightly incurved, obscurely 4-5-angleld, of the size of a man's finger, firm, not edible seeds many, black, tuberculate.

Common in northern Estern Ghats at high altitudes with a considerably high rainfall. Fl. & Fr.: April - July. Vern.: Tel.: *Chilaka arati, Kodhili, Goddala kodili.*

Maredumilli (EG), *BR & KNR* 10612; Near Dharawada (WG), *DN* 85522 (BSID); Valamuru (EG), *SS* 3475 (AU); Bodalanka (EG), *DN* 85561 (BSID); Valamuru, Maredumilli (EG), *MM* 100866 (MH); way to Sileru-Chintapalli (VSKP), *JLE* 37123 (MH); Chintapalli (VSKP), *GVS* 28207 (MH); Near Dabuguda, Laxmipuram (VSKP), *NRR* 83797 (MH), *NRR* & *DN* 84206 (BSID); Koinpur, below Mahendragiri (GJM), *HOS* & *MB* 3894 (RRL-B); Karlapat (KHD), *MB* & *Dhal* 7591 (RRL-B).

INDIA: Peninsular & E.India, E.Himalaya.

WORLD: Introduced and naturalized in Central and South America.

Musa shankarii Subba Rao & Kumari, Fl. Visakhapatnam Dist. 266. 2008.

Stoloniferous herbs, suckers not latent, rhizomes short. Pseudostem up to 7.5 m high, pale brown, blotched black, more or less waxy, up to 65 cm across at breast height. Juvenile leaf pale green, not blotched; midrib and petiole pale green, more or less waxy. Adult leaves ascending; lamina up to 2.4 x 0.6 m, elliptic or oblong, oblique at both ends, auricled at base, glabrous, green above, glaucous beneath, midrib yellowish-green; petioles up to 75 cm long, more or less waxy, pale-green, naviculate, gap up to 1 cm wide, wings of margin black; sheaths clasping firmly. Inflorescence pendulous, up to 1.35 m long, peduncle glabrous; 9-10 hands per inflorescence, 9-13 fingers per hand in two rows; dried bracts persistent till fruits mature. Fruits 6-9 x 2 -3 cm, negatively geotropic, waxy, pale-yellow, 4-5-angled, pulp scanty, skin up to 0.3 cm thick, tapering abruptly at both ends; pedicelss up to 2.5 cm long, neck up to 1 cm long, 4-angled, truncate; seeds many, black, up to 0.4 x 0.6 cm, subglobose, angled and warted. Male bud up to 17.5 cm long, subacute; bracts up to 24.5 x 11 cm, broadly ovate, obtuse, concave, 2-3 opening at a time; outer surface corrugated, waxy, dull reddish-brown with yellow-streaks; inner surface reddish-brown; tip some times pale-green, never recurved; gap between insertion of bracts and flowers and fruits nil. Male flowers up to 5 cm long; outer tepal up to 3 x 1 cm, oblong, obtuse, pale yellow with brown stripes, ultimately 5-fid; inner tepal up to 1.6 x 1 cm, obovate, translucent with brown stripes, shallowly 3-lobed at apex, middle lobe shortly acuminate; stamens 4-5, up to 3.1 cm long, filaments up to 1.1 cm long, anthers up to 2 cm long; pistillode up to 3.4 cm long; rudimentary ovary up to 1.4 cm long.

Found occasionally on hill slopes on rocky and laterite soil in open places in Araku to Korai road and Galikonda in Visakhapatnam district (Subba Rao and Kumari, 2008). Fl. & Fr.: March – August.

Araku (VSKP), *GVS* 22639 (holotype, MH); Galikonda hills (VSKP), *K.Prasad* 6424 (BSID).

INDIA: Andhra Pradesh. Endemic.

Ravenala madagascariensis Sonn. is cultivated in parks and gardens as an ornamental.

Heliconia stricta Huber and **H. rostrata** Ruiz & Pav. (Heliconiaceae) are cultivated in gardens of Yercaud and Kolli hills for its ornamental inflorescence.

CANNACEAE

CANNA Linnaeus

Canna indica L., Sp. Pl. 1. 1753; Saxena & Brahmam, Fl. Orissa 3: 1912. 1995. *C. orientalis* Roscoe, Monandr. Pl. t. 12. 1828; Fischer 3: 1496. 1928; Matthew, Fl. Tamilnadu Carnatic 3: 1620. 1983. *C. indica* var. *orientalis* (Roscoe) Hook.f., Fl. Brit. India 6: 260. 1892

Perennial erect herb, to 1.5 m. Leaves elliptic-lanceolate, 12-50 x 8-20 cm, glabrous, base obtuse, margin entire, apex caudate-acuminate. Flowers in terminal spikes scarlet, orange or yellow; sepals 3; more or less tubular below; stamens partly adnate to corolla tube, usually 1-5, one bearing single anther cell on the margin of a petaloid stamen, staminodes petaloid, one opposite the fertile stamen secured, other erect; ovary 3-celled, style flattened, stigma terminal. Capsules globose or ellipsoid, echinate, to 2 cm.

Commonly grown as ornamental. Also as an esape especially in moist places in villages, fields. Fl. & Fr.: Throughout the year. Tam.: *Kalvazhalai.*

Way to Koilur, Kolli hills (NMK), *P.G.Diwakar* 97083 (MH); Kiliyur water falls, Yercaud (SLM), *AVNR* 26962 (MH); Rogada (GJM), *JSG* 14004 (MH).

INDIA: Throughout India, Naturalized.

WORLD: Native to Tropical America. Cultivated through out the tropics.

BROMELIACEAE

ANANAS Miller

Ananas comosus (L.) Merr., Interpret. Herb. Amboin. 133. 1917; Matthew, Fl. Tamilnadu Carnatic 3: 1621. 1983. *Bromelia comosa* L., Herb. Amb. 21. 1754. *Ananas sativa* Schult. & Schult.f., Syst., 7: 1283. 1830; Fischer 3: 498. 1928.

Perennial herb, 0.9-1 m tall; stem thick, hidden by leaves. Leaves rigid, mostly densely tufted in a rosette at the base of the stem, linear or linear-lanceolate, 90-150 x 2.5-6.5 cm, margin spiny serrate. Inflorescence a head, generally from the centre of the rosette, terminal with numerous flowers. Flowers fleshy, each with a rough bract, pale blue with a sheen of faint purple. Cone-like syncarp formed by an aggregate of fleshy fruits. Syncarp cylindric, up to 20 cm long, dark green when unripe, dark orange when ripe with some mottling; fruit sometimes seedless under cultivation.

Cultivated in hills of Odisha, hills of Visakhapatnam district of Andhra Pradesh and in Kolli hills of Tamil Nadu for its fruits. Fl. & Fr.: October – June. Vern.: Tel.: *Anasapandu*, Ori.: *Sapuri*; Tam.: *Annasipazham*. English: Pineapple.

Billbergia pyramidalis (Sims.) Lindl. is often cultivated in Paderu Mandal in Visakhapatnam district – *Deepika Divya Kadiri* 21107 (BSID).

AMARYLLIDACEAE

1. Flower solitary, pink-rose; spathe tubular .... **Zephyranthes**
1. Flowers 2-many, umbellate, white; spathe bracteate:
    2. Filaments basally united by a coronal membrane into a toothed cup .... **Pancratium**
    2. Filaments not united by a coronal membrane .... **Crinum**

CRINUM Linnaeus

1. Perianth tube as long as lobes, pinkish-white .... **C. latifolium**
1. Perianth tube exceeding lobes, white:
    2. Leaves > 10 cm wide; umbels 6-12-flowered:
        3. Bulb with stoloniferous base; perianth lobes equaling the linear lobes, which considerably exceed the stamens .... **C. viviparum**
        3. Bulb not stoloniferous; perianth-lobes lanceolate, about as long as the stamens .... **C. amoenum**
    2. Leaves <5 cm wide; umbels 15-50-flowered .... **C. asiaticum**

Crinum amoenum Roxb., Fl. Ind. 2: 127. 1820; FBI 6: 282. 1892; Saxena & Brahmam, Fl. Orissa 3: 1920. 1995.

Large herb, bulbs globose, 5-7.5 cm diam., not stoloniferous. Leaves suberect, ensiform or lorate, 45-60 x 2.5-3.7 cm, apex acuminate, margins subscabrous. Scape 30-60 cm long, rather slender, subcylindric, umbels 6-12-flowered; spathe laceolate, 5 cm; flowers subsessile, white, perianth-tube straight, erect, 7.5 – 11 cm long, lobes lanceolate or linear-lanceolate, 5-7.5 cm long, stellately patent, about as long as the stamens; stamens spreading, filaments red.

Along river banks in Eastern Ghats of Odisha. Fl.: October.

Gurguria-Joshipur, Similipahar (MBJ), *SX* & *MB* 5312 (RRL-B).

Crinum asiaticum L., Sp. Pl. 292. 1753; FBI 6: 280. 1892; Fischer 3: 1504. 1928; Matthew, Fl. Tamilnadu Carnatic 3: 1624. 1983. *Amaryllis carnosa* Hook.f., Fl. Brit. India 6: 280. 1892.

A stout herb; stem with white bulbous root stock, 12 x 10 cm; neck 10-20 cm, clothed with old leaf-sheaths. Leaves linear-lanceolate, to 150 x 10 cm, glabrous, base narrow, margin entire, apex gradually tapering. Scape 0.5 – 1 m long, 1-2 cm across. Flowers white, 15-50 in an umbel; bracts hooded, ovate-lanceolate; perianth-tube white, salver-shaped, lobes oblong-linear, glabrous, cuspidate; stamens reddish, filaments slender, 4 cm, shorter than the lobes of the perianth, ovary 3-celled, style filiform, stigma minute, subcapitate. Fruit sub-globose, 2-5 cm in diameter, 1-(rarely 2)-seeded, dehiscing irregularly.

Throughout Eastern Ghats, up to 1300 m. Often grown as an ornamental. Fl. & Fr.: July – December. Vern.: Tel.: *Vishamungali, Kesarchettu*; Ori.: *Arsa*; Tam.: *Velimoongil; Vishamoongil.*

Isukaguam (KDP), *R.V.Reddy* 8462; Way to Puttagondhilanka-Tadepalli (EG), *NR* 84353 (MH); Rajavomangi (EG), *SS* 4840 (AU); Kedaripuram (VZN), *MV* 6838 (AU); Yanabedhahalla, Billigundala RF (SLM), *EV* 24119 (MH).

INDIA: Throughout India.

WORLD: Sri Lanka. Introduced into America and other countries.

Crinum latifolium L., Sp. Pl. 291. 1753; FBI 6: 283. 1892; Fischer 3: 1504. 1928; Matthew, Fl. Tamilnadu Carnatic 3: 1625. 1983.

A stout herb with large subglobose bulb, 12-15 cm in diameter; neck short, stout, 10 cm. Leaves numerous, 60-90 x 6-10 cm, lorate, acuminate, bright-green, the margins slightly scabrous. Scape inserted on the neck of the bulb, 40 cm long, stout, tinged with purple. Flowers white, fragrant, in 10-20-flowered umbels, bracts lanceolate, pedicels very short, perianth funnel-shaped, lobes white, with pinkish-reddish streaks without, oblong-lanceolate, base attenuate, apex obtuse, long cuspidate, stamens much shorter than the perianth lobes: ovary 3-celled, style longer than the stamens, filiform, ovary with 5-6 superposed ovules in each cell. Capsule globose, 4 x 3.5 cm.

Common in hill areas, up to 1800 m.Fl. & Fr.: June – August

Kekati RF (ATP), *TP 799*; Kalasamudram RF (ATP), *KRKS* 39511; Towards Tanjavanam (VSKP), *GVS* 47309 (MH); Geratli forest (DMP), *EV* 57975 (MH).

INDIA: Throughout India.

WORLD: Sri Lanka, Myanmar, China, Laos, Thailand, Vietnam.

Crinum viviparum (Lam.) R. Ansari & V.J.Nair, J. Econ. Taxon. Bot. 11: 205. 1987. *C. defixum* Ker-Gawl., J. Sci. Arts (London) 3: 105. 1817; FBI 6: 281. 1892; Fischer 3: 1504. 1928. *Amaryllis vivipara* Lam., Encycl. 1: 123. 1783. *A. coenosa* Hook.f., Fl. Brit. India 6: 282. 1892.

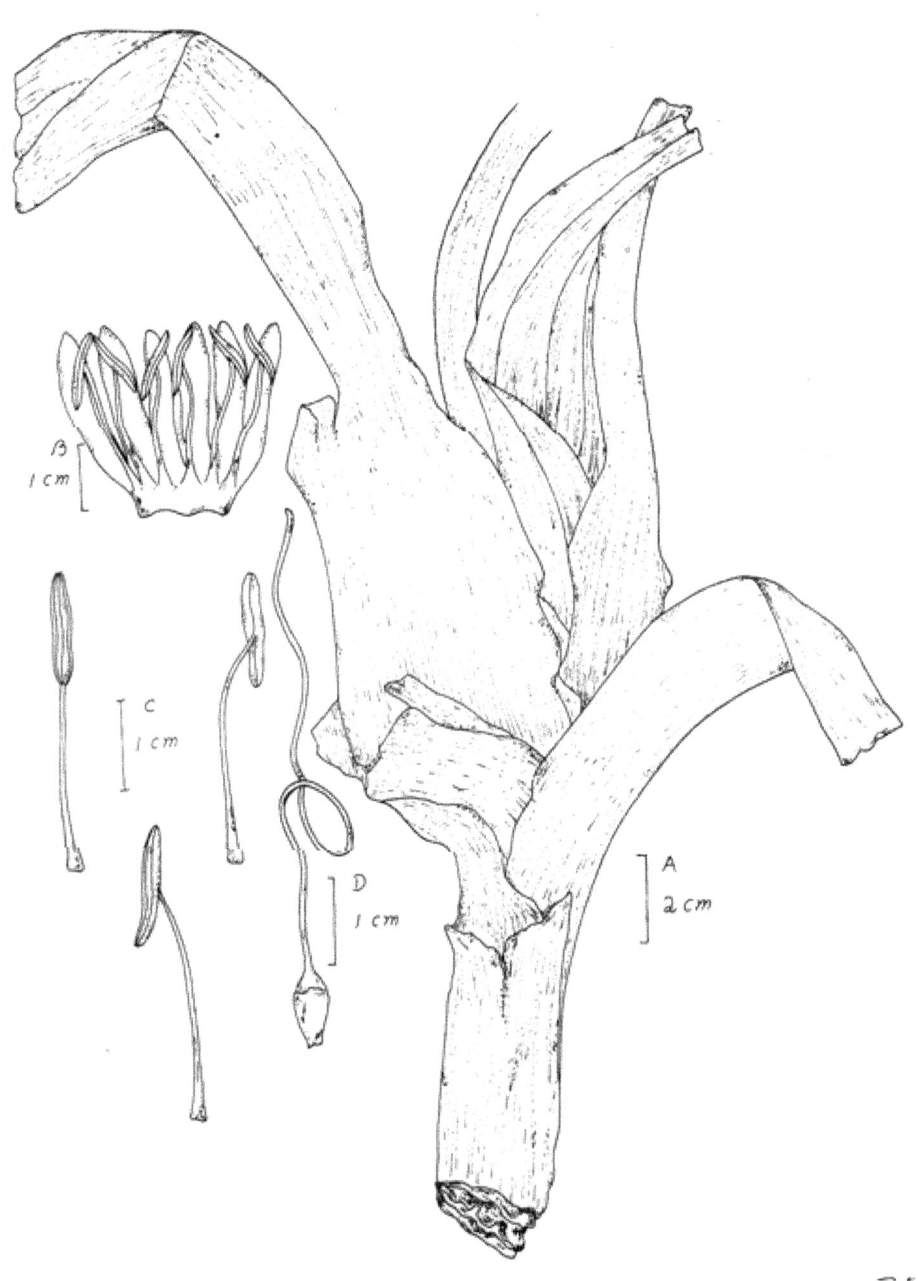

Figure 7. **Crinum latifolium** L.

A. Twig, B. Perianth split open with stamens, C. Stamen, D. Pistil.

Figure 8. **Crinum viviparum** (Lam.) R. Ansari & V.J .Nair
(Syn.: *C. defixum* Ker-Gawl.)
A. Twig, B. Perianth split open with stamens, C. Pistil.

A stout herb with bulbous rootstock; bulb ovoid, 8 x 6 cm, fusiform, stoloniferous base; neck 15 cm long, cylindric. Leaves 6-8 to a bulb, 60-90 cm long, deeply canaliculated, coriaceous, scaberulous, apex gradually tapering, scapes from the axils of the lowest leaves, 45-75 cm long, erect, cylindric. Umbels 14-20 cm long, 10 cm across. Flowers white, fragrant at night, 6-12-flowered, pedicels very short; bracts lanceolate, perianth salver-shaped; lobes white, oblong, glabrous, apex obtuse, cuspidate; stamens 6, bright-red; ovary 3-celled, oblong, style delicate, stigma simple. Capsule subglobose, seeds rugose.

Common along streams and back waters. Fl. & Fr.: November – January. Vern.: Tel.: *Kesarchettu*; Ori.: *Kondai*.

Kalasamudram (ATP), *BSS & KRKS* 39191; Kambagirikonda (KNL), *SS & AMR* 24427; Kothakota RF (ATP), *JSG* 20872 (MH); Mamandur (CTR), *GVS* 31974 (MH); Satyavedu (CTR), *MCB* 45253 (MH); Talakona (CTR), *DRC* 2150 (SVU); Hoganakkal (SLM), *KCJ* 18000 (MH); Sithamur (CPT), *S.India Flora* 11208 (MH); Borapuram (MBNR), *BSS & SKB* 32389 (BSID); Bhadrachalam (KMM), *RCS* 102446 (BSID); Purankotee, Satkosia Wild Life Sanctuary, *D.Hazra & D.Das* 18108 (BSID);

INDIA: Throughout India.

WORLD: Sri Lanka, Bangladesh, Myanmar, Nepal, Thailand, Vietnam.

PANCRATIUM Linnaeus

1. Bulbs with distinct neck; corona cylindrical or conical-trigonous .... **P. bhramarambae**
1. Bulbs without neck; corona infundibular or campanulate or obconic:
    2. Flowers solitary, sessile; spathe lanceolate .... **P. zeylanicum**
    2. Flowers 2-4; spathe ovate:
        3. Perianth tube less than 5 cm long:
            4. Leaves bifarious; scape less than 10 cm long; perianth tube as long as or shorter than lobes .... **P. telanganense**
            4. Leaves not bifarious; scape more than 15 cm long; perianth tube longer than the lobes .... **P. triflorum**
        3. Perianth tube more than 5 cm long .... **P. longiflorum**

Pancratium bhramarambae Sadas., Species 19: 133. 2018.

Perennial herbs; bulb globose, 2–4 cm dia., tunica membranous, pale brown, veined; neck distinct, 4–7 cm long. Leaves hysteranthus, 1-4, bifarious, linear, *c.* 15 x 1 cm, obtuse at apex. Scape terminal, solitary, cylindrical, greenish, thick, 8–12 cm high, with longitudinal grooves; umbels usually 3-4-flowered, rarely 1-2-flowered. Spathe single, tubular at base, hyaline, broadly ovate, *c.* 3 cm long, apex obtuse or rounded, veined. Flowers fade by afternoon. Pedicels sub-sessile to 1 cm long, triangular, greenish. Perianth tube 5.5–8 cm long, narrow and greenish-white below, widened and white above; lobes linear-lanceolate, 2.5–3 x 0.2–0.4 cm, white, light yellow on midrib below, recurved in the middle; corona cylindrical, 8–10 mm long,

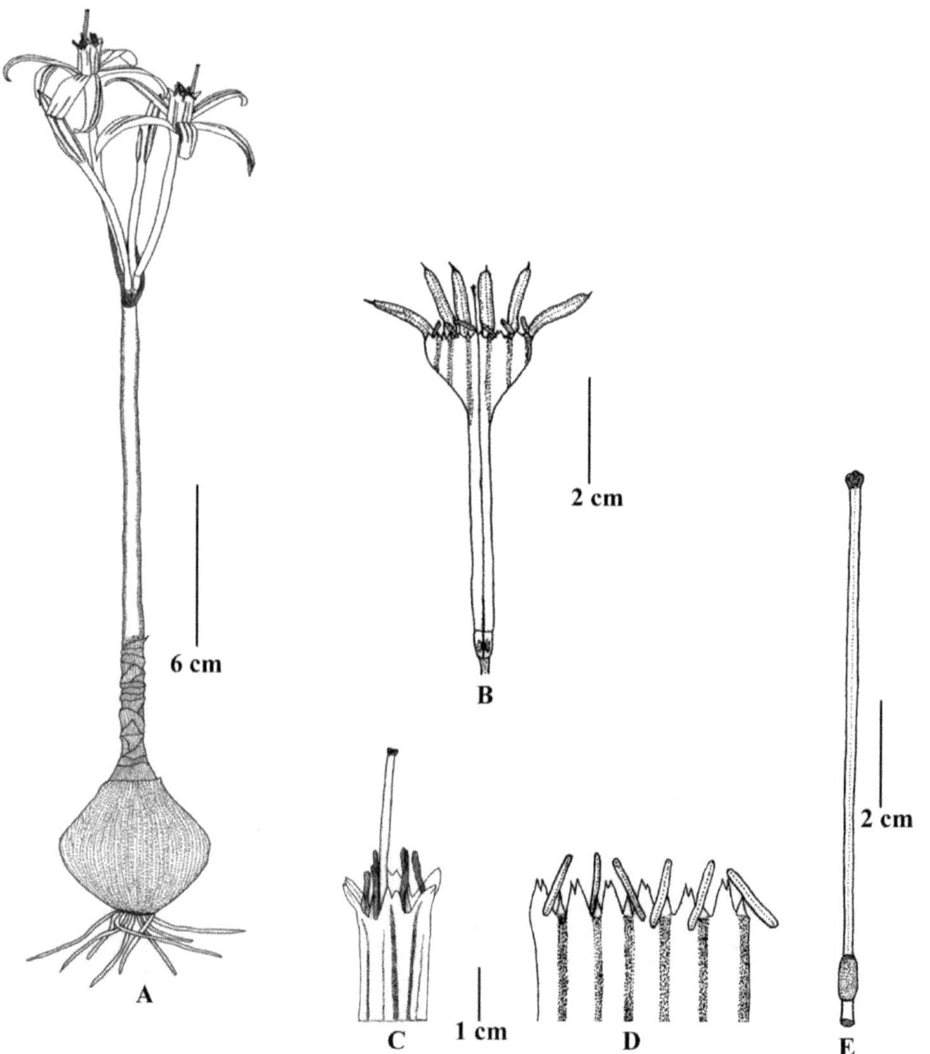

Figure 9. **Pancratium bhramarambae** Sadas.
A. Habit; B. Flower spilt open; C. Corona; D. Corona spilt open; E. Style.

rim 6-toothed; teeth triangular, *c.* 4 mm long, acute or bilobed at apex; filaments subulate, never exceed coronal teeth; anthers unequal, linear, *c.* 6 x 1 mm; ovary ellipsoid-trigonal, 6–9 x 4 mm, green; ovules 5–10 in each locule. Style longer than the tube, filiformis, 6.5–9 cm long; stigma capitate, 3-lobed.

Rare in moist deciduous forests at an altitude range of 700-900 m Nallamala hills (Telangana part) and Seshachalam hills (Andhra Pradesh). Plants sprout after first rains which commence at the end of the April and first week of while scapes with buds appear during first week of June. During second week of June blossoms with white flowers can be seen, even though other associates did not commence the leaf flushings. Before end of the June all the plants complete their flowering and fruiting, stipes become dried off. Then a few new leaves are produced and being fleshy and flexuous for at least a month. All other months, the plants perennate with bulbous underneath of soil in vegetative condition.

Nallamala hills, Mannanur range, near Farahabad (MBNR), *BSS & KP 2018* (holo. CAL!), 2045 (BSID).

INDIA: Telangana, Andhra Pradesh. Endemic.

Pancratium longiflorum Roxb., Fl. Ind. 2: 125. 1832; FBI 6: 286. 1892; Fischer 3: 1505. 1928.

A herb with globose bulb; 4-6 cm diameter, neck long cylindric. Leaves narrowly lanceolate, 20-30 x 1-2.5 cm, usually 1-fid. Scape compressed 1-(rarely 2-) flowered and much shorter than the leaves; spathe single, perianth pale green, lobes linear-lanceolate, staminal cup 2-toothed between the filaments, stamens 6; ovary oblong, 3-celled, ovules many per cell, 2-seriate, style filiform, stigma small, capitate. Capsule subglobose.

In hill areas of Visakhapatnam district. Fl. & Fr.: July – August.

INDIA: Peninsular, C. & E. India.

WORLD: Bangladesh and Malaysia.

Pancratium telanganense Sadas., Species 19: 136. 2018.

Perennial herbs; bulbs *c.* 5 cm dia., tunica membranous, pale brown, veined; neck distinct, 4–6 cm long. Roots fibrous, vermiform, curved upwards. Leaves remain small till the appearance of scapes, bifarious, linear, *c.* 30 x 1.5 cm, obtuse at apex. Scape terminal, solitary, erect, sub-triangular, 6–8 cm high, sulcate, greenish; umbels 2-flowered. Spathe single, hyaline, tubular at base, broadly ovate, *c.* 3.5 cm long, veined, apex bilobed, lobes acuminate. Flowers fade by afternoon; pedicels *c.* 1.4 cm long, triangular, greenish; perianth tube 3–4 cm long, narrow and green below, widened and faint greenish-yellow above; lobes lanceolate, 3–4 x 0.4–0.5 cm, white, midrib light yellow below, rugose near apex, awl shaped at apex; corona infundibular, 1.6–2.0 cm long, rim 12-toothed; tooth triangular-acuminate, 2.5–3.5 mm long; filaments much longer than the teeth, subulate, 1.2–1.8 cm long; anthers unequal, linear-falcate, 8–10 x 1.2 mm; ovary trigonus, 5–6 x 2–3 mm, green; ovules 4-6 in each locule; style filiform, 8–9 cm long, longer than the anthers, green at base, white above; stigma capitate, 3-lobed, whitish. Capsules subglobose, 3-angled, *c.* 1 x 1.2 cm, with persistent perianth; seeds angled, testa black, *c.* 5 x 5 mm.

Grass dominated areas in dry deciduous forests at an altitude range of 600-800 m. This species appear among the grasses in ephemeral nature for short period between June to July every year. Fl. & Fr.: June – July.

Nallamala hills, Mannanur range, near Vatuvarlapalli (MBNR), *BSS & KP* 2017 a (holo. CAL) 2017 b (iso. BSID), 2046 (BSID).

INDIA: Telangana. Endemic.

Pancratium triflorum Roxb., Fl. Ind. 2: 126. 1832; FBI 6: 285. 1892; Fischer 3: 1604. 1928; Matthew, Fl. Tamilnadu Carnatic 3: 1626. 1983; Sadasivaih *et al.*, Asian J. Plant. Sci. Res. 6: 27. 2016. *P. verecundum auct. non.* Ait. 1789; FBI 6: 286. 1892.

A herb with globose bulb, bulb 6 x 5 cm, neck 0 (or) short, 1 cm. Leaves linear-lanceolate, 12-28 x 1-2.5 cm, flat, thin coriaceous, glabrous, base narrow, apex subacute. Scape slender, 12-20 cm long. Flowers fragrant, 3-8 in an umbel, spathe ovate, acute; pedicel very short, perianth funnel-shaped, lobes 6, white, narrow, soon reflexed, stamens 6, filaments connate at base into a petaloid 2-dentate membranous cup; teeth c 7, ovary oblong, 3-celled, ovules many per cell, 2-seriate, style filiform, stigma small, capitate. Capsule subglobose.

In open forests and grasslands. Fl. & Fr.: July – Short lived. Vern.: Ori.: *Ku-kanda.*

Tyada RF (VSKP), *KNR* & *BR* 9894; Tirumalaiah Gutta (MBNR), *L. Paramesh* & *B. Sadasivaiah*, 43897; Similipahar (MBJ), *SX* & *MB* 4376 (RRL-B), *Dutta* & *SX* 371 (RRL-B); Purunakote, Satkosia Tiger Reserve, *KCM* 6567 (BSID).

INDIA: Odisha, Andhra Pradesh, Telangana, Maharashtra.

WORLD: Bangladesh.

Pancratium zeylanicum L., Sp.Pl. 290. 1753; FBI 6: 285. 1894; Chandramohan *et al.*, Indian Forester 143: 1239. 2016.

Tuberous herb, up to 30 cm high; bulb globose, smooth, 4.5 x 3.2. cm, neck 0. Leaves radical, 4-6, thin, linear to lanceolate, contemporary with flowers (synanthous), glossy green, bifarious, 21-24 x 2.1-2.4 cm, slightly curved, apex acute to acuminate. Scape glabrous, subtereete, 12-14 cm long, shorter than leaves, axillary, 1-flowered. Spathe single, membranous, 2-nerved, apically bilobed, lanceolate and tubular at base. Flowers short-pedicelled, white, fragrant; staminal cup broad, up to 3.1 cm wide, perianth tube 3.1-4.3 cm long, light green, shorter than corolla lobes, corolla lobes 6, linear to lanceolate, subequal, recurved at apex, greenish white at back of the lobes; stamninal sheath broadly funnel-shaped, bilobed between long filaments, lobes triangular or deltoid, acuminate at apex; stamens 6, with 2.8 cm long filaments, much longer than the broad shallow 12-toothed staminal cup, slightly curved upwards, anthers light yellow, linear, 0.5 cm long; style 6 cm long; stigma capitates. Capsule ellipsoid, obscurey 3-lobed.

Undergrowth in moist, shady localities and alluvial soils along water courses in northern Eastern Ghats. Fl. & Fr.: April – June.

Karaka (VSKP), *CAB* 1654 (MH); Satkosia Wildlife Sanctuary – Purunakote forest, *KCM* 6568 (BSID).

INDIA: Andhra Pradesh, Odisha, Kerala and Tamil Nadu.

WORLD: Sri Lanka, the Maldive Islands, Borneo, Java, Maluku, Sulawesi and the Philippines.

ZEPHYRANTHES Herbert

Zephyranthes carinata Herb., Bot. Mag. 52. t. 2594. 1825; Matthew, Fl. Tamilnadu Carnatic 3: 1626. 1983.

Bulbous herbs. Leaves linear, arising from ground level. Flowers solitary, pink-rose, tubular, lobes spreading; spathe notched at apex; perianth funnel-form, with a prominent tube; lobes obovate; stamens 6, unequal; ovary inferior; ovules many. Capsule globose, loculicidal; seeds balck, compressed.

Widely cultivated as an ornamental and become naturalized. Fl. & Fr.: Monsoon seasons.

INDIA: Throughout.

WORLD: Native of Mexico and spread everywhere.

Zephyranthes citrina Baker (yellow rain lily) is also cultivated in gardens and home yards.

AGAVACEAE

1. Ovary superior:
   2. Stem present, woody .... **Dracaena**
   2. Stem absent or very short; leaves thick and leathery .... **Sansevieria**
1. Ovary inferior:
   3. Stamens exserted .... **Agave**
   3. Stamens inserted .... **Furcraea**

AGAVE Linnaeus

1. Leaves narrow, linear-oblong, marginal spines hooked, curved upwards .... **A. cantula**
1. Leaves distinctly broader at or above the middle, marginal spines not as above:
   2. Leaves pendant, at least basal leaves .... **A. vera-cruz**
   2. Leaves erect always:
      3. Spine to 1.8 cm, black .... **A. angustifolia**
      3. Spine to 3 cm, grey green .... **A. americana**

Agave americana L., Sp. Pl. 323. 1753; FBI 6: 277. 1892; Fischer 3: 1505. 1928.

A perennial stemless, scapigerous herb. Leaves at least in their upper halves, distinctly exsculptate between the marginal spines, not with horny margins, erecta patent, lower ones partly decumbent, larger ones very often with recurved, rarely incurved upper halves, leaves not combined into dense squarrose crown, lanceolate, with very strong widely patent or recurved marginal spines and 2-5 cm long apical one, thick, rigid, glaucous, very often with longitudinal white or yellow streaks or bands. Panicle (inclusive of peduncle) 4-8 m long; branches numerous, widely patent, sigmoid, apically repeatedly branched, flowers densely crowded, with a strong evil smell; perianth funnel-shaped, segments oblong, yellowish green, stamens exerted. Capsule oblong-clavate, beaked, *c.* 4 cm long; seeds shining, robust.

Grown as a hedge plant in fields and in dry forests, often as escape in wastelands. Fl. & Fl.: September – January. Tel.: *Kithanara*; Ori.: *Murba, Muruga, Barabarasia, Birhot okumari.*

Hampapuram (ATP), *KNR* 10635; Peravali (ATP), *TP & NY* 529; Diguvametta (PKM), *P.Venkata Krishnaiah* 43313; Ambakampalli (KDP), *KRKS & BSS* 40468; Bommalapuram (PKM), *RKM* 0791 (CAL); Bogapuram (VZN), *MV* 3951 (AU).

INDIA: Naturalized in Peninsular Indian plains.

WORLD: Native of Mexico, naturalized in the Mediterranean region, India and Pakistan.

Agave anugustifolia Haw., Syn. Pl. Succ. 72. 1812; var. **angustifolia**. *A. wightii* Drumm, & Prain, Bengal Agric, Bull. 8: 15. 1906 & Agric. Ledger 12(7): 91. 1906; Fischer 3: 1505. 1928. *A. vivipara* Wight, Icon. Pl. Ind. Orient. t. 2024. 1853; FBI 6: 277. 1892.

A perennial, somewhat stout herb; stem 30-45 cm long; woody, simple, with a rosette of leaves. Leaves erect, narrowed at both ends, constricted at base with a spinescently dentate margin, and an aciculate tip with 1-2 cm long greyish green spine. Spines stout, upcurved. Scape stout, scaly, 2-4 m high. Flowers in terminal panicles; perianth 4-5 cm long, perianth segments oblong-lanceolate, green or brown, thin, with a hardened apex, apically with minute hairs; stamens 6, filaments subequl. Capsule oblong, shortly beaked; seeds numerous, dull. Bulbils often produced.

Common in waste lands, along road sides, forest clearings and often planted as hedge plant. Fl. & Fr.: July – September. Vern.: Tel.: *Kithanara*; Ori.: *Baramasi.*

Rampachodavaram (EG), *GVS* 24479 (MH); Towrds Kinchipda (VSKP), *GVS* 19763 (MH).

INDIA: Naturalized in Peninsular India.

WORLD: Native of Mexico and Central America.

Agave cantula Roxb., Fl. Ind. 2: 167. 1832; Fischer 3: 1505. 1928. *A. americana* sensu FBI 6: 277. 1892 (non L. 1753).

A perennial, somewhat stout, scapigerous herb. Stem very short, woody, simple with a crown of rosette leaves. Leaves patent, recurved in the upper part, remotely spinescently dentate with upwardly hooked spines, spines 1-2 cm long, sharp, dark brown; apical spine robust, 1-2.5 cm long. Scapes stout, scaly, up to 5 m high, patently branched at the top. Flowers greenish, in panicles; flowers shortly pedicelled, perianth 6-8 cm long, tube widened upwards, with 6 longitudinal grooves; segments oblong, on the innerside of the top with a tuft of hairs; filaments brown-blotched. Bulbils present.

Common in waste places road sides, often planted as hedge plant. Fl. & Fr.: July – September.

Polavaram (EG), *CAB* 2037 (CAL).

INDIA: Naturalized in many parts of India.

WORLD: Native of Mexico. Planted and naturalized in Tropical Asia.

Agave vera-cruz Mill., Gard. Dict. ed. 8: 7. 1768. *A. vera-crucis* Haw., Syn. Pl. Succ. 72. 1812; Jacobi, Vers. Syst. Ord. Agav. 111. 111. 1865. *A. vernae* A. Berger, Agaven 245. 1915. *A. breviscapa* A. Berger ex Roster, Bull. Soc. Tosc. Ortic. 4(1): 36. 1916. *A. cyanophylla* Jacobi, Hamburger Garten- Blumenzeitung 22: 175. 1866. *A. haworthiana* M. Roem., Fam. Nat. Syn. Monogr. 4: 290. 1847. *A. lepida* D. Dietr., Syn. Pl. 2: 1192. 1840. *A. lurida* Aiton, Hort. Kew. 1: 472. 1789. *A. magni* Desf., Tabl. École Bot. ed. 2: 33. 1815. *A. mexicana* Lam., Encycl. 1: 52 1783. *A. polyphylla* K.Koch, Wochenschr. Vereines Beförd. Gartenbaues Königl. Preuss. Staaten 3: 38. 1860.

Stem reduced, covered with leaf bases; rhizome thick, cylindrical, erect, up to 1 m long. Leaves forming a lax rosette, arching upwards, the ends drooping, 1.5-2 m, linear-oblong, widest just above the middle, neck hardly constricted; apical spine 1-1.5 cm, almost black; marginal spine broad, spreading or pointing downwards, blackish-brown. Leaves green, often glaucous sometimes with a whitish bloom, shallowly channelled. Scap stout, up to 7 m long.

Rarely cultivated in land margins and sometimes leaf fiber is extracted. Fl. & Fr.: July – October. Tam.: *Railkatralai.*

Punagaadu (NA), *EV* 54578 (MH).

INDIA: Tamil Nadu.

WORLD: Native of Mexico, introduced to Europe, North West Africa, Mauritius, Pakistan and Sri Lanka.

DRACAENA Linnaeus

Dracaena terniflora Roxb., Fl. Ind. 2: 159. 1824; FBI 6: 328. 1892; Fischer 3: 1521. 1928. *D. terniflora* var. *heyneana* Hooker.f., Fl. Brit. India 6: 329. 1892.

A slightly branched glabrous straggling shrub, stem scarcely more than 0.6 cm thick, sometimes rooting at the base. Leaves somewhat crowded, elliptic-lanceolate, 10-20 x 3-5 cm, thinly coriaceous, apex acute or acuminate, base narrowed into the petiole; petioles 2.5-6 cm long, with widened amplexicaul base. Flowers white, often 2-3 together on simple 10-20 cm long raceme, jointed in the middle; bracts scarious, ovate, acute, perianth lobes white, linear, obtuse; stamens 6, adnate to the base of the perianth tube; ovary 3-celled, single ovule in each cell. Berry globose, red, 1-3-seeded; seeds globose.

Occasional in hills. Fl. & Fr.: May – September. Vern.: Tel.: *Bhoda dishti.*

Tirumala (CTR), *SSR* 16182, *GVS* 31910 (MH); Maredumilli RF (EG), *BR & KNR* 10609; Tirumalai (CTR), *GVS* 31910 (MH); Rampa hills (EG), *JSG* 15951 (MH); Vankachinta (VSKP), *GVS* 42606 (MH); Minumuluru towards Paderu (VSKP), *GVS* 30064 (MH); Bokoi hill, Lakshmipuram (VSKP), *NRR* & *DN* 84235 (BSID); Similipahar (MBJ), *SX* & *MB* 5200 (RRL-B); Jenabil, Similipahar (MBJ), *A.R.K.Sastry* & *G.P.Singh* 12373 (BSID); Dhuanalai, Puri forest division, *Yoganarasimhan* 077 (RRL-B); Berbera forest, Puri, *Yoganarasimhan* 072 (RRL-B); Kwadoli, Raigoda, Satkosia Tiger Reserve, *KCM* 6717 (BSID).

INDIA: Peninsular & N.E.India.

WORLD: Bangladesh, China, Malaysia, Thailand.

FURCRAEA Ventenat

Furcraea foetida (L.) Haw., Suppl. Pl. Succ. 73. 1819; Matthew, Fl. Tamilnadu Carnatic 3: 1631. 1983. *Agave foetida* L., Sp. Pl. 323. 1753. *Furcraea gigantea* Vent., Bull. Sci. Soc. Philom. Paris 1: 65. 1793; Fischer 3: 1505. 1928.

A perennial herb; superterraneous stem short or long, woody. Leaves densely crowded, squarrose, lanceolate or linear-lanceolate, acute, tough-fibrous, spinous-dentate or entire. Flowers in terminal, large, widely branched panicles; flowers along the rachises of the inflorescence fascicled or solitary, shortly pedicelled, pendulous; perianth constricted above the ovary, chorophyllous, tepals 6, erecto-patent, subequal, the inner ones broader; stamens 6, inserted at the base of the tepals, erect, filaments in the middle with a spongy thickening, anther medifixed; ovary 3-celled, cells numerous-ovuled, style basally much thickened and triquetrous-trilobed, higher up filiform-subulate, stigma small. Capsule ellipsoid, trigonous, loculicidally 3-valved, seeds numerous, flat.

Occasional in dry tracts. Fl. & Fr.: August – October. Vern.: Tel.: *Morga, Kithali.*

Bakarapeta (CTR), *KNR* & *DAM* 9841; Anjodigedda (VSKP), *TP* & *EC s.n.;* Anantagiri (VSKP) *GVS* 32823 (MH); Near Dabuguda, Lakshmipuram (VSKP), *NRR* & *DN* 84208 (BSID); Chulliyar shola, Pannaikadalu (DGL), *K.Ravi Kumar* 92521 (MH); Chinnakalrayan (SLM), *N.Venugopal* 14575 (RHT, MH); Kodagarai (DMP), *TRS* 84121 (MH, BSID); Anchetty forest (DMP), *EV* 57932 (MH):

INDIA: N. to E. India.

WORLD: Colombia,Venezuela north to Costa Rica and the Caribbean.

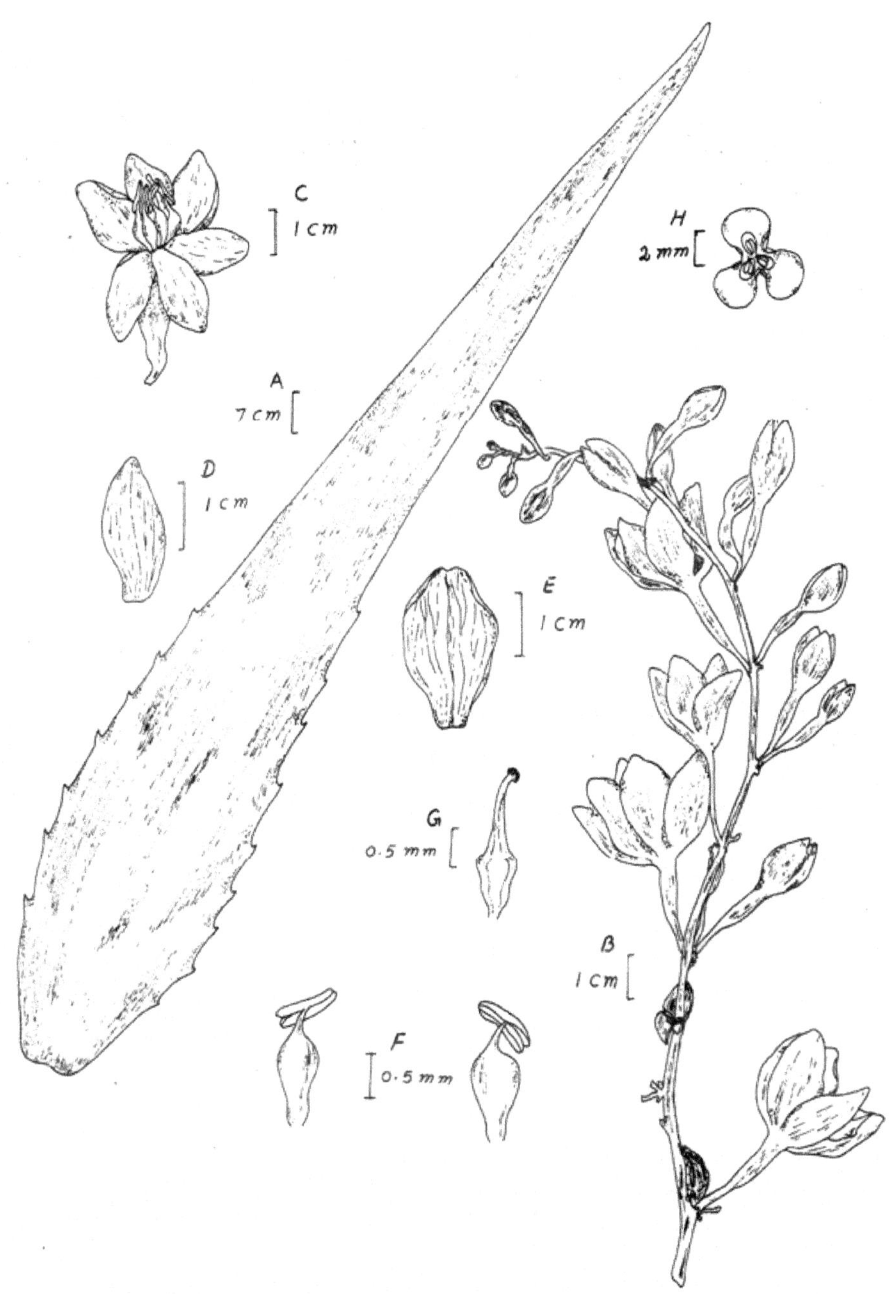

Figure 10. **Furcraea foetida** (L.) Haw.

A. Leaf, B. Inflorescence, C. Flower, D. Outer tepal, E. Inner tepal, F. Stamens, G. Pistil, H. Fruit.

SANSEVIERIA Thunberg *nom. cons.*

Sansevieria roxburghiana Schult. & Schult. f., Syst. 7: 357. ff. 12 D & E. 1829; FBI 6: 271. 1892; Fischer 3: 1520. 1928; Matthew, Fl. Tamilnadu Carnatic 3: 1631. 1983. *S. zeylanica* Roxb., Pl. Coromandel. t. 184. 1805,(non (L.) Willd. 1799).

A herb with very short stem or stemless; root stock creeping. Leaves about 8 or 9 in a tuft, 75-90 x 2-5 cm, towards the middle, suberect, rigid, pale green with transverse bands of dark green, thick, with spine-like tip. Flowers greenish-white tinged with violet, in fascicles of 3-6 on 30-60 cm long racemes, bracts membranous: perianth-tube slender, lobes 6, long, narrow; stamens 6, on the perianth tube, filaments filiform, anthers dorsifixed; ovary 3-celled, ovules solitary in each cell, erect, style filiform, stigma simple. Fruit memberanous, indehiscent; seeds 1-3, large, globose, fleshy, ripening outside the pericarp.

Occasional throughout Eastern Ghats both in forests and waste lands. Fl. & Fr.: June – August. Vern.: Tel.: *Nela kithala, Yerra jaga;* Ori.: *Murga.*

Pennahobilam (ATP), *TP* 1276; Vengalammacheruvu RF (ATP), *TP & NY* 2120; Bukkarayasamudram (ATP), *KNR* 10641; Erramalai hills (KNL), *TP & RVR* 1827; Palakonda hills (KDP), *CS* 9395; Kukkalakunta (CTR), *PANR* & *AJR* 20427; Peddamantanala (PKM), *RVK* 15835; Macherla RF (GNT), *VRK* 3820; Thumburatheertham (CTR), *J.Swamy* & *S.Nagaraju* 7428 (BSID); Tirumala (CTR), *GVS* 31922 (MH); Near Yeguvaguda, Donubai (SKLM), *NRR* 83608 (BSID); Purunakote Satkosiaa Tiger Reserve, *KCM* 5211 (BSID); Hogainakkal forest (KGR), *EV* 20607 (MH); Yercaud (SLM), *SKK* 26872 (MH).

INDIA: Southern and Eastern India.

WORLD: Pakistan.

HYPOXIDACEAE

1. Ovary produced upwards into a filamentous rostrum........................ **Curculigo**
1. Ovary not or hardly produced upwards into a rostrum:
    2. Leaves sessile .... **Hypoxis**
    2. Leaves petiolate .... **Molineria**

CURCULIGO Gaertner

Curculigo orchioides Gaertn., Fruct. 1: 63. t. 16. f. 11. 1788; FBI 6: 279. 1892; Fischer 3: 1502. 1928; Matthew, Fl. Tamilnadu Carnatic 3: 1626. 1983.

A plant with stout root stock, short or elongated (sometimes 30 cm long) with fleshy root fibres. Leaves lanceolate, 5-20 x 0.8-1.5 cm, membranous, plicate, glabrous or sparsely softly hairy, the tips sometimes rooting and reaching the ground, base sheathing, margin entire, apex acuminate; petiole usually short or often absent; scapes 3-5 or more, slender, hidden by leaf-sheaths. Flowers bright-yellow, in 8 cm long raceme, distichous, spathaceous, glabrous, densely imbricating; perianth 6-lobed to the base, lobes yellow, oblong-elliptic, outer ones pilose without, inner ones sparsely pilose along nerves; stamens 6, filaments filiform; ovary 3-celled,

Figure 11. **Sansevieria roxburghiana** Schult. & Schult. f.

A. Habit, b. Perianth split open with stamens, C. Perianth lobe with stamen, D. Stamens, E. Pistil.

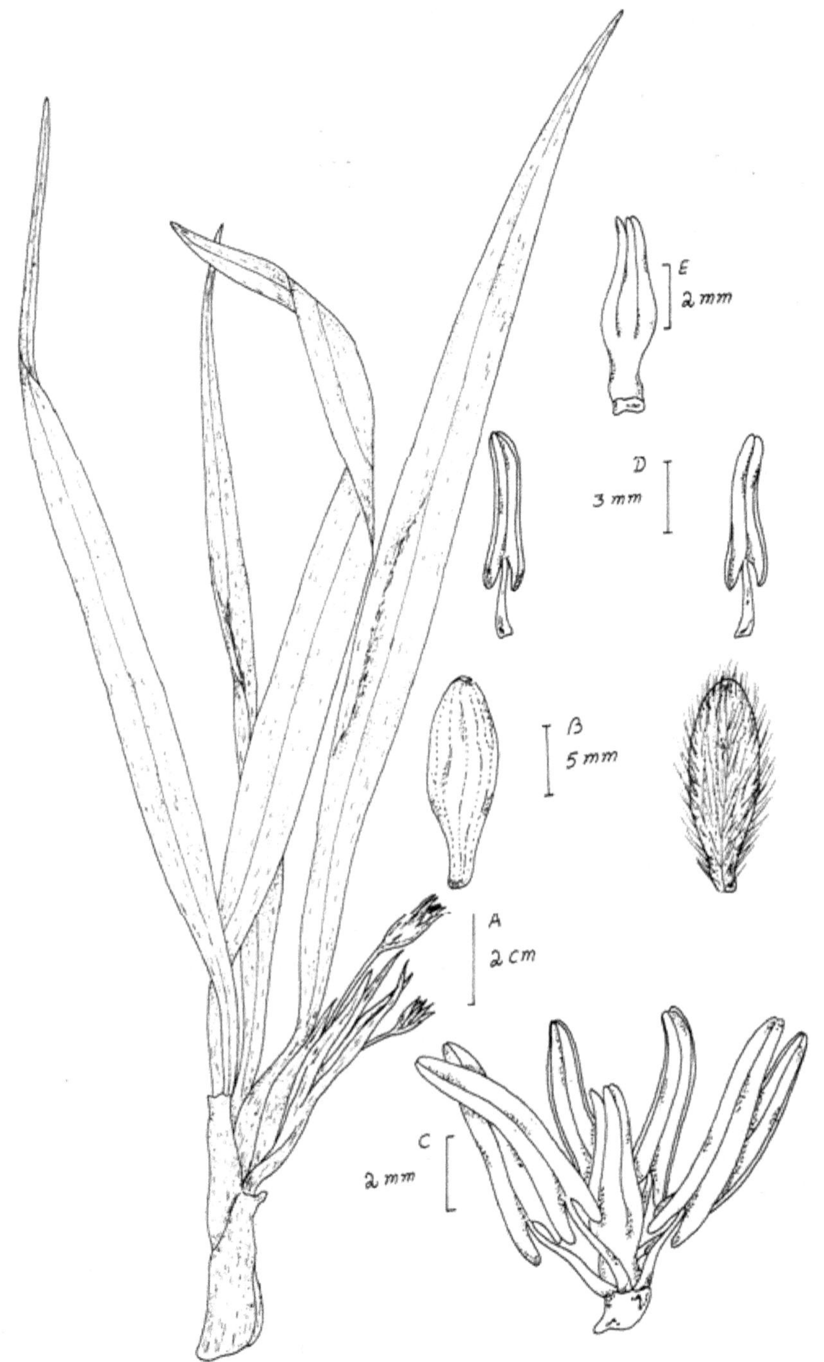

Figure 12. **Curculigo orchioides** Gaertn.

A. Twig, B. Perianth lobes (outer and inner), C. Stamens with pistil, D. Stamens, E. Pistil.

oblong, stigmas 3, lobes elongate. Berry oblong, glabrescent, pericarp membranous; seeds *c.* 8, globose, glossy, beaked.

Abundant on forest floor, moist shady places. Fl. & Fr.: Throughout the year. Vern.: Tel.: *Neella thati;* Ori.: *Talmuli, Mania kanda, Mainikuakenda.*

Kalasamudram (ATP), *KRKS* & *BSS* 39183; Thalakona (CTR), *KNR* & *PSPB* 9850; Penchalakona (NLR), *SSY* 16236; Chinnarutla (PKM), *RVK* 15801; Nallamalais (KNL), *CAB* 8050 (MH); Chelama (KNL), *JLE* 16708 (CAL); Talakona RF (CTR), *JHFB* 112960 (BSID); Thotamamidi, near Devarapalli (EG), *NRR* 76816 (BSID); Chintapalli (VSKP), *GVS* 28146 (MH); Donubai (SKLM), *GVS* 62459 (MH); Vatavarlapalli (MBNR), *VBH* 84956 (BSID); Ratham Hutta hills (KMM), *RCS* 104232 (BSID); Sirumalai (DGL), *S.Indian Flora* 9023 (MH); Guindy deer park (MDS), *ANH* 45627 (MH); Gingee RF (SA), *KRM* 13064 (MH); Shanikulam (SA), *CAB* 906 (MH); Melpat (SA), *CAB* 946 (MH); Koradiyur via Nagalur, Yercaud (SLM), *AVNR* 26786 (MH); Pampasar RF, Satkosia Tiger Reserve, *KCM* 5114 (BSID); Satkosia Wild Life Sanctuary, *D.Hazra* & *D.Das* 18155 (BSID); Mahendragiri (GJM), *VNS* 5577 (MH); BR Hills (Mysore dist.), *VBH* 96825 (MH).

INDIA: Peninsular & N.E.India, Andaman & Nicobar Islands, Subtropical Himalayas.

WORLD: Pakistan, Myanmar, China, Cambodia, Indonesia, Japan, Laos, Papua New Guinea, Philippines, Thailand, Vietnam.

HYPOXIS Linnaeus

Hypoxis aurea Lour., Fl. Cochinch. 1: 200. 1790; FBI 6: 277. 1892; Fischer, 3: 1502. 1928.

A herb with a tuberous root stock or a coated corm; root stock subglobose or elongate and erect, crowned with the fibrous remains of old leaves; leaves 6-12, narrowly linear, 10-25 x 0.5 cm; subcoriaceous, acute, keeled; scapes 1-4, filiform, 2-10 cm long, 1-2-flowered, nearly glabrous or sparsely clothed with pale brown hairs; perianth-segments yellow, thinly hairy, elliptic-lanceolate; anthers sagittate; ovary broadly clavate, clothed with golden-brown, short, shining hairs. Capsule with thin walls, oblong or clavate, ultimately 3-valved, crowned with the erect perianth-segments; seeds black, finely tuberculate.

Rare in hills above 100 m, in shady localities. Fl. & Fr.: June – September.

Ellivada (EG), *SS* 8353 (AU); Mahendragiri (GJM), *VNS* 5753 (MH), *SX* & *MB* 2154 (RRL-B).

INDIA: Throughout India.

WORLD: Pakistan, Myanmar, Nepal, Korea, Laos, Papua New Guinea, Philippines, Thailand, Vietnam.

MOLINERIA Colla

1. Racemes lax-flowered, erect .... **M. trichocarpa**
1. Racemes subcapitate .... **M. capitulata**

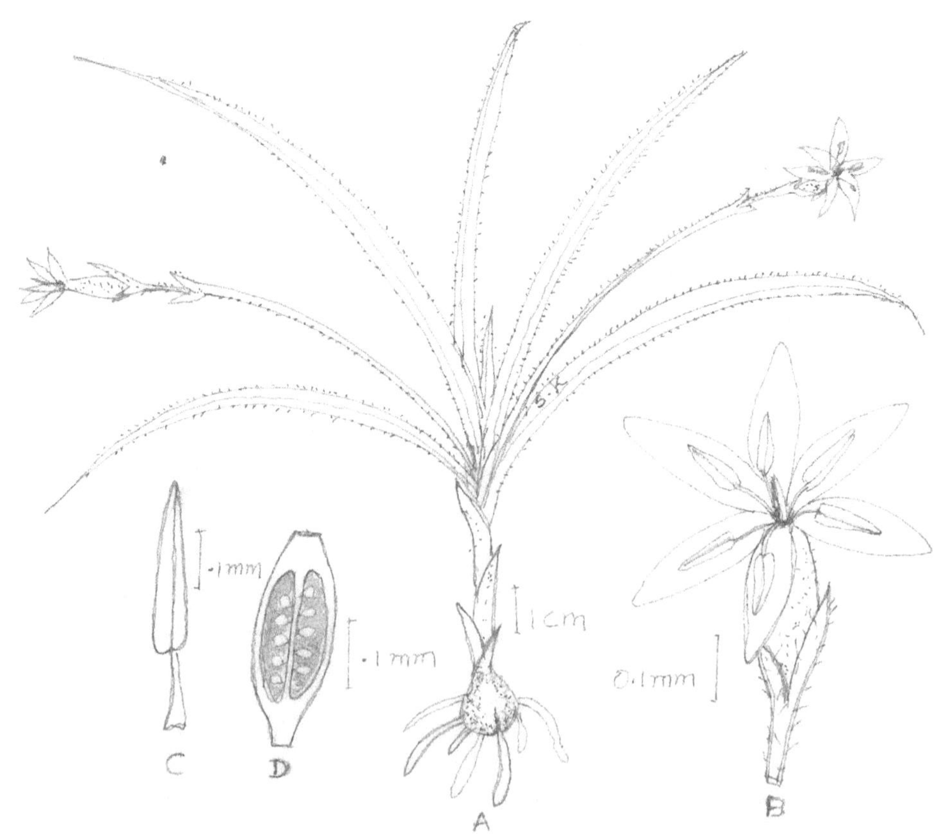

Figure 13. **Hypoxis aurea** Lour.

A. Habit; B. Flower-entire; C. Stamen; D. Ovary-l.s.

Molineria capitulata (Lour.) Herb., Amaryllidaceae 84. 1837. *Leucojum capitulatum* Lour., Fl. Cochinch. 199. 1790. *Curculigo capitulata* (Lour.) Kuntze, Revis. Gen. Pl. 2: 703. 1891; Saxena & Brahmam, Fl. Orissa 3: 1935. 1995. *C. recurvata* Dryand in Aiton Hort. Kew. ed. 2. 2: 253. 1811; FBI 6: 278. 1892.

An erect, large herb, root stock tuberous. Leaves large, lanceolate, plicate, palm like, nerves hairy beneath, 60-90 x 7.5-19 cm; petiole 30-60 cm long, channeled and hairy below. Flowers yellow, 1.5-1.8 cm across, in dense heads, villous; bracts ovate-lanceolate, acuminate, more or less hairy; pedicels 6.2 mm; perianth-lobes up to 1 x 0.5 cm, ovate, pilose without; stamens 6, adnate to the base of the perianth-lobes, filaments short, anthers cohering; ovary turbinate, scarcely beaked, stigma capitate. Berry globose, 6.2-7.5 mm diam., hairy; seeds black, deeply grooved.

Common near streams, occurring only in northern Eastern Ghats. Fl. & Fr.: October – January.

Bokai hills near Laxmipuram (VSKP), *NRR* 83698 (MH, BSID); Ampodar valley, (KHD), *HFM* 3208 (DD).

INDIA: Tropical Himalaya, E.India, Andhra Pradesh, Andaman & Nicobar Islands.

WORLD: Sri Lanka, Pakistan, Bangladesh, Myanmar, Nepal, China, Malaysia, Japan, Indonesia, Papua New Guinea, Philippines, Thailand.

Molineria trichocarpa (Wight) N.P. Balakr., J. Bombay Nat. Hist. Soc. 63: 330. 1967; Matthew, Fl. Tamilnadu Carnatic 3: 1628. 1983. *Hypoxis trichocarpa* Wight, Icon. Pl. Ind. Orient. t. 2045. 1853. *Curculigo trichocarpa* (Wight) Bennet & Raiz., Indian J. Forest. 4: 68. 1981; Saxena & Brahmam, Fl. Orissa 3: 1936. 1995. *Molineria finlaysoniana* Wall. ex Baker, J. Linn. Soc. 17: 121. 1878. *Curculigo finlaysoniana* (Wall. ex Baker) Hook. f., Fl. Brit. India 6: 279. 1892.

A perennial herb with erect rhizome, elongate. Leaves lanceolate, 8-24 x 1.5-3 cm, flat-plicate, thin-coriaceous usually glabrous, rarely pilose, base narrow, apex gradually acuminate; petiole usually very elongate, 10-25 cm, rarely reduced, leaf-sheath membranous. Raceme 5-8-flowered, bracts linear, elongate, densely pilose; pedicel filiform, curved. Flower usually bisexual, sometimes male ones at the top; perianth lobes 6, yellow, oblong-elliptic, persistent; stamens 6, ovary 3-celled, ovules numerous per cell, style columnar, stigma capitate. Berry ellipsoid; seeds globose.

Hills above 1000 m, in moist, shady localities. Fl. & Fr.: June – July.

Batrepalli (ATP), *BSS* & *KRKS* 39603; 13th Km from Maredumilli towards Chinturu (EG), *GVS* 68509 (MH); Bodalanka (EG), *DN* 85563 (BSID); Maredumilli (EG), *MM* 105012 (MH, BSID); Forest near Sunkarimetta (VSKP), *NPBK* 10918 (MH); Anantagiri (VSKP), *GVS* 21726 (MH); Thadiguda (VSKP), *GVS* 44259 (MH); Kurupam (VZN), *MV* 2082 (AU); Kuthadya hill (GJM), *VNS* 5911 (MH); Similipahar (MBJ), *SX* 3786 (RRL-B), *SX* & *MB* 4610 (RRL-B).

INDIA: Peninsular India.

WORLD: Sri Lanka.

TACCACEAE

TACCA Forster *nom. cons*

Tacca leontopetaloides (L.) Kuntze, Revis. Gen. Pl. 3: 311. 1893. *Leontice leontopetaloides* L., Sp. Pl. 313. 1753. *Tacca pinnatifida* Forst., Char. Gen. 70. f. 35. 1776; FBI 6: 287. 1892; Fischer 3: 1506. 1928.

A plant with root stock globose, up to 30 cm in diameter; rootlets superficial. Leaves 30-90 cm in diameter, circular in outline, 3-partite, the segments variously pinnatifid, margins undulate; petioles 30-90 cm long, terete, with pale and dark green stripes. Flowers pedicellate, drooping, green, tinged with purple, involucral bracts 6-12, oblong-lanceolate, acuminate, recurved, striped with purple, bracteoles filiform, longer than the bracts; perianth subglobose, greenish, lobes margined with purple, connivent; stamens 6, adnate to the perianth tube, filaments short; ovary, inferior, 3-angular, 1-celled, ovules many, on 3 parietal placentas. Fruit globose, ovoid or oblong, 3-6-ribbed, forming an indehiscent berry; seeds numerous, ovoid, compressed, longitudinally striate.

More or less common throughout Eastern Ghats. Fl. & Fr.: June – November. Vern.: Tel.: *Pedda kanda gadda*; Ori.: *Dhoi*.

Balugram RF (KNL), *TP & RVR* 2754; Rollapenta (PKM), *RVK* 15829; Maredumilli (EG), *KSK* 23450; Way to Vishnunandi-Mahanandi (KNL), *JLE* 25440 (MH); Gundlabrahmeswaram (KNL), *JLE* 16939 (MH); Bison Hills (EG), *CAB* 5029 (MH); Tiger camp, Maredumill (EG), *MM* 102573 (BSID); Sesharayi (EG), *SS* 9981 (AU); Barnakonda RF (VSKP), *KCJ* 17167 (MH); Kurupam (VZN), *MV* 2010 (AU); Tulka RF, Satkosia Tiger Reserve, *KCM* 5915 (BSID).

INDIA: Peninsular, C. & E.India, Andaman & Nicobar Islands.

WORLD: Widely spread in tropical areas, either as a native plant or naturalized, from Africa, through Asia to Australia and the Pacific.

DIOSCOREACEAE

DIOSCOREA Linnaeus

1. Leaves compound:
    2. Fertile stamens 6 .... **D. hispida**
    2. Fertile stamens 3:
        3. Leaves glabrous .... **D. pentaphylla**
        3. Leaves tomentose .... **D. tomentosa**
1. Leaves simple:
    4. Stems twining to left:
        5. Tubers numerous, stalked .... **D. esculenta**
        5. Tubers 1-3, sessile .... **D. bulbifera**
    4. Stems twining to right:

6. Stems densely pubescent .... **D. pubera**
6. Stems glabrous:
    7. Axis of the male spikes zig-zag:
        8. Leaves lanceolate to ovate; bulbils 0; stem angled, not alate, wild .... **D. hamiltonii**
        8. Leaves broadly ovate; bulbils large; stems 4-angled or 4-alate, cultivated .... **D. alata**
    7. Axis of the male spikes not zig-zag:
        9. Bases of leaves acute-rounded .... **D. oppositifolia**
        9. Bases of leaves cordate:
            10. Secondary nerves of leaves irregular .... **D. belophylla**
            10. Secondary nerves of leaves regular .... **D. wallichii**

Dioscorea alata L., Sp. Pl. 3: 1033. 1753; FBI 6: 296. 1892; Fischer 3: 1512. 1928. *D. globosa* Roxb., Fl. Ind. 3: 797. 1832; FBI 6: 296. 1892.

Stout twiner, stem twining to the right, tuber large, shallow or deep underground but without long stalks, stem compressed or strongly 4-angled at the base and sometimes with scattered prickles below, and 4-many-winged above. Bulbils large, germinate, often many on special branches, mostly oblong, attaining 7.5 x 2.5-5 cm, brown with a tessellated or longitudinally cracked corky surface. Leaves mostly opposite, glabrous, lower very broadly ovate-cordate with very broad sinus, up to 20 x 15 cm, suddenly cuspidate, 9-costate with lower costa forked, transverse secondary nerves subscalariform, upper leaves smaller and narrower, those on the flowering branches often lanceolatee, to 7.5 x 3 cm or less, acuminate, 5-costate of which 3 reach the apex, base cordate; petiole bases sometimes with scattered prickles. Male spikes with zigzag winged rachis, 1-1.8 cm long, 2-3-nate or subverticillate on axillary branches, one flower at each angle of the spike. Flowers subglobose, 1.2 mm, outer perianth lobes elliptic-oblong, rounded and concave, 1.2 mm long, inner shorter, obovate; stamens 6 perfect, short, in central column around small pistillode. Female spikes 10-20 cm long, axillary, solitary, sometimes forming large brachiate panicles by suppression of upper leaves; outer perianth lobes to 3 mm long, thick, boat-shaped, but sharply convex, rather than keeled outside, concave within, inner perianth lobes broadly obovate or subquadrate, very thick and fleshy. Young fruits shortly, stoutly beaked and wings with thickened margins. Capsules broadly obcordate, 2.5-3.7 cm wide.

Commonly cultivated. Fl. & Fr.: October – December. Vern.: Tel.: *Pendalamu, Dukka pendalam, Kavili gadda;* Ori.: *Khambo alu, Kham alu.*

On the way to Yeguvaguda, Seedhi (SKLM), *NRR* 83601 (MH),

INDIA: Cultivated.

WORLD: Cultivated in tropics.

Dioscorea belophylla (Prain) Haines, For. Fl. Chota Nagpur 530. 1910. *D. nummularia* var. *belophylla* Prain, Bengal Pl. 2: 1065, 1067. 1903. *D. glabra* auct non Roxb.: FBI 6: 294. 1892; Fischer 3: 1512. 1928.

A vine; stem unarmed, enlarged into a small rhizome at base emitting long fleshy tuber-bearing fibres. Leaves variable, usually ovate or tapering to an acute point, 7-9-ribbed, veins rather regular, close and parallel, petioles 2-4 cm long. Male flowers in short spikes in the axils or on leafless axillary shoots; sepals ovate-oblong;, petals cuneately obovate; pistillode minute. Capsules sub-orbicular, to 5 x 4 cm, slightly broader than long.

Occasional in northern Eastern Ghats. Fl. & Fr.: October – December. Vern.: Tel.: *Naratega;* Ori.: *Bhat kando, Parlu, Korondi alu, Kunda alu, Tar kanda, Malara.*

Maredumilli (EG), *BR & KNR* 10608, *GVS* 67582 (MH); Gidikuppa-Godavari river (EG), *CAB* 5342 (MH); Araku (VSKP), *GVS* 21573 (MH).

INDIA: Peninsular & E. India, Himalaya.

WORLD: Pakistan.

Dioscorea bulbifera L., Sp. Pl. 1033. 1753; Matthew, Fl. Tamilnadu Carnatic 3: 1634. 1983. var. **bulbifera**; Fischer 3: 1511. 1928. *D. sativa* Thunb., Fl. Jap. 151. 1784 (non L. 1753); FBI 6: 295. 1892.

A vine; tubers solitary or 2-3, large, globose; stem terete, unarmed, glabrous, bulbiferous in leaf axils; bulbils warted. Leaves broadly ovate-sub-orbicular, 8-20 x 4-12 cm, dark green, base deeply cordate, margin entire, apex acuminate to shortly caudate. Male flowers variable in size, green or purplish in 3-10 cm long spikes, close or scattered on long, pendulous, axillary panicles; bracts ovate, acuminate; perianth sessile by a broad base; stamens 6, filaments short, anthers minute, didymous, pistillode 3-lobed. Female flowers sessile, axillary, solitary or fascicled, pendulous spikes, perianth sessile; staminodes 6; bracts below the ovary minute, ovate, acuminate, style short, conical, stigmas 3, very short. Capsules recurved, quadrately oblong, slightly widened upwards, seeds winged at the base only.

Common in deciduous forests. Fl. & Fr.: September – December. Vern.: Tel.: *Malakakaya pendalamu, Chedupaddu dumpa;* Ori.: *Pita kanda, Pita alu;* Tam.: *Vethalavalli.*

Krishna Nandi (KNL), *KNR* 9863; Papanasanam (CTR), *BR* & *BSS* 30486; Rollapenta (PKM), *RVK* 15830; Modugamamisala (KNL), *JLE* 42195 (MH); Way to Mantur from Devipatnam (EG), *GVS* 67529 (MH); Jidikuppa-Godavari river (EG), *CAB* 5343 (MH); Rajavommangi (EG), *SS* 742 (AU); Araku valley (VSKP), *NPBK* 10750 (MH); way to Javapuram konda (VSKP), *GVS* 42516 (MH); Srungavarapukota (VJN), *GVS* 21810 (MH); Murugampadu RF (KMM), *RCS* 102451 (BSID); Nayakkanur, Javadi hills (NA), *MBV* 996 (MH); Jamnamarathur (NA), *MBV 886* (MH); Curangi (GJM), *VNS* 5791 (MH); Boragharo (GJM), *VNS* 5950 (MH); Purunkote RF, Satkosia Tiger Reserve, *KCM* 5180, 5604 (BSID).

INDIA: Peninsular & E. India, Andaman & Nicobar Islands.

WORLD: Africa, Oceania, Myanmar, Bhutan, China, Cambodia, Japan, Korea, Thailand, Vietnam.

Dioscorea esculenta (Lour.) Burkill, Gard. Bull. Straits Settlem. 1: 396. 1917; Fischer 3: 1511. 1928. *Oncus esculentus* Lour., Fl. Cochinch. 1: 194. 1790. *Dioscorea acubata* sensu Hook. f., FBI 6: 296. 1892 (non L., 1753). *D. spinosa* Roxb. ex Hook. f., Fl. Brit. India 6: 291. 1892.

A vine, tubers very large, edible, stalked; base of the stem with long woody rigid fibres bearing 1.5 cm long spines. Leaves orbicular or reniform, 10-20 x 8-16 cm, acuminate or cuspidate, 5-7-nerved, rather membranous, basal lobes rounded. Male flowers often in very dense cymules, sessile or shortly pedicelled; bracteoles very broad; perianth segments remote from the large oblong pistillode; perfect stamens 6, anthers large. Female flowers in short racemes. Capsules oblong, slightly narrowed below, apex retuse; seeds broadly winged.

Cultivated, also as an escape. Fl. & Fr.: September – December.

Way to Punyagiri (VZN), *BR & KNR* 9882.

INDIA: Throughout India.

WORLD: Native to India, Malaysia, Papua New Guinea and Thailand. Cultivated in Tropical Asia.

Dioscorea glabra Roxb., Fl. Ind. 3: 803. 1832; FBI 6: 294. 1892; Fischer 3: 1512. 1928.

Prickly, glabrous, climbing vine, stem terete, swollen at nodes. Leaves ovate-oblong, up to 9.5 x 5.5 cm, glabrous, base deeply cordate to sagittate, apex acuminate; petiole about 7 cm long, slender, canaliculate. Male flowers scattered in short spike, ca 2 cm long, showy, female flowering axis solitary or 2 together; sepals 3, minute, ovate-oblong, concave; petals 3, up to 1.5 mm long, obovtate; stamens 6, basifixed, pistillode minute, inconspicuously winged. Capsules obtuse at base.

A frequent climber along the foot hills associated with other bushes, usually near moist regions in northern Eastern Ghats. Fl. & Fr.: August – December. Vern.: Tel.: *Gintiga theega;* Tam.: *Naaruvalli.*

Tadepalli (EG), *NRR* & *DN* 84319, 84319, 84320 (MH); Jidikuppa (EG), *CAB* 5342 (MH); Maredumilli to Kakur (EG), GVS 67582 (MH); Paderu (VSKP), *NRR* & *DN* 84225, 84318, 84319, 84320 (MH); Sunkarimetta (VSKP), *GVS* 21668 (MH); Araku (VSKP), *GVS* 21573 (MH); Purunakote section, Satkosia Tiger Reserve, *KCM* 51821 (BSID); Tulka RF, Satkosia Tiger Reserve, *KCM* 5957 (BSID).

INDIA: Throughout India.

WORLD: Myanmar, Bhutan, China, Cambodia, Laos, Malaysia, Indonesia, Thailand, Vietnam.

Dioscorea hamiltonii Hook. f., Fl. Brit. India 6: 295. 1892; Fischer 3: 1512. 1928.

A vine, stem twining to the right; tubers long-stalked, deep underground; stem winged or regularly angled, quite glabrous; whole plant reddish when dry. Leaves ovate or lanceolate, 4.5 – 10 x 2-5 cm, base truncately to deeply cordate, margin entire, apex acuminate; petiole 3-6 cm long. Male flowers in spikes, rachis very slender, zig zag; sepals broad; petals cuneate-obovate; stamens very short, pistillode obscure. Capsules reniform, retuse.

Occasional in forests of Vizianagaram and East Godavari districts. Fl. & Fr.: September – December.

Vathangi (EG), *SS* 806 (AU); Sunki (VZN), *MV* 6734 (AU); Similipal RF (MBJ), *A.R.K.Sastry* & *G.P.Singh* 12226 (BSID).

INDIA: S., E. & N.E.India, C. to E. Himalaya.

WORLD: Nepal, Bhutan, China, Myanmar, Thailand, Vietnam.

Dioscorea hispida Dennst., Schluess. Hort. Malab. 15. 1818; Fischer 3: 1511. 1928. *D. daemona* Roxb., Fl. Ind. 3: 805. 1832; FBI 6: 289. 1892.

A vine, stem twining to the left; tubers lobed, biennial. Leaves 3-foliolate, on 10-20 cm long, prickly petiole, leaflets broadly cuneate-obovate, 10-20 x 5-7 cm, all petiolulate, cuspidately caudate-acuminate, villous when young, glabrous with age, sometimes reticulately veined, base tapering, lateral leaflets very oblique. Male flowers in dense cylindric pedunculate spikes, arranged in clusters along the more or less prickly, pubescent or villous rachis of a 15-40 cm long raceme; bracts shorter than the flowers; sepals orbicular-ovate, membranous, shorter than the coriaceous incurved petals; stamens 6, all antheriferous; anthers sub-sessile, pistillode very low, broad. Female flowers in spikes, distant, perianth like that of the male. Capsules quadrately oblong, ends truncately rounded, smooth; seeds winged at the base only.

More or less common throughout Eastern Ghats. Fl. & Fr.: October – December. Vern.: Tel.: *Shanda gadda, Puli dumpa, Tella gini gedda;* Ori.: *Bainya alu*; Tam.: *Vaelikizhangu.*

Sivapuram (KNL), *BSS* & *SKB* 29440; Lankamalai RF (KDP), *SRSR* 13049; Chelama (KNL), *JLE* 16756 (MH); Pulusumamella (KNL), *VBH* 83930 (BSID); Near Mallavaram – Tadepalli (EG), *NRR* 84360 (MH); Vatigadda (EG), *SS* 806 (AU); Araku (VSKP), *GVS* 19733 (MH); Purunkote RF, Satkosia Tiger Reserve, *KCM* 5604, 5181 (BSID).

INDIA: Throughout India.

WORLD: Bhutan, China, Indonesia, Thailand.

Dioscorea oppositifolia L., Sp. Pl. 1033. 1753; Matthew, Fl. Tamilnadu Carnatic 3: 1634. 1983; var. **oppositifolia**; FBI 6: 292. 1892; Fischer 3: 1512. 1928. *D. lanceolata* Heyne ex Hook. f., Fl. Brit. India 6: 292. 1892.

A vine; rhizomes branched with many long cylindric roots as thick as swan's quill; stem slender, unarmed, glabrous. Leaves oblong or ovate-suborbicular, 3-12 x 1.5-9 cm, base obtuse-rotund, margin entire, apex sub-acute-obtuse, mucronate, glabrous. Male flowers yellowish-green, slighty scented, in dense shortly pedunculate spikes, which are fascicled in the leaf axils or along a slender axillary rachis; bracts linear, bracteoles scarious, triangular, subequal; perianth-lobes oblanceolate; stamens 6. Female flowers distant, in solitary or fasciculate, axillary,

10-20 cm long spikes; flowers distant, perianth-lobes ovate. Capsules obovoid, stipitate; seeds orbicular, winged throughout.

Common throughout Eastern Ghats in deciduous and semievergreen forests. Fl. & Fr.: August – November. Vern.: Tel.: *Adavi dumpa*; Ori.: *Bainya alu*; Tam.: *Kaatuvalli, Verrolaivalli, Malayankizhangu.*

Batrepalli (ATP), *B.Aswarthappa* 340 (BSID); Krishna Nandi (KNL), *TP & KNR* 9871; Ahobilam (KNL), *TP & KNR* 2589, 9857; Lankamalai RF (KDP), *SRSR* 8720; Thalakona (CTR), *KNR & PSPB* 9849; Balapalli (KDP), *JLE* 14915 (MH & CAL); Malakondapenta (PKM), *VBH* 83975 (BSID); Towards Akasaganga (CTR), *GVS* 46877 (MH); Horsley konda (CTR), *JSG* 15179 (MH & DD); Penchalakona (NLR), *PV* 109938; Jaddangi RF (EG), *GVS* 68654 (MH); Gundlabramheswaram (KNL), *JLE* 16926 (MH); Kondapalli RF (KSN), *PV* 5320 (MH); Tadepalli (EG), *NRR & DN* 84321 (BSID); Araku (VSKP), *NPBK* 10716 (MH); Malleswarakonda (SKLM), *NRR* 79496 (BSID); way to Yeguvaguda, Seedhi (SKLM), *NRR* 83603 (MH, BSID); Mannanur RF (MBNR), *VBH* 86609 (BSID); Siddharam (KMM), *PVS* 84079 (BSID); Thattilanga forest (KMM), *RCS* 102418 (BSID); Nilagiri view point (MBNR), *S.R. Srinivasan* 109724 (BSID); Kambukudi (NA), *KS* 6083 (MH), *EV* 51982 (MH); Tippukadu RF (NA), *KRM* 176579 (MH); Melpat (SA), *CAB* 979, 1030 & 1031 (MH); Swamimalai RF (SA), *S.R.Raja* 17911 (MH); Near Pereri, Chitteri (DMP), *TRS* 95533 (MH, BSID); Hogainakkal forest (KGR), *EV* 21913(MH): Way to Semmedu, Kolli hills (NMK), *A.A.Ansari* 47905 (MH); Sanyasimalai (SLM), *K.M.Sebastine* 14405 (MH); Kannimarthirai, Thandikudi hills (DGL), *K.Ravikumar* 92375 (MH, BSID); Yercaud (SLM), *G.Bidie s.n.* (MH); Nedungundram, near Vandalur (CPT), *DN* 1012 (MH); Boragharo (GJM), *VNS* 5976 (MH); Rukeekonda (GJM), *VNS* 60650 (MH); Thargaikonda (KPT), *T.K.Paul* (BSID); Similipal RF (MBJ), *A.R.K.Sastry* & *G.P.Singh* 12328 (BSID);

INDIA: S., E. & N.E. India.

WORLD: Sri Lanka.

Dioscorea pentaphylla L., Sp. Pl. 1032. 1753; FBI 6: 292. 1892; Fischer 3: 1511. 1928; Matthew, Fl. Tamilnadu Carnatic 3: 1635. 1983. *D. jacquemontii* Hook.f., Fl. Brit. India 6: 281. 1892.

A gregarious tuberous vine; tubers oblong, 50-100 cm long; stem slender, glabrous, prickly towards the base, often bulbiferous in the leaf axils. Leaves alternate, 3-5-foliolate, glabrous or sparsely pubescent beneath, terminal one elliptic ovate, 4-8 x 1.5-3 cm, lateral ones inequilateral, 3-7 x 2-3 cm, thin coriaceous, 3-5-nerved at base, rusty-pubescent below, sparsely so above, base subacute-attenuate, margin entire, apex acute, mucronate; petiole grooved. Male flowers pale greenish, fragrant, in very slender shortly pedunculate, 2-4 cm long racemes which are solitary or in fascicles along the hairy branches, bracteoles connate, orbicular-ovate, woolly. Female flowers in axillary pendulous pubescent spikes, flowers distant. Capsules oblong, deeply angled, winged, glabrous; seeds 4-6, subquadrate, apically winged.

Common throughout Eastern Ghats in deciduous and semievergreen forests. Fl. & Fr.: February – September. Vern.: Tel.: *Duka pendalam, Adavi ginusu teega*; Ori.: *Pittalo kanda, Karwa, Karba, Koraba alu, Mundi kando*; Tam.: *Nalvaelikizhangu.*

Kalasamudram RF (ATP), *EC* 2930; Ahobilam (KNL), *TP & KNR* 9858 & 9860; Lankamalai RF (KDP), *SRSR* 13050; Venkatagiridurgam (NLR), *SSY* 12404; Pedarutla (PKM), *RVK* 15821; Chelema (KNL), *JLE* 16789 (MH); Balapalli (KDP), *JLE* 14947 (MH); Marthinayani cheruvu (CTR), *GVS* 45297 (MH); Horsley hills (CTR) *JSG* 20978 (DD); Penchalakona (NLR), *PV* 109928 (BSID); Bison hill (EG), *CAB* 5175 (MH); Maredumilli (EG), *GVS* 68535 (MH); Nulakamaddi (EG), *VNS* 331 (CAL); Minumuluru (VSKP), *GVS* 42590 (MH); Araku (VSKP), *GVS* 21572 (MH); Junaiparai, Andiappanur (NA), *MBV* 1157 (MH); Kottaiyur RF, Yelagiri hills (NA), *MBV* 534 (MH); Vellimalai slope, Kalrayans (SA), *KRM* 79636 (MH); Mottampatti RF (SA), *KRM* 79668 (MH); Takarai RF (SA), *KRM* 53813 (MH); Vandalur (CPT), *S.Indian Flora* 11572 (MH); Tickepalli -Linepada (GJM), *sine coll. s.n.* (MH); Kolli hills (NMK), *P.G.Diwakar* 97082 (MH); Tulka RF. Satkosia Tiger Reserve, *KCM* 5966 (BSID).

INDIA: Throughout India.

WORLD: Africa, Bangladesh, Myanmar, Nepal, China, Laos, Malaysia, Philippines, Vietnam, Australia.

Dioscorea pubera Blume, Enum. Pl. Jav. 1: 21. 1827. *D. anguina* Roxb., Fl. Ind. 3: 803. 1832; FBI 6: 293. 1892; Fischer 3: 1513. 1928.

A vine, stem twining to the right, unarmed, woody, densely pubescent; tubers long, cylindric. Bulbils axillary, potato-like, 2.5-5 cm across, with a greenish or grey-brown skin. Leaves opposite, broadly ovate or suborbicular, 6-18 x 5-12 cm, base cordate, margin entire, apex acuminate-cuspidate, pubescent on the nerves beneath. Male flowers in 20-30 cm long, axillary, panicled, pubescent spikes; bracts ovate, reaching nearly halfway up flower, deflexed; outer perianth-lobes broadly oblong or ovate-oblong, inner periath lobes ovate; stamens 6, perfect, adnate to the inner perianth lobes. Female spikes densely pubescent, 5-10 cm long, axillary; perianth 1.2 mm long, outer perianth-lobes ovate-rounded, inner rather smaller; ovary densely pubescent or tomentose, not beaked. Capsules subcordate, subglabrous; seeds orbicular, winged all round.

In hill forests. Fl. & Fr.: July – September. Vern.: Ori.: *Kosa kanda, Dang alu, Chekka alu, Garh kanda, Kosa alu, Saiga alu, Kasi theega.*

Penubakam (NLR), *A.S.Rao* 415 (BSID); Vathangi (EG), *SS* 4926 (AU); Way to Kodavala lanka (EG), *NRR & DN* 84359 (MH); Palakonda Range (VSKP), *KCJ* 17239 (MH); near Seedhiguda, Donubai (SKLM), *NRR* 83604 (MH);

INDIA: S., E. & N.E.India, Tropical Himalaya.

WORLD: Myanmar, Nepal, Bhutan, Indonesia.

Dioscorea tomentosa J. Koenig ex Spreng., Pl. Pugil. 2: 92. 1815; FBI 6: 289. 1892; Fischer 3: 1511. 1928; Matthew, Fl. Tamilnadu Carnatic 3: 1636. 1983.

A vine, stem twining to the right, densely pubescent or tomentose, unarmed; tubers elongate; branchlets tomentose, rarely sparsely prickled. Bulbils axillary, potato like, 2.5-5 cm across, with thin greenish or grey-brown skin. Leaves 3-5-foliolate, ovate-broadly elliptic or lanceolate, laterals oblique at base; leaflets 4-12 x 1.5-7 cm, coriaceous, 3-5-nerved, margin entire, apex round-acute, mucronate. Male flowers in axillary and/or terminal, 10-20 cm long spikes; bracteoles ovate; perianth-lobes coriaceous, tomentose; stamens 3, staminodes 3; pistillode short. Female flowers axillary, paired, 5-12 cm long spikes, perianth-lobes ovate, ovary oblong, 3-celled, ovules 2 per cell, stigma capitate. Capsules oblong, apex broad, tomentose, glabrescent later; seeds obovoid, apically winged.

Common in hill forests throughout Eastern Ghats. Fl. & Fr.: March – November. Vern.: Tel.: *Nuetiga chettu, Burdi gadda, Tegamdumpa, Nallatiga;* Ori.: *Taraga;* Tam.: *Vallikizhangu, Peruvallikizhangu, Nalvallikizhangu.*

Batrepalli (ATP), *KRKS* 39580; Peccheruvu RF (KNL), *RVR* 2190; Lankamalai RF (KDP), *SRSR* 10100; Rollapenta (PKM), *RVK* 15828; Anantagiri – Galikonda (VSKP), *TP* & *EC* 7386; Gundlabrahmeswaram (KNL), *JLE* 16905 (MH); Dalgutta (KSN), *PV* 5203; Jaddangi (EG), *GVS* 68657 (MH); Pingerkonda (EG), *SS* 462 (AU); Forest near Araku valley (VSKP), *NPBK* 10785 (MH); Palakonda Range (VSKP), *KCJ* 17231 (MH); Sunki (VZN), *MV* 6702 (AU); way to Samarelle (SKLM), *PVS* & *NRR* 76891 (MH); Guthirayanbetta, Kodagaai (DMP), *TRS* 84137 (MH, BSID); Way to Kulivalur, Kolli hills (NMK), *A.A.Ansari* 78866 (MH); Jamnamarathur, Javadi hills (NA), *EV* 51975 (MH), *MBV* 872 (MH); Kakka shola, Shevaroy hills (SLM), *EV* 82558 (MH); Singarapettai hills (NA), *MBV* 1430 (MH); Way to Kambukudi (NA), *KS* 6477 (MH); Melpat (SA), *CAB s.n.* (MH), Pakamalai RF, Gingee (SA), *KRM* 53525 (MH);

INDIA: S. & E. India.

WORLD: Bangladesh, Sri Lanka.

Dioscorea wallichii Hook. f., Fl. Brit. India 6: 295. 1892; Fischer 3: 1511. 1928. *D. aculeata* L., Sp. Pl. 1033. 1753. *p.p.;* FBI 6: 296. 1892.

A vine; stem stout, twining to the right, often prickly below, the tubers growing out direct from the base of the stem. Leaves opposite, ovate or orbicular, 6-15 x 4-12 cm, base cordate, 7-nerved, secondary nerves regular and nearly parallel. Male flowers in 2.5 cm long spikes in axillary and terminal spreading paniclels; flowering bracts ovate-acuminate, minute; outer perianth-lobes ovate-oblong, concave, larger than the inner ones, inner ones suborbicular; stamens 6, short, pistillode large, globose. Capsules broadly obovate, to 5 x 4 cm, apex emarginated; seeds orbicular, broadly winged.

Occasional in forest of North Coastal and Kurnool district. Fl. & Fr.: August – December. Vern.: Tel.: *Chirangi, Kandamal;* Ori.: *Phalokando, Tunga alu, Pita alu.*

Way to Diguvametta (PKM), *JLE* 22156 (MH); Gundlabrhmeswaram (KNL), *JLE* 32656 (MH); Gundlakamma (KNL), *KNS* 3025 (DD); Bison hill (EG), *CAB* 5176

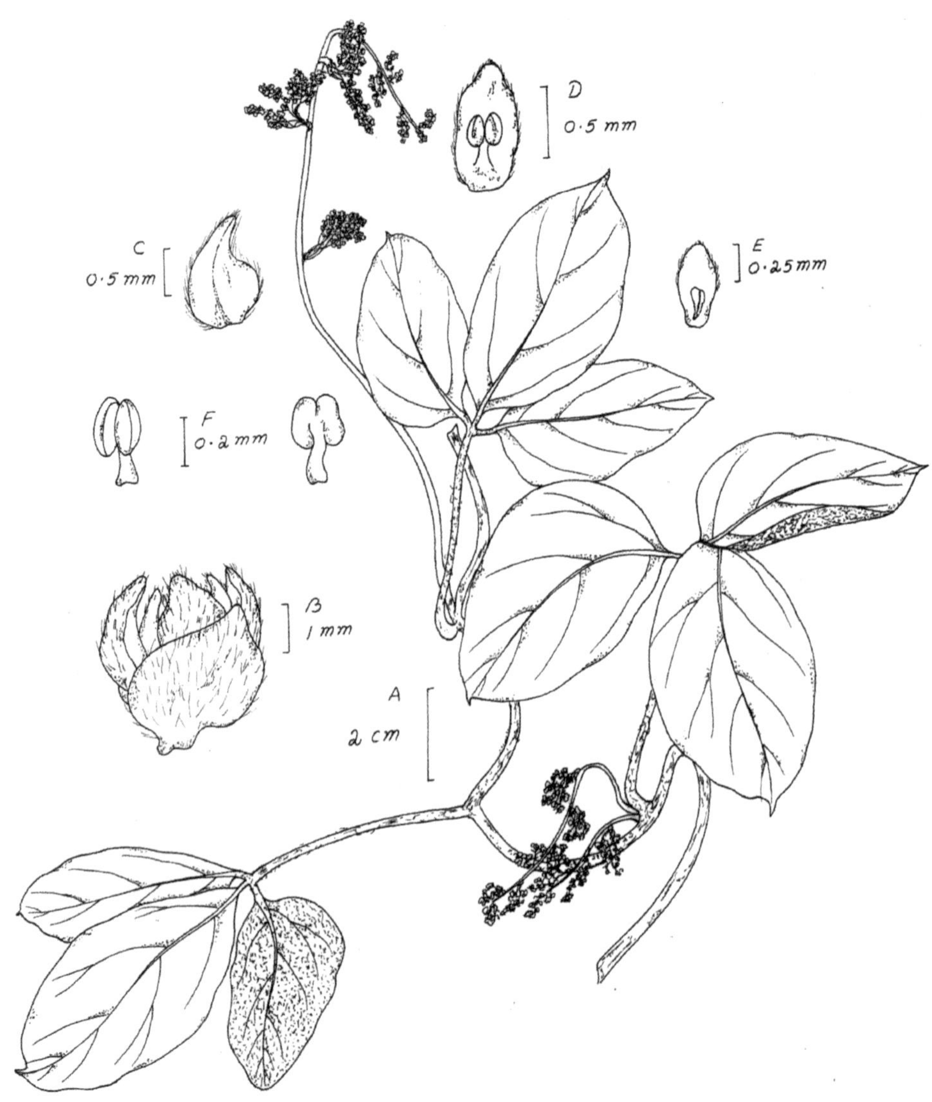

Figure 14. **Dioscorea tomentosa** J. Koenig ex Spreng. Male
A. Twig, B. Male flower, C. Bracteole, D. Perianth lobe (outer), E. Perianth lobe (inner), F. Stamens.

(MH); Way to Sesharayi, Rampa (EG), *VNS* 611 (CAL); Galikonda area (VSKP), *GVS* 21651 (MH); Near Dabuguda, Lakshmipuram (VSKP), *NRR* & *DN* 84204 (BSID).

INDIA: In hotter parts of India.

WORLD: Bangladsh, Myanamr, China, Malaysia, Thailand.

STEMONACEAE

STEMONA Loureiro

Stemona tuberosa Lour., Pl. Cochinch. 2: 490. 1790; FBI 6: 298. 1892; Gamble 3: 1514. 1928; Matthew, Fl. Tamilnadu Carnatic 3: 1637. 1983; var. **minor** (Hook.f.) C.E.C.Fischer, Fl. Madras, 3: 1514.1928. *S. minor* Hook.f., Fl. Brit. India 6: 298. 1892.

A plant with twining stem, leafing and flowering at the same time. Leaves ovate, membranous, 4-7.5 x 3-5 cm, opposite, rarely alternate, 9-nerved, glabrous, base shallow-cordate, margin entire, apex acuminate-caudate. Flowers in 1-3-flowered raceme, perianth segments lanceolate, many-nerved, greenish with purple; stamens large, filaments red, stout, deeply grooved with crenulate margins, connective green; ovary free, compressed, stigma small, ovules 2 or more. Capsules ovoid-oblong, 5-8-seeded.

Occasional in open forests and thickets. Fl.: February - April. Vern.: Ori.: *Koelakand, Kelya kanda.*

Maredumilli (EG), *BR & KNR* 10610, *GVS* 24410 (MH); Cuddapah hills (KDP), *RHB* 42 (MH); Balapalli (KDP), *JLE* 17932 (MH); Tirumala – Dharmagiri (CTR), *DRC* 1484 (MH); Rampa hills (EG), *JSG* 51681 & 16062 (MH); Maredumilli (EG), *GVS* 24410 (MH); Karaka (VSKP), *CAB* 1699 (MH); Sunkarimetta (VSKP), *GVS* 21695 (MH); Minumuluru towards Ontally (VSKP), *GVS* 30021 (MH); S. Ghat (VZN), *MV* 6829 (AU); Damadua to Andaba (GJM), *CAB* 1480 (MH); Mahendragiri (GJM), *JSG* 13921 (CAL, MH), *SX* & *MB* 2218 (RRL-B); Dudhari (KPT), *SX* & *MB* 6573 (RRL-B); Maliput (KPT), *SX* & *MB* 6658 (RRL-B); Kambakkam hills (CPT), *sine coll. s.n.* (MH);

INDIA: S. to E.India.

WORLD: Bangladesh, Mynmar, China, Laos, Cambodia, Phlippines, Thailand, Vietnam.

LILIACEAE

1. Leaves minute, often spinescent scales, bearing axillary tufts of needle like or slightly flattened cladodes .... **Asparagus**
1. Leaves well developed:
    2. Climbing plants .... **Gloriosa**
    2. Erect undershrubs or herbs:
        3. Leaves very thick, cartilaginous or fleshy .... **Aloe**
        3. Leaves not very thick, neither cartilaginous nor fleshy:

4. Undershrubs or stout herbs with definite above ground leafy stem.. **Disporum**
4. Herbs without distinct above – ground stems:
   5. Root stock small with fleshy or tuberous roots:
      6. Ovary inferior.. **Ophiopogon**
      6. Ovary superior:
         7. Leaves semi-terete, fistular.................................**Asphodelus**
         7. Leaves flat, not fistular.................................. **Chlorophytum**
   5. Root stock a bulb or corm:
      8. Flowers several-many, racemose on a simple, naked scape:
         9. Outer perianth series larger than inner and recurved ....**Dipcadi**
         9. Both perianth series equal, not recurved:
            10. Flowers distant, usually appearing before leaves, dingy – brown .... **Drimia**
            10. Flowers close, appearing with leaves, greenish purple ....**Ledebouria**
      8. Flowers solitary or few corymbose or umbel; leaves scattered on stem:
         11. Plant stout, more than a meter tall; flower 10 cm across .... **Lilium**
         11. Plant slender, less than 15 cm tall; flower 1 cm across ....**Iphigenia**

ALOE Linnaeus

Aloe vera (L.) Burm. f., Fl. Ind. 83. 1768; Fischer 3: 1520. 1928; Matthew, Fl. Tamilnadu Carnatic 3: 1639. 1983. *A. perfoliata* L. var. *vera* L., Sp. Pl. 320. 1753.

A stoloniferous herb. Leaves dense, aggregated, succulent, ensiform, 40-60 x 3-8 cm, narrowed from base to apex, pale green, with distant horny prickles on the margins. Scape 60-90 cm long, simple or branched. Raceme dense, 15-30 cm long, bracts lanceolate, flowers reddish-yellow, bisexual; perianth-tube terete, campanulate, somewhat curved, lobes 6, reddish-yellow or orange, oblong, 3-nerved; stamens 3 + 3, filaments inserted in a pit in the connective; ovary 3-celled, ovules numerous per cell on axile placentae, style elongated, stigma obscurely lobed. Capsule ellipsoid-oblong.

Common in open wastelands and field hedges. Fl. & Fr.: September – December. Vern.: Tel.: *Chinna kalabanda*; Tam.: *Sothukathalai.*

Peddachoragiri – Hottabetta RF (ATP), *TP* & *NY* 770; Hampapuram (ATP), *KNR* 9856; Kalvabugga (KNL), *RVR* 3102; Palakonda hills (KDP), *CS* 9326; Ethipothala falls (GNT), *VRK* 3572; Guvvalacheruvu (KDP), *KS* 6826 (MH & CAL); way to

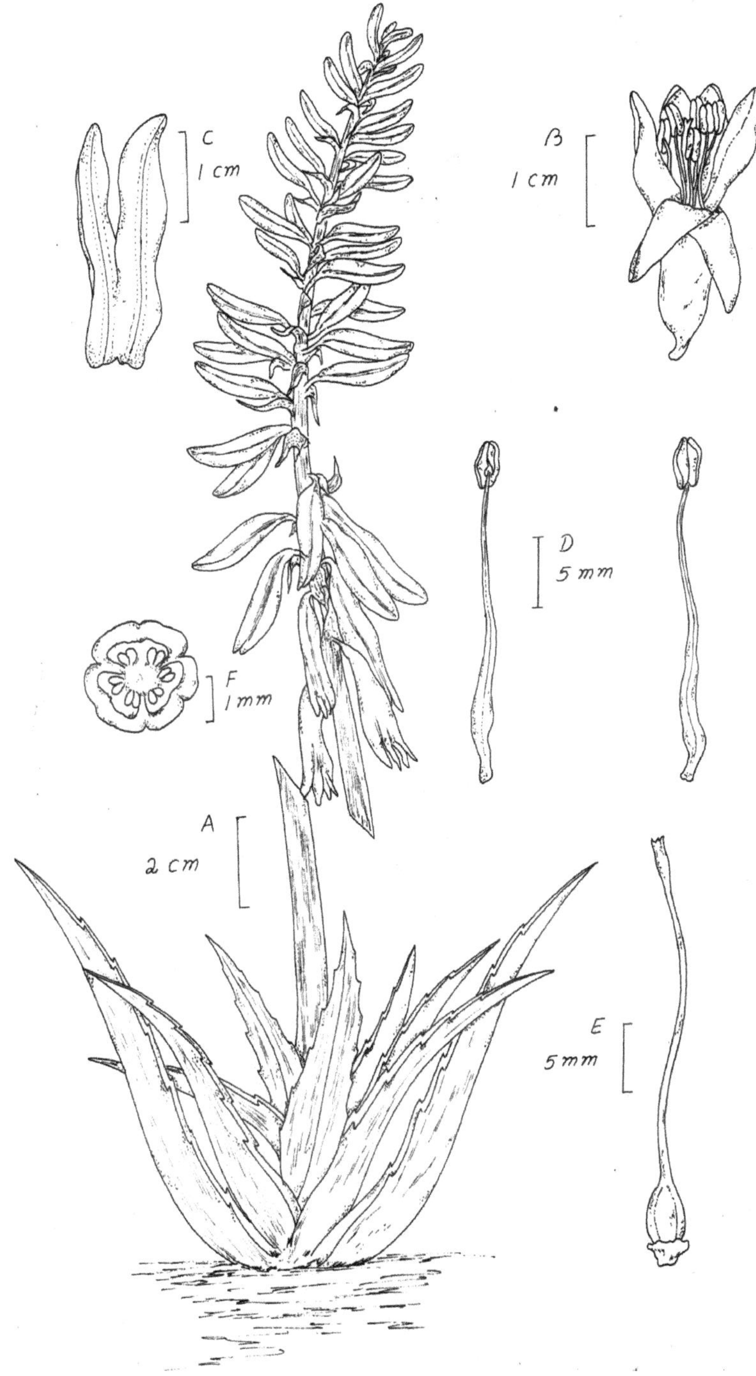

Figure 15. **Aloe vera** (L.) Burm.f.
A. Habit, B. Flower, C. Perianth lobes (outer and inner), D. Stamens, E. Pistil, F. T.s. of ovary.

Kodavalalanka, Tadepalli (EG), *NRR & DN* 84357 (MH); way from Kavalur to Komattiyur (NA), *KS* 6465 (MH); Bathikanaru, Natrapalayam (DMP), *EV* 57971 (MH);

INDIA: Introduced and naturalized.

WORLD: Native of Mediterranean region, cultivated and naturalized elsewhere.

ASPARAGUS Linnaeus

1. Cladodes 2–5 cm .... **A. laevissimus**
1. Cladodes 0.5–2 cm:
   2. Cladodes triquetrous or very slightly compressed .... **A. racemosus**
   2. Cladodes flat, falcate-ensiform, 2-6-nate, acute or acuminate .... **A. gonoclados**

Asparagus gonoclados Baker, J. Linn. Soc. 14: 627. 1875; FBI 6: 318. 1892; Fischer 3: 1517. 1928.

A much – brached subscandent armed undershrub; main stem smooth, terete; branches firm, green, 3-quetrous. Leaves squrred at the base with hard spines, 0.8-1.25 cm long. Cladodes flat, falcate-ensiform, 2-6-nate, acute or acuminate, 1-2 cm long, narrowed to both ends, finely spinous-pointed. Raceme 2.5-8 cm long, often fascicled; flowers white; bracts ovate boat-shaped; perianth lobes spreading, the outer linear-oblong, the inner subspathulate; anthers minute. Berry globose.

Occasional in Kambakkam forest and Tirumala hills of Chittoor district, in thickets of deciduous forests. Fl. & Fr.: August – January.

Papanasanam (CTR), *BR & BSS* 30415 (BSID); Kambakkam (CTR), *DRC* 2379 (SVU); Tirumala hills (CTR), *DRC* 2434 (SVU).

INDIA: Peninsular India.

WORLD: Sri Lanka.

Asparagus laevissimus Steud. ex Baker, J. Linn. Soc 14: 623. 1875; FBI 6: 317. 1892; Fischer 3: 1517. 1928.

Perennial, subscandent under shrub, to 1.5 m long. Stems shiny, branches striate. Leaves with conical stout spurs; cladodes 3–6-nate, spreading, triquetrous. Flowers in racemes, pedicels short; bracts large, cymbiform; tepals linear-oblong, obtuse, white; stamen 6, as long as perianth; style short, stout; stigmas 3. Fruit berry.

Rare in forests in Anantapuram district. Fl. & Fr.: June–September.

Batrepally (ATP), *KRKS & BSS* 39579.

INDIA: Peninsular India.

Asparagus racemosus Willd., Sp. Pl. 2: 152. 1799; FBI 6: 316. 1892; Fischer 3: 1517. 1928; Matthew, Fl. Tamilnadu Carnatic 3: 1639. 1983. *A. racemosus* Willd. var. *zeylanicus* Baker, J. Linn. Soc. Bot. 14: 623. 1875. *A. zeylanicus* (Baker) Hook.f., Fl. Brit. India 6: 317. 1892.

Key to Varieties

1. Racemes terminated by cladodes .... var. **javanica**
1. Racemes not terminated by cladodes .... var. **racemosus**

Asparagus racemosus Willd. var. **racemosus**

An extensivey scandent spinous much-branched under shrub; root stock tuberous; branchlets angular. Leaves scaly, triangular, 4-6 mm, stiff-acuminate, cladodes 3-6, linear, 1-2 x 0.5-0.7 cm, base narrow, margin entire, apex acuminate. Flowers bisexual, white, fragrant, in simple (rarely branched), 5-15 cm long racemes; peduncle much reduced; bracts triangular; perianth-lobes 6, white, oblong-obovate, obtuse; stamens 6, adnate to the base of perianth; ovary obovoid, 3-celled, ovules 2 per cell on axile placentae, style terminal, apex 3-fid, stigmas 3, recurved. Berry globose; seeds 3-6, globose or angled.

Common throughout Eastern Ghats in deciduous forests. Fl. & Fr.: September – October. Vern.: Tel.: *Satamula, Devdani, Pilligaddalu, Challa gadda*; Ori.: *Chhataori, Satawari, Gaichero, Deobadini,Hateri kanda.*

Muchukota RF (ATP), *TP* & *NY* 209, *KNR* 10624; Racherla RF (KNL), *RVR* 1363; Chelama RF (KNL), *KNR & DAM* 10632; Palakonda hills (KDP), *CS* 8237; Macherla RF (GNT), *VRK* 3817; Anantagiri (VSKP), *KSK* 23426, *GVS* 21745 (MH); Gooty (ATP), *JSG* 16487 (MH); Horsley hills (CTR), *JSG* 15074 (MH); Kokulakonda RF (NLR), *KCJ* 83345 (MH); Near Isakgundam (PKM), *VBH* 83999 (BSID); Bison hills (EG), *CAB* 5161 (MH); Araku valley (VSKP), *NPBK* 10839 (MH); Edrallapalli (KMM), *RCS* 102404 (BSID); Umamaheswaram temple (MBNR), *S.R.Srinivasan* 109153 (BSID); Koosuguttai, Yelagiri hills (NA), *MBV* 586 (MH); Shanikulam (SA) *CAB* 764 (MH); Kaduvanoor (SA), *KRM* 50615 (MH); Gingee RF (SA), *KRM* 13017 (MH); Vandalur (CPT), *sine coll,* 14121 (MH); Yercaud (SLM), *AVNR* 18247 (MH); Hogainakkal (KGR), *EV* 20718 (MH); Marandhalli RF (DMP), *EV* 57843 (MH); Mahendragiri (GJM), *VNS* 57210 &5672 (MH); Rukeekonda (GJM), *VNS* 6066 (MH); Simonbadi to Darugambadi (GJM), *CAB* 1330 (MH); Pati varu (KPT), *T.K.Paul* 1433 (BSID); Chahala, Similipal RF (MBJ), *A.R.K.Sastry* & *G.P.Singh* 12216 (BSID); Tikarpura, Satkosia Wild life sanctuary, *D.Hazra* & *D.Das* 19668 (BSID).

INDIA: Throughout Inda.

WORLD: Australia, Malaysia, Myanmar, Bhutan, Sikkim, Nepal, Kashmir, Pakistan, Africa.

Aspragus racemosus Willd., Sp. Pl. 2: 152. 1799; var. **javanica** (Kunth) Baker, J. Linn. Soc. Bot. 14: 624. 1875; FBI 6: 317. 1892.

Scandent undershrubs, spinous, much branched, root stock tuberous, branches angular. Leaves 0.43-0.63 cm long, linear- subulate, straight or slightly curved, 0.33-0.43 cm long at base. Cladodes spinous-pointed. Flowers white in 5-15 cm long racemes, pedicels jointed above middle. Berries *ca* 0.43 cm in diam., globose, ripens red.

Rare in moist deciduous forests in Lankapakalu in Visakhapatnam district (C.S. Reddy *et al.*, 2001). Fl. & Fr.: August – January. Vern.: Tel.: *Pilli teegalu.*

INDIA: Peninsular India.

WORLD: Tropical Africa to N. Australia.

Note: If differs from the typical variety in racemes terminated by cladodes.

ASPHODELUS Linnaeus

Asphodelus tenuifolius Cav., Anales Cl. Nat. 3: 46. t. 27. 1081; FBI 6: 332. 1892; Fischer 3: 1524. 1928.

An annual herb with slender root-fibres. Leaves 15-30 cm long, terete, fistulous, acute, sheathing at the base, finely puberulous. Scapes several, 30-60 cm long, glabrous or sparsely minutely puberulous. Flowers white, distant, laxly racemose, solitary in each bract; jointed at or below the middle; bracts broadly ovate, acute, scarious, with a strong brownish keel, perianth segments oblong, obtuse, with a strong conspicuous brownish costa; ovary 3-celled, ovules 2 in each cell. Capsule globose, erect, the valves deeply wrinkled; seeds sharply trigonous, acute, black.

A weed of fields and gardens, rare. Fl. & Fr.: December – March. Vern.: Tel.: *Neeru malli.*

Tokapalli (KNL), *sine coll., s.n.* (MH)

INDIA: Throughut plains of India.

WORLD: North Africa, S.W. Europe, S.W. Asia, Pakistan.

CHLOROPHYTUM Ker - Gawler

1. Scape branched .... **C. nimmonii**
1. Scape unbranched:
    2. Flowers remotely scattered on racemes solitary, or in pairs............ **C. laxum**
    2. Racemes dense-flowered:
        3. Leaves linear, not or rarely narrowed into a petiole:
            4. Scape longer than leaves ....**C. tuberosum**
            4. Scape shorter than leaves ....**C. malabaricum**
        3. Leaves narrow-lanceolate or oblanceolate narrowed into petiole:
            5. Anthers longer than the glabrous filaments; scape as long as the leaves .... **C. arundinaceum**
            5. Anthers shorter than the papillose filaments; scape longer than the leaves ....**C. glaucum**

Chlorophytum arundinaceum Baker, J. Linn. Soc. Bot. 15: 323. 1876; FBI 6: 333. 1892; Fischer 3: 1526. 1928.

An erect herb with a short hard root stock producing fascicled roots, often thick and fleshy and tuber like. Leaves narrow, lanceolate or oblanceolate, acuminate

15-50 cm long (including the petiole). Scape 30-60 cm long, naked, usually as long as the leaves. Flowers racemose on simple or shortly branched scapes, usually fascicled in the axils of small scarious or large membranous bracts; jointed usually below the middle; perianth-lobes white, petaloid, lanceolate; stamens 6; ovary oblong, 3-celled, ovules 4 or more per cell, stigma simple. Capsule depressed-globose; seeds suborbicular, flat, black.

Common in humus or rocky places. Fl. & Fr.: June – September.

Madhuvitipadu (EG), *SS* 8301 (AU); Gundlabrahmeswaram (KNL), *JLE* 16920 (MH); Muzumamidivalasa (EG), *MM* 105013 (BSID); way to Seedhiguda (EG), *PVS & NRR* 79410 A; Sirivaka (EG), *Bourne* 3372 (MH); Donubai (VSKP), *KCJ* 17221 (MH); Paderu (VSKP), *GVS* 28086 (MH); Elvinpet (VZN), *MV* 2047 (AU); Way to Seedhiguda (SKLM), *PVS & NRR* 79410 A (BSID); Zaring forest (KHD), *K.Safui* 13381 (CAL); Gunpur (KHD), K.*Safai* 13580 (CAL); Puthur (GJM), *VNS* 6035 (MH); Labangi, Satkosia Wild life sanctuary, *D.Hazra & D.Das* 18156 (BSID); Tikarpara, Satkosiaa Wild life sanctuary, *D.Hazra & D.Das* 18157 (BSID); Pampasar RF, Satkosia Tiger Reserve, *KCM* 5146 (BSID).

INDIA: Peninsular, E. & NE. India

WORLD: Nepal, Bhutan.

Chlorophytum glaucum Dalz., Hooker's J. Bot. Kew Gard. Misc. 2: 142. 1850; FBI 6: 334. 1892; Fischer 3: 1526. 1928. *C. orchidastrum* sensu Hook. f., FBI 6: 336. 1892.

A herb with cylindric root-fibres. Leaves 6-8, membranous, 30-45 x 2-5 cm, narrowly oblanceolate, acute, glabrous, attenuated into a short broad petiole. Scape erect, simple, 30-60 cm long, clothed with many narrowly lanceolate sheaths. Flowers white, in simple dense 15-30 cm long racemes; bracts persistent, forming coma before flowering, lanceolate, acuminate, pedicels ascending, slender, jointed at or above the middle; perianth lobes white, linear-lanceolate, sub-obtuse, 5-nerved. Capsule globose, emarginate, acutely 3-winged, the cells 2-4-seeded; seeds orbicular, compressed, dull black.

Rare in deciduous forests in Papi hills. Fl. & Fr.: June – September.

Papikonda RF (WG), *DCSR* 462 (CAL).

INDIA: Goa, Madhya Pradesh, Maharashtra, Karnataka, Andhra Pradesh. Endemic.

Chlorophytum laxum R. Br., Prodr. 277. 1810; FBI 6: 336. 1892; Fischer 3: 1526. 1928.

A low plant rarely exceeding 30 cm high; root-fibres usually with small, oblong tubers hanging from them. Leaves 6-12, grass like, 15-30 cm, subdistichous, usually spreading and recurved, longer than the scape, narrowly linear, falcate, flat, acute, glabrous, with numerous conspicuous nerves. Scape very slender, flexuous, 2-6 cm long, naked. Flowers few, greenish-white, very distant, in lax very slender imple or forked racemes, 10-25 cm long; bracts ovate, acuminate, pedicels jointed,

Figure 16. **Chlorophytum laxum** R. Br.

A. Habit; B. Flower-entire; C. Stamen; D. Ovray-l.s.; E. Capsule; F. seed.

about the middle, short, but becoming longer and drooping in fruit; perianth lobes greenish-white, oblong, obtuse, obscurely 3-nerved; stamens alternately short and long. Capsule broadly obcordate, 3-winged, the cells 1-4-seeded; seeds irregularly angled, black, minutely papillose.

Rare in shady localities of forest plains. Fl. & Fr.: May – June.

Somulavandlapalli (ATP), *BSS & KRKS* 41907; Eswramala hills (ATP), *B. Aswarthappa* 606 (BSID); Ahobilam (KNL), *TP & KNR* 9861; Horsley hills (CTR), *SKW* 6356 (BLAT); Y.Palem (PKM), *BR* & *BSS* 30682; Isakagundam (PKM), *VBH* 83997 (BSID); Rampachodavaram (EG), *MM* 102656 (BSID); Tulka, Satkosia Tiger Reserve, *KCM* 8321 (BSID); Sirumalai hills (DGL), *SKS* 828 (SGH).

INDIA: Peninsular India.

WORLD: NE Tropical Africa, Arabian Peninsula, India to W. Malesia, N. Australia.

Chlorophytum malabaricum Baker, J. Linn. Soc. Bot. 15: 331. 1876; FBI 6: 335. 1892; Fischer 3: 1526. 1928; Matthew, Fl. Tamilnadu Carnatic 3: 1641. 1983.

A perennial herb with tuberous roots. Leaves linear, 4.5-20 x 3-0.6 cm, canaliculate, keeled, ca 20-nerved, base sheathing, margin scaberulous, apex obtuse; scape shorter than the leaves. Raceme simple or branched, flowers distant, paired or solitary, bracts lanceolate, bracteoles linear, flowers 4-8 cm across, perianth-lobes white, oblong, subequal; stamens 6, anthers shorter than the filaments; ovary 6-lobed. Capsule globose, 3-winged; seeds ovoid, warty.

In moist shallow soil in Horsley hills of Chittoor district. Fl. & Fr.: June – November.

Horsley hills (CTR), *CECF* 4371 (CAL).

INDIA: Peninsular India.

Chlorophytum nimmonii Dalzell, Hooker's J. Bot. Kew Gard. Misc. 2: 142. 1850; Dalzell & Gibson, Fl. Bombay 252. 1861; Henry *et al.*, Fl. Tamil Nadu Analysis 3: 39. 1989; Karthik. *et al.*, Fl. Ind. Enum. Monocot. 92. 1989. *Anthericum nimmonii* J.Graham ex Nimmo, Cat. Pl. Bombay 220. 1839. *Phalangium oligospermum* Wight, Icon. Pl. Ind. Orient. t. 2038 1853. *Chlorophytum orchidastrum* Lindl., Sketch Veg. Swan R. pl. 813 sensu Baker, J. Linn. Soc., Bot. 15: 323. 1876; FBI 6:333. 1894.

Perennial geophytes, 35 – 65 cm tall, grows solitary, rhizome reduced to disc, covered with fibrous remnant of older leaves; roots numerous, 12 – 20 x 0.5 – 0.75 cm, fasciculated, fleshy, evenly thickened, non tuberous. Leaves all rosette, pseudopetiolated, lanceolate to elliptic, 45 –75 x 3 –6 cm, glaucous green beneath, 21–25 veined, gradually narrowing at base, pseudopetiole winged 5–9 cm, apex acute; margin undulate, hyaline. Scape solitary or 2 –3, branched, 35 –70 cm, naked, upper 2/3 part sparsely flowered panicle, longer than leaves; sterile bracts absent; branching of panicle guarded by large foliaceous bracts, 6 –9 x 0.5 1.5 cm, elliptical, apex acute to acuminate, floral bracts reduced, linear to lanceolate, persistent, green,

1.5 –5.5 x 0.5–1.5 cm, 5 –9-nerved, apex acute–acuminate, margin broad, translucent; bracteoles 2–4 per floral bract, covering 2 -3 floral buds, translucent, papery, 3-veined, brown at margin. Flowers drooping, pedicellate, white, 1.2 –1.7 cm across, usually in alternate or rarely subopposite, 5 –9-flowered clusters, pedicels 0.6 – 1.2 cm long, cylindrical, greenish white, smooth, jointed at the below middle, pedicel portion below the joint 0.2– 0.5 cm long; pedicel portion above the joint 0.4 –0.7 cm long. Perianth segments 6, in two whorls of 3 each; outer perianth segments 0.7 –0.9 x 0.3 –0.4 cm, white, linear-lanceolate, 5–nerved, apex acute, margins hyaline; inner perianth segments 0.7 –1.0 x 0.4 –0.6 cm, white, ovate to elliptical, 5–nerved, apex obtuse, margins hyaline. Stamens 6, 0.8 –1.0 cm long, connate, erect; filaments 0.2 –0.5 cmlong, white, smooth; anther 0.4 –0.6 cm long, much longer than filaments, yellow, dehiscing by longitudinal slits; ovary 0.2–0.3 cm in diam., sessile, green, globose; style 0.5 – 0.8 cm long, straight, white; stigma minutely papillose. Capsule green, triquetrous, 3 – sulcate, *c.* 0.3 x 0.5 cm, obcordate, pedicel elongated; seeds 1 –2 each cells, one cell abortive, *c.* 0.2 cm across, orbicular, discoid, black, minutely papilose.

It occurs near the water streams under the shades of trees and bushes in northern Eastern Ghats (Adsul, 2015). Fl. & Fr.: July – November.

Addatigala (EG), *SKW* 3944 (BLAT); Tulka, Satkosia Wild Life Sanctuary, *D.Hazra* & *D.Das* 18178 (BSID).

INDIA: Maharashtra, Goa, Karnataka, Kerala, Tamil Nadu, Andhra Pradesh, Odisha. Endemic.

Chlorophytum tuberosum (Roxb.) Baker, J. Linn. Soc. Bot. 15: 332. 1876; FBI 6: 334. 1892; Fischer 3: 1526. 1928; Matthew, Fl. Tamilnadu Carnatic 3: 1642. 1983. *Anthericum tuberosum* Roxb., Fl. Ind. 1: 149. 1832.

A herb with cylindric root-fibres and ellipsoid tubers hanging from them. Leaves 6-12, membranous, sessile, 15-30 x 1.5-2.5 cm, shorter than the scape, usually falcate, recurved, acuminate, the margins undulate. Scape terete, naked, 8-25 cm long. Flowers white, in simple or shortly branched recemes, peduncle stout, angular, bracts elongate, pedicel 1-2 cm, jointed at or below-middle; perianth lobes white, oblong-obovate; stamens 6; ovary 3-celled, ovules 4 or more per cell. Capsule oblong, triquetrous, strongly nerved, glabrous; seeds subquadrate, irregularly pitted.

Common in fallow lands throughout Eastern Ghats, up to 500 m. Fl. & Fr.: June – November.

Thippareddipalli- Tadapatri (ATP), *TP* 423; Pothurajukalva (ATP), *TP & KNR* 9852; Guvvalacheruvu (KDP), *RVR* 8464; Pulivendula Ghat (KDP), *KRKS* 39670; V.P. South (GNT), *VRK* 6833; Chelama (KNL), *JLE* 16742 (MH); Lukke (KNL), *JLE* 42223 (MH); Horsley hills (CTR), *JSG* 15112 (MH); Bata, Atmakur (NLR), *A.S.Rao* 6591 (BSID); Near Rollapenta (PKM), *VBH* 83963 (BSID); Lakshmipuram (VSKP), *CAB* 1719 (MH); Karakakonda (VSKP), *CAB* 1708 (MH); Elwinpet (VZN), *MV* 6828 (AU); Mannanur (MBNR), *BR* & *BSS* 30652 (BSID); Mannanur RF (MBNR), *VBH* 86617 (BSID); Pullajalamala (MBNR), *VBH* 84982 (BSID); Similipahar (MBJ), *SX* & *MB*

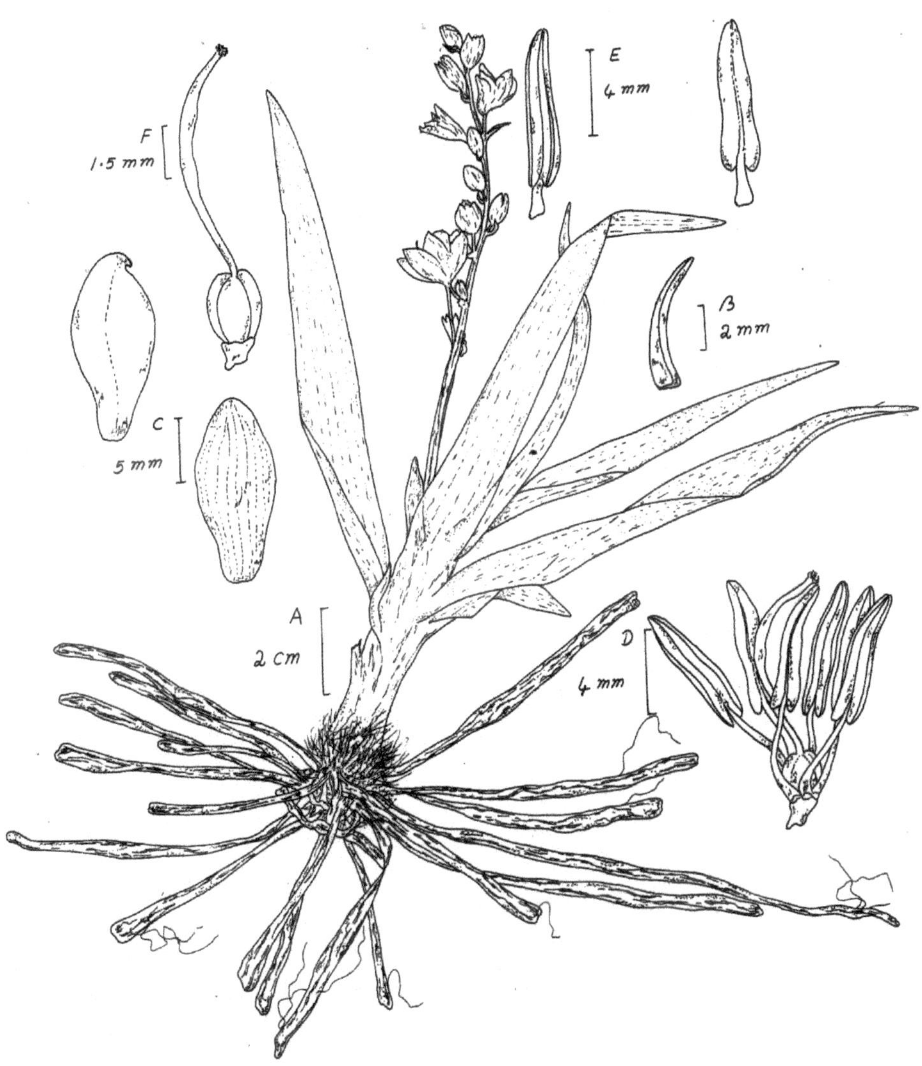

Figure 17. **Chlorophytum tuberosum** (Roxb.) Baker
A. Habit, B. Bract, C. Perianth lobes (outer and inner), D. Stamens with pistil, E. Stamens, F. Pistil.

4576, 4681 (RRL-B); Kambukudi, Poovankulam RF (NA), *KS* 6057 (MH); Kanankadu RF (SA), *KRM* 50633 (MH); Gingee RF (SA), *KRM* 13037 (MH); Toranakallu (Bellary dist.), *sine coll., s.n.* (MH).

INDIA: Throughout plains of India.

WORLD: Tropical Africa, Nepal, Sri Lanka, Myanmar.

DIPCADI Medikus

Dipcadi montanum (Dalz.) Baker, J.Linn. Soc. 11: 398. 1871; FBI 6: 346. 1892; Swamy *et al.*, J. Econ. Taxon. Bot. 41: 61. 2017 var. **montanum.** *Uropetalum montanum* Dalz. in Hook., Kew J. Bot. 2: 142. 1850. *'Uropetalon'*.

A bulbous scapose herb, 18 – 25 cm high; bulbs ovoid, white glabrous, 1.2 – 2 x 1-1.5 cm. Leaves in a rosette, 3-8 per bulb, linear, 14-24 x 0.2-0.3 cm long, deeply channelled, green, slightly broader and white at base, margins entire, apex narrowly acute, glabrous. Flowers in 15-35 cm long raceme, 7-15-flowered, pedicellate, greenish white, glabrous, 1.2 – 1.3 cm long; pedicel 0.1-0.3 cm; bracts 0.5-1.0 x 0.3-0.4 cm, much longer than pedicel, ovate - lanceolate, acuminate; perianth tube 4.8 – 5 mm long, sparsely pubescent at upper part; outer perianth lobes elliptic to oblong, 0.7 - 0.8 x 0.2 – 0.25 cm long, glabrous, obtuse to rounded at apex, glandular hairs at apex; inner lobes constricted in the middle, 6.8– 7.2 x c. 2.5 mm long, coherent to form a flask-shaped structure with apical parts spreading, round to obtuse, glandular hairs at apex; stamens 6, 5.2 – 5.4 x *c.* 3 mm long, filaments 5 – 5.4 mm long, originating at the mouth of the perianth tube; anthers *c.* 3 mm long, yellow; ovary stipitate, 3.8 – 4 x 1.6 – 1.8 mm, trilocular, obovoid; style 9-13 mm; stigma trifid, glabrous, but appearing simple to begin with. Capsule distinctly 3-lobed, elliptic, green coloured at mature stage, 3-5 seeds per locule.

Occasional in dry, gravely soils in open places among grasses in hills of Anantapur district, Seshachalam Biosphere reserve in Andhra Pradesh and Eastern Ghats of Odisha. Fl. & Fr.: October – January.

Moulatat Dargah, Kadapa-Renigunta highway (KDP), *J.Swamy* 8801 (BSID); Khandual Mali (KHD), *HFM* 3470 (DD).

Key to Varieties

1. Leaves 2; raceme short with about 7-flowered........................ var. **madrasiacum**
1. Leaves 8; raceme long with about 15-flowered...........................var. **montanum**

Dipcadi montanum (Dalz.) Baker var. **madrasicum** (E.Barnes & C.E.C.Fisch.) Deb & S. Dasgupta, J. Bombay Nat. Hist. Soc. 75: 59. 1978 & Fasc. Fl. India 7: 7. 1981; Saxena & Brahmam, Fl. Orissa, 3: 1963. 1995; Swamy & Ahmedullah, Plant Sci. Res. 38: 109. 2016; E. Barnes & C.E.C.Fisch., Kew Bull. 1940: 301. 1941; Matthew, Fl. Tamilnadu Carnatic 2: 1642. 1983.

Small bulbous herbs, *c.* 46 cm high, bulbs ovoid, white, glabrous, 2.5-3 x 2-2.5 cm. Leaves 2 per bulb, linear, 10-18 x 0.2 cm, green, slightly broader and white at base, entire, acute at apex, glabrous. Scapes *c.* 38 cm long, terete, glabrous. Inflorescence *c.* 8 cm long, racemose, loose 7-flowered. Flowers *c.* 12 x 3.5 mm, pedicellate, gree or greenish white, glabrous; pedicel 3-4 mm long, filiform, glabrous; bracts deltoid, 5-8 mm long, acuminate at apex, 6-nerved; perianth 3+3, unequal, green, mildly perfumed, glaucous or glabrous, slightly hooded, sparsely glandular pubescent at subapex; outer perianth united up to one third of its length, coherent to form a flask-shaped structure with apical parts spreading; lobes 3-4 x 2-2.2 mm, obtuse at apex; stamens 6, *c.* 4.1 mm long, filaments arising from the base of the perianth, adherent throughout the tube, slightly free above; gynoecium *c.* 9 mm long, glabrous, ovary subsessile, obovoid, style 4.5-5 mm long, stigma trifid, glandular-pubescent.

Occasional in Eastern Ghats of Odisha, Narsimhakonda in Andhra Pradesh and Shevaroy hiils in Tamil Nadu. Fl.: June – July.

Narasimhakonda(NLR), *J.Swamy* & *S.Nagaraju* 790, 7902 (BSID); Tambaram (CPT), *E.Barnes* 1801 (Lecto - K), *E. Barnes* 2805 (para – K); Shevaroy hills (SLM), *sine coll., s.n.* (MH); Sanyasimalai, Shevaroy hills (SLM), *KP* 6407 (BSID); Pottangi, Sunki (KPT), *SX* & *MB* 6683 (RRL-B).

DISPORUM Salisbury

1. Perianth-lobes spurred at the base .... **D. calcaratum**
1. Perianth segments not spurred at the base .... **D. cantoniense**

Disporum calcaratum D. Don, Trans. Linn. Soc. 18: 516. 1841; FBI 6: 359. 1892; Fischer 3: 1522. 1928.

A herb with erect angular leafy stem, 30-60 cm, arising from a creeping root stock. Leaves 5-8 cm long, oblong-lanceolate, acuminate, base rounded. Flowers in sessile or shortly peduncled umbels, 1.2-1.8 cm long, usually narrow, white greenish or purplish, perianth-lobes spurred at the base, lobes lanceolate, acute, minute ciliate; stamens 6; ovary 3-celled, ovules 2-6 in each cell. Berry fleshy, pisiform, black; seeds few, subglobose, testa brown.

Rare in Visakhapatnam district at Lochili, 1300 m (A.W. Lushington – Fischer). Fl. & Fr.: April – September.

INDIA: Temperate Himalaya, N.E.India, Andhra Pradesh.

WORLD: Myanmar, Nepal, Bhutan, China, Thailand, Vietnam.

Disporum cantoniense (Lour.) Merr., Trans. Hort. Soc. London 1: 331. 1812. *D. pullum* Salisb., Trans. Hort. Soc. 1: 330; FBI 6: 360. 1892; Fischer 3: 1522. 1928.

A herb with erect, angular, leafy stem arising from a creeping root stock. Leaves narrow to ovate-lanceolate, base rounded, margin entire or papillose, apex acuminate, sessile or shortly petioled, 6-10 cm long. Flowers in axillary, few-flowereed umbels, usually peduncled; pedicels up to 1.1 cm long in flower, elongating in fruit up to 3.5 cm, papillose; perianth-lobes 6, spathulate or lanceolate, acute or acuminate, white or dull purple, perianth segments not spurred at the base; stamens 6; ovary 3-celled, glabrous, ovules 2-6 per cell, stigmas 3, divided halfway down. Berry fleshy; seeds few, subglobose, rusty-brown, rugose.

Common in moist places in northern Eastern Ghats. Fl. & Fr.: February – May. Vern.: Ori.: *Launiki.*

Dharwada (EG), *SS* 10194; Maredumilli to Kakur (EG), *GVS* 67591 (CAL); Sunkarimetta (VSKP), *NPBK* 677 (CAL); Satwa, near Araku (VSKP), *NPBK* 10853 (MH); Cherukonda (VSKP), *GVS* 28149 (MH); Barakutni, Deomali (KPT), *SX* & *MB* 6597 (RRL-B); Garasika forest, Keonjhar, *MB* & *Dhal* (RRL-B); Kwadoli, Raigoda, Satkosia Tiger Reserve, *KCM* 6712 (BSID); Chaha, Similipal RF (MBJ), *A.R.K.Sastry* & *G.P.Singh* 12214 (BSID).

INDIA: E. & N.E.India.

WORLD: Myanmar, Bhutan, Nepal, China, Laos, Thailand, Vietnam.

DRIMIA Jacquin ex Willdenow

1. Flowers appearing before the leaves:
    2. Flowers dingy brown or greenish brown:
        3. Flowers dingy brown .... **D. indica**
        3. Flowers greenish brown .... **D. raogibikei**
    2. Flowers white with green shade:
        4. Leaves 8-16; raceme 15-30 cm long .... **D. nagarjunae**
        4. Leavs 2-3; raceme 30-60 cm long .... **D. polyantha**
1. Flowers appearing with the leaves .... **D. wightii**

Drimia indica (Roxb.) Jessop, J. South Afr. Bot. 43: 272. 1977. *Scilla indica* Roxb., Fl. Ind. 2: 147. 1832. *Urginea coromandeliana* (Roxb.) Hook. f., Fl. Brit. India 6: 347. 1892; Fischer 3: 1527. 1928. *Scilla coromandeliana* Roxb., Fl. Ind. 2: 148. 1832. *Urginea wightiana* Hook. f., Fl. Brit. India 6: 347. 1892. *U. indica* (Roxb.) Kunth., Enum. Pl. 4: 333. 1843.

A herb with tunicate bulb; bulb the size of an apple, pale, 5-10 cm long, ovoid, thick. Leaves appearing after the flowers, 15-45 x 1-2 cm, nearly flat, sub-bifarious, linear, acute. Scape erect, brittle, 30-45 x 0.4-0.6 cm at the base. Flowers dingy brown, very distant, in slender, laxy flowered racemes, 15-25 cm long; bracts minute, soon falling, slender, spreading or decurved; perianth campanulate, lobes oblong-lanceolate, obtuse; stamens 6, adnate at or near the base of the perianth lobes; ovary 3-celled, ovules numerous in each cell. Capsule ellipsoid, tapering to both ends, the cells 6-9-seeded; seeds elliptic, 6 mm, flattended, black.

Common throughout Eastern Ghats in shady localities of forest. Fl. & Fr.: March – April. Vern.: Tel.: *Nakkavulligadda, Adavithellagadda*; Ori.: *Kendai, Ban uli, Terkodhan, Ban piajo*; Tam.: *Narivengayam.*

Kalasamudram (ATP), *KRKS* 39903; Vempalli (KDP), *KRKS* & *BSS* 39087; Palakonda hills (KDP), *CS* 9391; V.P.South (GNT), *VRK* 6825; Jatharevurasta (KNL), *VBH* 84976 (BSID); Mahanandi (KNL), *JLE* 25501 (MH); Horsley hills (CTR), *CAB* 15466 (MH); Nerabylu to Talakona (CTR), *GVS* 31990 (MH); Addatigala (EG), *SS*

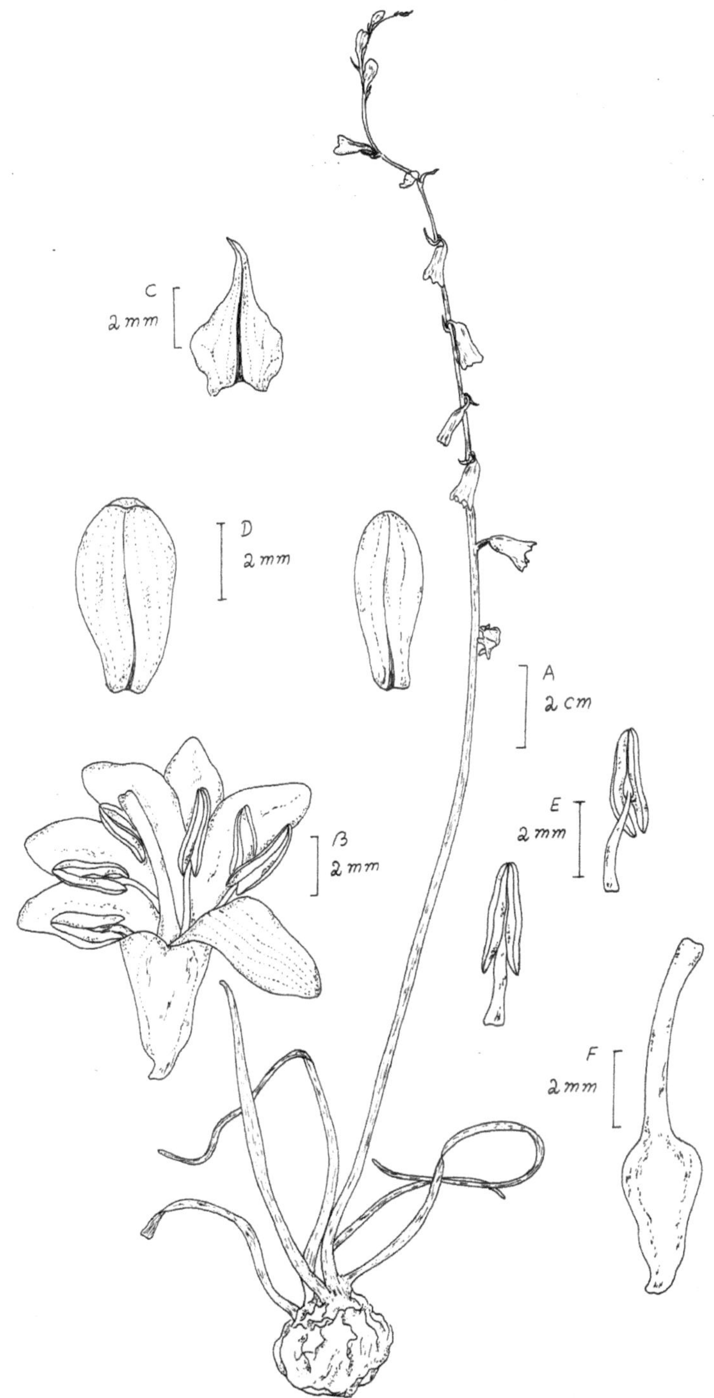

Figure 18. **Drimia indica** (Roxb.) Jessop
A. Habit, B. Flower, C. Bract, D. Perianth. E. Stamens, F. Pistil.

3569 (AU); Tirumalayapale RF (EG), *GVS* 24507 (MH); near Vizag (VSKP), *CAB* 1536 (MH); Ragupalem (VSKP), *CAB* 1536 (MH); Punyagiri (VZN), *GVS* 19404 (MH); Raigoda, Satkosia Tiger Reserve, *KCM* 7601, 7760 (BSID); Gingee RF (SA), *KRM* 13009 (MH); Shanikulam (SA), *CAB* 787(MH); Addapallam RF (DMP), *EV* 57966 (MH); Krishnagiri (KGR), *CAB* 14891 (MH); Hogainakkal RF (KGR), *EV* 20579 (MH).

INDIA: Throughout India plains.

WORLD: Tropical Africa south to South Africa, Pakistan, Myanmar, Vietnam.

Drimia nagarjunae (Hemadri & Swahari) Anand Kumar, J. Econ. Taxon. Bot. 5: 962. 1984. *Urginea nagarjunae* Hemadri & Swahari, Ancient Sci. Life 2: 105. 1982.

A scapigerous herb with perennial tunicated subglobose bulb. Leaves appear after flowering, 8-16-whorled at base, ensiform oblong-lanceolate, 25-45 x 3.5-5.5 cm, green, glaucous on both sides, base narrowed, apex acute. Scape solitary, 25-50 cm long. Raceme 15-30 cm long, many-flowered, flowers 2-2.2 cm long, closely arranged one per bract, bracts broadly ovate, acute-acuminate, auriculate, extending at the base into a spur; perianth lobes 6, in two whorls, oblong-lanceolate, to 2.1 cm, white with greenish shade along the two very closely arranged midribs; stamens 6, filaments white, flesh, dorsally compressed, anther versatile, yellow; ovary ovoid-ellipsoid, to 1 x 0.5 cm. Capsule ellipsoid, to 2 x 1.1 cm, trilocular, glabrous; seeds many, black, to 1.2 x 1 cm, having brownish papery, transparent wing.

Occasional in the Coastal districts in waste lands. Fl. & Fr.: May – July. Vern.: Tel.: *Adaviulligadda.*

Bhata village (Near Udayagiri hill fort) (NLR), *Hemadri,* 3001 A (Holotype) (Regional Research Centre Experimental Garden, Vijayawada).

INDIA: South India. Endemic.

Drimia polyantha (Blatt. & McCann) Stearn, Ann. Mus. Goulandris 4: 209. 1978. *Urginea polyantha* Blatt. & McCann, J. Bombay Nat. Hist. Soc. 32: 735. 1928; Chorghe *et al.*, Nelumbo 59: 70. 2017.

Bulbs ovoid or globose, white, *c.* 5 x 7 cm with 8-10 roots at base, neck *c.* 1.5 cm long. Leaves 2 or 3, linear-oblong, 40-53 x 1-2.5 cm, acuminate at apex, ashy green coloured. Inflorescences up to 60 cm long. Flowers about 1 cm long; pedicels filiform, *c.* 6 mm long; bracts triangular, *c.*1mm long; perianth rotate, yellowish-green; lobes obtuse, 5–6 x 3 mm; stamens *c.* 2.5 mm long; ovary oblong, 6 grooved; style short, stout; stigma trigonus.

Common on the rocky crevices of dry deciduous forests. Fl. & Fr.: February-May.

Penchalakona (NLR), *L. Rasingam, M. Sankara Rao & Alok Chorghe* 2977 (BSID).

INDIA: Andhra Pradesh and Maharashtra. Endemic.

Drimia raogibikei (Hemadri) Pullaiah, Fl. Andhra Pradesh 1930. 2018. *Urginea raogibikei* Hemadri, Med. Flora Pragati Resorts 386. 2006.

Scapigerous, perennial herb with tunicate, compressedly subglobose-ovoid, 4-5 cm long and 4-6 cm broad bulb with a neck of 2-4 cm long, sometimes longer than broad if grown amongst stones; outer scales dirty white, inner white. Leaves 3-7, appearing after flowering (*i.e.*, during the months of July-August and retain till December – January), 20-30 x 1.1-2 cm, green, ashy both sides, linear-oblong with acute apex. Flowering scape solitary, (sometimes a second one at a later stage), appearing during April-May (-June), 15-20 cm high including 9-12 cm long raceme, 0.4-0.6 cm thick at base, cylindrical, greenish-brown. Inflorescence a raceme of 20-30 flowers, ending with a few ill-formed buds. Bracts ovate, 0.2-0.3 x 0.15 -0.2 cm, the lowest 2-3 bracts deltoid and spurred. Flowers 0.75-0.9 cm long, closely arranged; pedicel 0.3-0.4 cm long; perianth lobes 6 in two whorls, connate at base, non-reflexed (non revolute) when fully opened, 0.7-0.8 x 0.35-0.4 cm, elliptic-oblong, light yellow with scarious margins, brownish-buff coloured along the midrib; stamens 6, arising from the base of perianth segments; filaments about 0.2 cm long, gynoecium 0.6-0.7 cm long including 0.3 cm long style; ovary 0.3 cm thick, green. Fruit an ovoid, triquetrous capsule of 1-1.2 x 0.8-1 cm with 0.4-0.5 cm long stalk; seeds many, 0.08-0.1 x 0.5-0.7 cm, black and broadly winged.

Occasional in the scrub jungles situated along the Udayagiri-Bhata village road, Nellore district. Fl. & Fr.: May – June. Vern.: Tel.: *Adavi ulligadda* (This name is common for all the Indian species of *Drimia*).

INDIA: Andhra Pradesh, Telangana. Endemic.

Drimia wightii Lakshmin., Kew Bull. 58: 507. 2003. *Urginea congesta* Wight, Icon. Pl. Ind. Orient. t. 2064. 1853; FBI 6: 348. 1892; Fischer 3: 1527. 1928. *Drimia congesta* (Wight) Ansari & Raghavan, J. Bombay Nat. Hist. Soc. 17: 174. 1980 illegitimate.

A small bulbous herb; bulb ellipsoid, 2.5 cm diameter. Leaves linear, 8-15 cm, filiform, grooved above, leaves appearing with the flowers. Scape 15 cm, flexuous. Flowers racemed, 5-8 cm, bracts minute, deltoid; perianth white and purplish; stamens 6; ovary 3-celled, ovules numerous in each cell. Capsule subglobose, 8 mm broad, cells 3-4-seeded.

Rare in the hills of East Godavari district. Fl. & Fr.: April – May

Addatigala (EG), *SS* 3571 (AU).

INDIA: Peninsular India.

WORLD: Sri Lanka.

GLORIOSA Linnaeus

Gloriosa superba L., Sp. Pl. 305. 1753; FBI 6: 358. 1892; Fischer 3: 1519. 1928; Matthew, Fl. Tamilnadu Carnatic 3: 1643. 1983.

A herbaceous, tall, glabrous branching climber, root stock tuberous, solid, fleshy-white, cylindric, 15-25 x 2.5-4 cm, pointed at each end, V-shaped, producing a new joint at the end of each branch; roots fibrous; stems annual, 3-6 m long, given off from the angles of the young tubers, herbaceous. Leaves alternate, opposite or whorled, oblong-lanceolate, sessile, 5-12 x 1-3 cm, flat, glabrous, base cordate, margin entire, apex tendrillar. Flowers large, 8-10 cm across, bisexual, solitary or some what subcorymobose at the ends of branches; perianth lobes 6, oblong-lanceolate, with crisply waved margins, greenish at first, then yellow, passing through orange and scarlet to crimson, flower exhibits herkogamy; stamens 6, spreading; ovary oblong, ovules many per cell on axile placentae, stigmas 3. Capsule ellipsoid-oblong, torulose; seeds numerous, globose, dorsally compressed, warty.

Common throughout Eastern Ghats, often climbing on hedge-row plants and forests. Fl. & Fr.: August – September. Vern.: Tel.: *Adavi-nabhi*; Ori.: *Ognisikha, Meheria phulo, Karihari, Dasara phulo, Na nangalia.*

Eswaramala hills (ATP), *B.Aswarthappa* 502 (BSID); Rangapuram RF (KNL), *RVR* 1711; Palakonda hills (KDP), *CS* 7679; Kalikiri (CTR), *KNR & DAM* 9835; Venkatagiridurgam (NLR), *SSY* 12415; Peddamantanala (PKM), *RVK* 15791; Macherla (GNT), *VRK* 3998; Anantagiri (VSKP), *KSK* 23431; Near Jeerangi RF (Gajapathy distr.) *K. Sampath Kumar* 23480; Balapalle (KDP), *JSG* 18169 (MH); Palamaner RF (CTR), *KCJ* 454 (MH); near Chittoor, *RHB* 1855 (MH); Mallammagonda (EG), *GVS* 67553 (MH); Sunkarimetta (VSKP), *NPBK* 10902 (MH); Chintapalli (VSKP), *GVS* 19782 (MH); Simalaguda (VZN), *MV* 2744 (AU); Edrallapalli (KMM), *RCS* 98759, 102405 (BSID); Kambukudi, Poovankulam RF (NA), *KS* 6059 (MH); Tippukadu RF (NA), *KRM* 17641 (MH); Rengappanur (SA), *KRM* 79662 (MH); Kaduvanoor (SA), *KRM* 50612 (MH); Theerthamalai RF (DMP), *EV* 51808 (MH); Kakkambadi, Yercaud (SLM), *SKK* 26860 (MH); Doddasampige, B.R.Hills (Mysore distr.), *VBH* 96899 (MH); Hogainakkal RF (KGR), *EV* 2926 (MH); Vandalur (CPT), *CAB* 14122 (MH); Rukeekonda (GJM), *VNS* 6064 (MH); Raxi (Sundargarh Dist.), *D.Namhata* 2416 (BSID); Purunakote, Satkosia Wild life sanctuary, *D.Hazra* & *D.Das* s.n. (BSID).

INDIA: Throughout India.

WORLD: Tropical Africa, Sri Lanka, Myanmar, Nepal, China, Cambodia, Laos, Indonesia, Thailand, Vietnam.

IPHIGENIA Kunth *nom. cons.*

1. Leaves acicular; perianth filiform .... **I. mysorensis**
1. Leaves not acicular; perianth elliptic-linear .... **I. indica**

Iphigenia indica (L.) A. Gray ex Kunth, Enum. Pl. 4: 213; 1843; FBI 6: 357. 1892; Fischer 3: 1528. 1928; Matthew, Fl. Tamilnadu Carnatic 3: 1643. 1983. *Melanthium indicum* L., Mant. Pl. 226. 1771.

A herb with small tunicate corm, corm subglobose, 0.5-1 cm long, 1.5 cm across, narrowed into short neck, tunicate with pale brown sheaths, stem 8-15 cm high, rigid or flexuous. Leaves alternate, few, sessile, the lower 8-15 x 0.2-0.4 cm, the upper gradually smaller, all narrowly linear or linear-lanceolate, glabrous, margin entire, apex acuminate; leaf sheath 4 cm, tubular. Flowers purple, few or many in

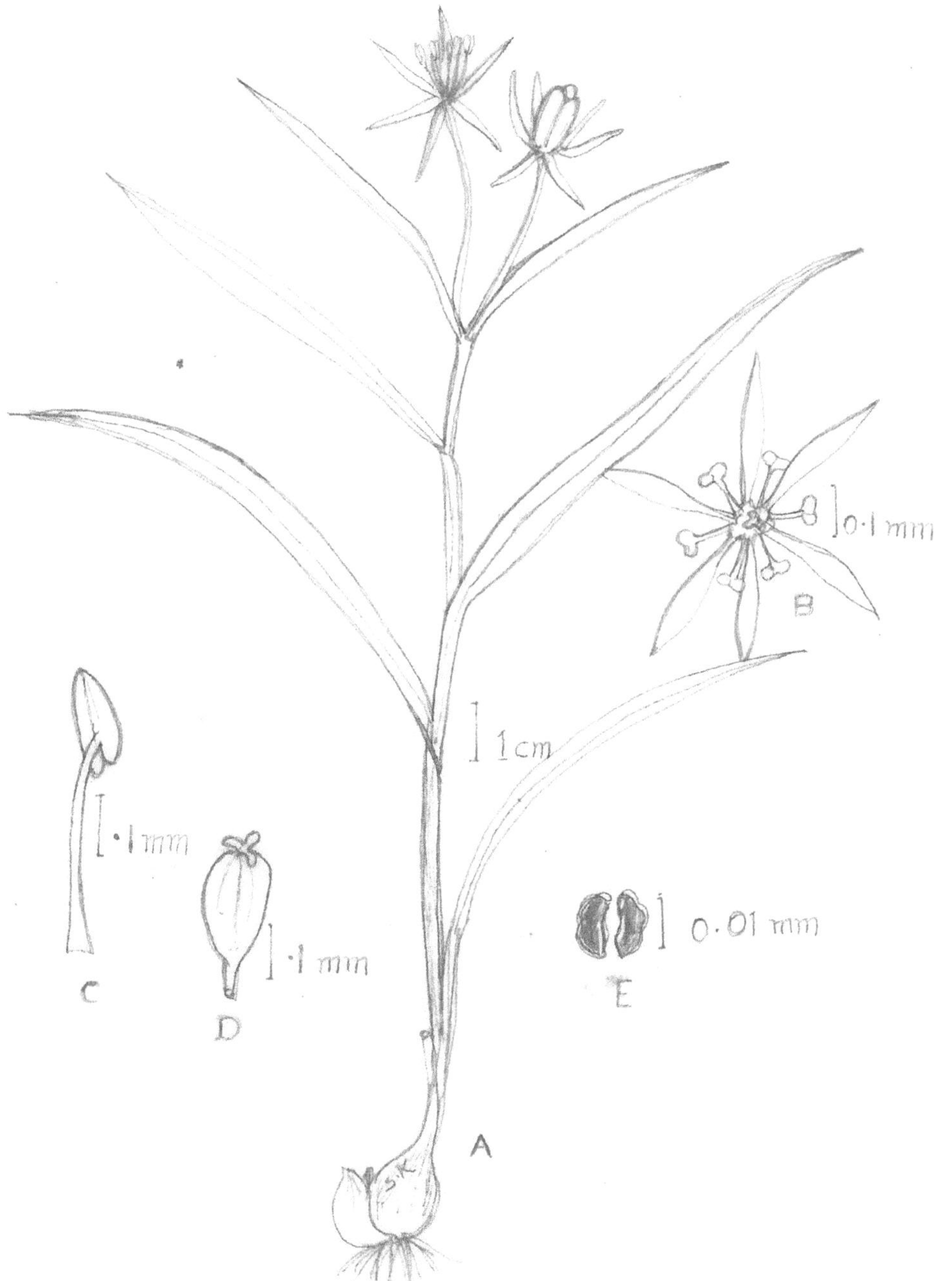

Figure 19. **Iphigenia indica** (L.) A. Gray ex Kunth

A. Habit; B. Flower-entire; C. Stamen; D. Pistil; E. Seeds.

a terminal erect raceme; bracts linear or subulate, the lower leafy; perianth lobes 6, deep purple, linear-oblong, spreading; stamens oblong; ovules many per cell on axile placentae. Capsule oblong, loculicidal; seeds ovoid, warty, brown.

Occasional in deciduous forests throughout Eastern Ghats. Fl. & Fr.: July – September. Ven.: Tam.: *Neerpanai.*

Kalasamudram (ATP), *KRKS* 39959; Peccheruvu (KNL), *BR* & *BSS* 30378; Nichenala kanuma (KDP), *BR* & *BSS* 29563; Mahanandi (KNL), *JLE* 25501 (MH); Muskanbavi (KNL), *VBH* 83944 (BSID); Kakula Dibba (CTR), *GVS* 46897 (CAL); Ankirandi kuntala (CTR), *GVS* 46033 (MH); Venkataagiri (NLR), *BSN* 5287 (BSID); Similipahar (MBJ), *SX* & *MB* 4499, 4741 (RRL-B); Mahendragiri (GJM), *SX* & *MB* 3600 (RRL-B); Mahendragiri (GJM), *VNS* 5581 (MH); Purunkote RF, Satkosiaa Tiger Reserve, *KCM* 5184 (BSID); Vandalur (CPT), *CAB* 11562 (MH); Royakota (SLM), *CAB* 13934 (MH); Way Ariyur shola, Kolli hills (NMK), *A.A.Ansari* 97010 (MH).

INDIA: Throughout India.

WORLD: Sri Lanka, Myanmar, Nepal, China, Cambodia, Philippines, Thailand, Vietnam, N. Australia.

Iphigenia mysorensis Arekal & Swamy, Bot. Not. 125: 220. 1972.

Perennial, slender, corm bearing plant; corm 0.9-1 cm in diam., bulb-like with pale brown sheath and somewhat flattened on the side, stem up to 20 cm high. Leaves aciular, up to 13 x 0.15cm, thick. Flowers dark purple in terminal, 2-4-flowered raceme; perianth lobes 6, free, filiform, lobes 0.6-1 cm long; stamens 6; styles 3. Capsule obovoid; seeds several, trigonous.

Rare in forests in Anantapur district. Fl. & Fr.: September-October.

Kalasamudram RF (ATP), *EC* 2902.

INDIA: South India. Endemic.

LEDEBOURIA Roth

1. Leaves spreading with dark purplish blotches, bulbils with apex..**R. revoluta**
1. Leaves erect without purplish blotches, bulbils absent........**R. hyderabadensis**

Ledebouria hyderabadensis M.V.Ramana *et al.,* Kew Bull. 67(3); 1. 2012; Kaliamoorthy & Saravanan, Indian J.Forestry 41: 57. 2018.

Bulbous herb, scapigerous, up to 15 cm high including scape; bulb ovoid, 1.5 x 1.5 – 2 cm, tunicated; tunic dark brown. Leaves radical, 2 – 6; lamina bright green, fleshy, devoid of blotches on both surfaces, oblong, linear-lanceolate, 5 – 10 x 0.5 – 1.0 cm, attenuate and sheathing at base, entire at margin, acute at apex, glabrous, devoid of bulbils. Scape slender, unbranched, longer than leaves, 8 – 14 cm long, 10 – 25 flowered, racemose; flowers pinkish green, campanulate, nodding, 5 – 8 mm long, rotate when fully opened, bracteate; bract 1, lunate with a semicircular notch above, *c.* 2 mm; pedicels 3 – 4 mm long; perianth lobes divided near to the base; lobes 6, biseriate, oblong, ovate, acute, *c.* 6 x 2 mm valvate, pinkish purple with green veins, reflexed from middle in mature flowers, persistent; stamens 6; filaments free, adnate at base to perianth, pinkish, 4 mm long; anthers oblong, 2-celled, dorsifixed and versatile, dehiscing longitudinally; ovary tricarpellary, stipitate, shallowly but distinctly 3-lobed, 3-celled, 1 x 1 mm; lobes more foliar; ovules 6, 2 per cell, attached at base; style simple, erect, pinkish, *c.* 4 mm long; stigma triquetrous, penicillate. Capsule obovoid, membranous, 1 – 2 lobed, rarely 3 –lobed, *c.* 5 x 7 mm with persistent perianth and staminal filaments; seeds 1 per cell, attached at base, smooth, obovoid, 4 mm, brownish, albuminous.

Growing in grasslands in open sunlight on steep and rocky slopes in associationwith *Ledebouria revoluta* (L.f.) Jessop, *Brachystelma elenaduense* Satyan. and *Iphigenia indica* (L.) A.Gray ex Kunth, between 1150 and 1490 m in Yercaud in Shevaroy hills. Fl. & Fr.: May - August.

Manjakuttai, Yercaud (SLM), *S.Kaliamoorthy* & *T.S.Saravanan* 109606 (BSID).

INDIA: Telangana and Tamil Nadu.

Ledebouria revoluta (L.f.) Jessop, J. S. African Bot. 36: 255. 1970. *Hyacinthus revoluta* L.f., Suppl. Pl. 204. 1782. *Scilla hyacinthina* (Roth) Macbr., Contr. Gray. Hrb. 56: 14. 1918; B.R.P. Rao in Pullaiah, Fl. Andhra Pradesh 3: 992. *Ledebouria hyacinthina* Roth, Nov. Pl. Sp. 195. 1821. *Scilla indica* (Wight) Baker in Saunders, Refug. Bot. 3(App.): 12. 1870; (non Roxb. 1832); FBI 6: 348. 1892; Fischer 3: 1527. 1928. *Barnardia indica* Wight, Icon. Pl. Ind. Orient. t. 2041. 1853.

A herb with tunicate bulb; bulb ovoid or globose, 2-6 x 1.5-3 cm, apex narrowing into a short neck. Leaves radical, 4-15 x 2-3.5 cm, variable, oblong to lanceolate or oblanceolate, chartaceous, occasionally with purplish blotches, glabrous, base gradually attenuate and sheathing, apex obtuse-subacute, mucronate. Scape (s) 1-3, racemes 5-12 cm, 30-60-flowered; flowers greenish-purple, bisexual, bracts scarious, bracteoles 2, adnate to the base of pedicel; perianth deeply 6-lobed to the base, greenish-purple, oblong, subequal; stamens 6, adnate to the base of perianth; ovary 3-lobed, 3-celled, ovules 2 or more per cell. Capsule globose, loculicidal, thin-walled.

Occasional in forests and waste lands. Fl. & Fr.: June – September. Vern.: Tel.: *Adavi thellagadda;* Tam.: *Narivengayam.*

Gani RF (KNL), *RVR* 1509; Rudrakod (KNL), *BR* & *BSS* 29566; Palakonda hills (KDP), *CS* 9393; Punganur (CTR), *BR* & *MVS* 31165; Pottepalem (NLR), *PMR* 16486; near Kurnool, *JSG* 17723 (MH); Guvvalacheruvu (KDP), *KS* 6433 (MH); Balapalle (KDP), *JLE* 14265 (MH); Thalakona (CTR), *GVS* 3199 (MH); Venkatagiri (NLR), *BSN* 6100 (BSID); Chinnaganjam (PKM), *KR* & *RG* 88123 (BSID); Mougalathala RF (PKM), *RKM* 0832 (CAL); Gollapalem (KSN), *PV* 5801(MH); Thirumalayapalem (EG), *GVS* 24507 (MH); near Vatavarlapalli (MBNR), *VBH* 84957 (BSID).

INDIA: Peninsular India.

WORLD: Tropical Africa, Southern Africa, Sri Lanka.

LILIUM Linnaeus

Lilium wallichianum Schultes & Schultes f. var. **neilgherrense** (Wight) H. Hara, Fl. E. Himal. 3rd. Rep.: 132. 1975; Matthew, Fl. Tamilnadu Carnatic 3: 1644. 1983. *L. neilgherrense* Wight, Icon. Pl. Ind. Orient. 6: 20, t. 2031. 1853.

An erect, unbranched, perennial herb, to 1 m high, bulb globose. Leaves alternate, lanceolate, 7-11 x 1.5-2 cm, scattered, stalkless, pointed at tip, glabrous. Flowers white, 1-3 in terminal umbels; perianth funnel-shped 15-18.5 x 1-3 cm, lobes 3 + 3, broadly ovate, unequal; stamens 6, filaments 15 cm long, included, free; ovary 3-celled, 6-furrowed, ovules many, stigma obscurely 3-lobed. Capsule oblong, to 4 x 0.5 cm, 3-volved, seeds numerous, compressed, black.

Scarce in Shevaroy and Sirumalai hills. Fl. & Fr.: November – February.

Sirumalai hills, Vellimalai peak, Agastiyapuram (DGL), *SKS* 711 (SGH).

INDIA: Peninular India.

WORLD: Nepal.

OPHIOPOGON Ker-Gawler *nom. cons*

Ophiopogon intermedius D. Don, Prodr. Fl. Nepal 48. 1825; FBI 6: 269. 1892; Fischer 3: 1499. 1928; Matthew, Fl. Tamilnadu Carnatic 3: 1645. 1983. *O. parviflorus* (Hook. f.) Hara, J. Jap. Bot. 40: 21. 1965 & Fl. East. Himal. 409. 1966. *O. intermedius* var. *parviflorus* Hook.f., Fl. Brit. India 6: 269. 1892. *Flueggea grifffithii* Baker, J.Linn. Soc. Bot. 17: 502. 1879. *Ophiopogon griffithii* (Baker) Hook.f., Fl. Brit. India 6: 270. 1892. *O. wallichianus* Hook.f., Fl. Brit. India 6: 268. 1892.

A scapigerous herb, stem short from a short root stock. Leaves 12-40 x 1-2 cm, linear, apex acute or acuminate, dark green above, paler beneath. Flowers white, rather distant, in lax racemes, with slender 10-25 cm long scapes, bracts narrowly lanceolate; perianth campanulate, segments elliptic-oblong, obtuse; stamens 6; ovary inferior, 3-celled, the crown flat or depressed, ovules 2 in each cell. Seeds subglobose, crowded round and almost concealing the small withering pericarp.

Common in forest in higher hills. Fl. & Fr.: August – October.

Mahendragiri (GJM), *VNS* 5646 (MH); Kuthadiya hill (GJM), *VNS* 5912 (MH); Singaraju parbat (GJM), *S.Misra* Acc. No. 3281 (RRL-B); Shevaroy hills, Yercaud (SLM), *AVNR* 26777 (MH);

INDIA: S.India, Temperate Himalaya, N.E.India.

WORLD: Sri Lanka, Bangladesh, Myanmar, China,

SMILACEAEAE

SMILAX Linnaeus

1. Stem unarmed or nearly so; branches unarmed; leaves <15 cm long, membranous; umbels solitary .... **S. lanceifolia**
1. Stem prickly, branches prickly or not; leaves >10 cm long; umbels rarely solitary:

Fiure 20. **Lilium wallichianum** Schultes & Schultes f.

A. Habit; B. Bulb; C. Stamen; D. Pistil; E. Ovary-l.s.; F. Ovary-c.s.

2. Umbels sessile .... **S. aspera**
2. Umbels not sessile:
   3. Umbels 1-3; petiolar sheath narrow, not auriculate .... **S. zeylanica**
   3. Umbels more than 7; petiolar sheath broad; auriculate .... **S. perfoliata**

Smilax aspera L., Sp. Pl. 1028. 1753; FBI 6: 306. 1892; Fischer 3: 1518. 1928.

A large shrub, branches armed or not. Leaves ovate-deltoid, hastate-lanceolate or broadly cordate, sometimes a little broader than long, acute or acuminate, base more or less deeply cordate, margins and veins below sometimes prickly, 4-10 x 2.5-8 cm, petioles sometimes armed, hardly sheathed at the very base; tendrils from near the base. Flowers dioecious, umbellate, umbels sessile on an axillary peduncle, sometimes forming a terminal panicle by the suppression of the upper leaves. Male flower: perianth-lobes 6, free, subequal, outer one broad, inner one narrow; stamens 6, adnate to the base of perianth, filaments free, erect, anthers oblong; pistillode O. Female flower: staminods 3 or 6, filiform, ovary 3-celled, 3-gonous; ovules 1-2 in each cell. Berry globose, ca 3-seeded.

Rare in forests in Visakhapatnam district at Ventala, 1300 m (A.W. Lushington cf. Fischer). Fl. & Fr.: February – April, Fruits April onwards.

INDIA: Throughout India.

WORLD: E and N Africa, SW Asia, C and S Europe, Myanmar, Bhutan, Nepal. China.

Smilax lanceifolia Roxb., Fl. Ind. 3: 792. 1832; FBI 6: 308. 1892; Saxena & Brahmam, Fl. Orissa 3: 1971. 1995.

Slender climber; stem usuallay unarmed, some times prickly, branches unarmed. Leaves oblong, elliptic-oblong or ovate-oblong, up to 7.5 x 3.5 cm, apex obtusely acuminate, base obtuse or acute, glossy on both sides; petiole up to 1.2 cm long, tendrils from the sides of the older petioles. Flowers in axillary umbels, peduncle 5-7.5 cm long; male umbel 1.8-2.2 cm dim., tepals linear, 3-6.2 mm long; female umbels subsimilar, staminodes 3; ovary obtusely trigonous, stigmas obtuse. Berry to 7.5 mm diam.

Rare in Mahendragiri hills in Odisha. Fl.: April. Vern.: Ori.: *Ramdantuni.*

Mahendragiri hills (GJM), *SX* & *MB* 2224 (RRL-B).

Smilax perfoliata Lour., Fl. Cochinch. 2: 622. 1790; Matthew, Fl. Tamilnadu Carnatic 3: 1647. 1983. *S. prolifera* Wall. ex Roxb., Fl. Ind. 3: 795. 1832; FBI 6: 312. 1892; Fischer 3: 1519. 1928.

A vine of 8-10 m long; branchlets glabrous; prickles recurved. Leaves broadly ovate or elliptic-lanceolate, 7-16 x 4-12 cm, coriaceous, 5-nerved, glabrous, base rotund-subacute, margin undulate, entire, apex obtuse, sometimes retuse; abruptly cuspidate; leaf-sheath broad, auriculate, amplexicaul. Male flowers: umbels *c.* 7, axillary, alternate or verticillate; bracts ovate, cuspidate, bracteoles scarious; perianth

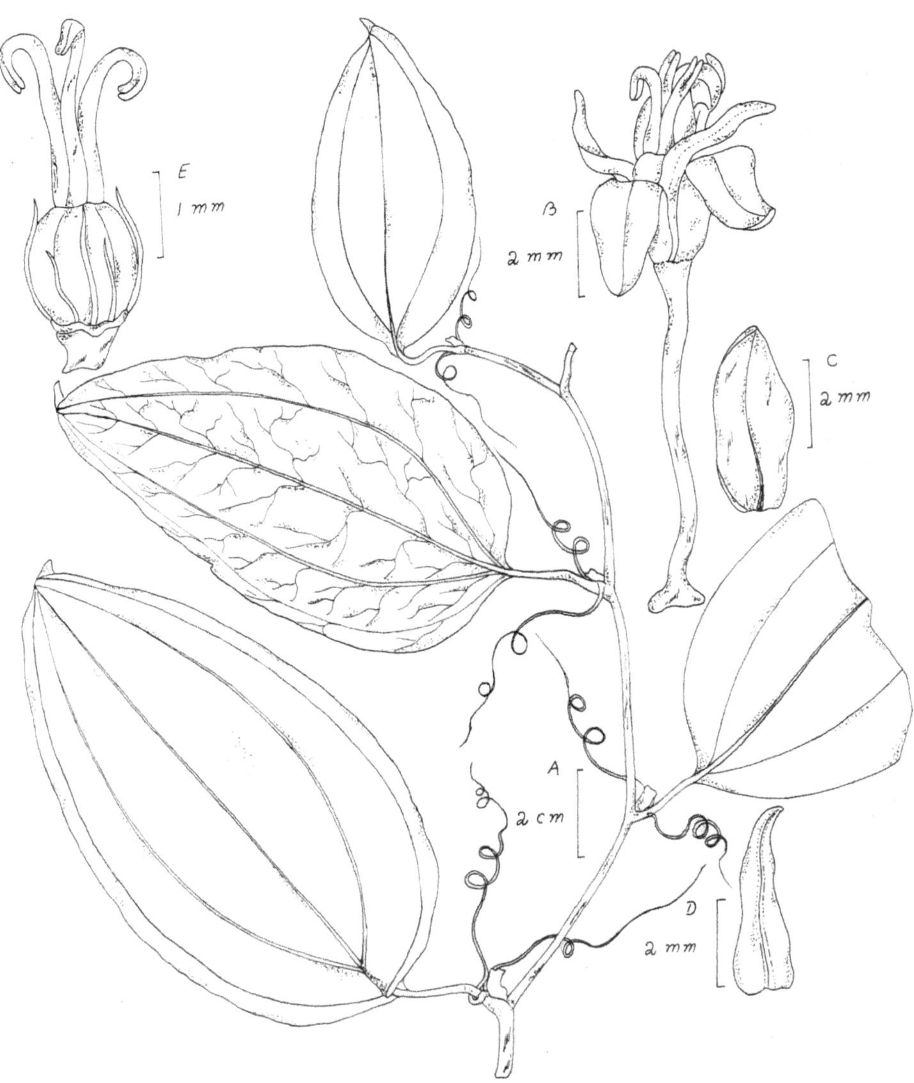

Figure 21. **Smilax perfoliata** Lour.
A. Twig, B. Female flower, C. Perianth lobe (outer), D. Perianth lobe (inner), E. Pistil and staminodes.

lobes greenish, oblong, outer ones broad, apiculate, falcate, inner ones narrow, obtuse; stamens 6. Female flower: umbels axillary, alternate or verticillate; ovary 3-celled, 3-gonous, ovules 1-2 in each cell. Berry globose.

Common in forests. Fl.: February – April. Fr.: April onwards. Vern.: Ori.: *Ramdatani, Ramdatun, Mutri, Mothuri.*

Talakona (CTR), *KNR* & *PSPB* 9851; Thumbura Theertham, *BSS* & *MVS* 30394 (BSID); Cuddapah forest, *CBDM* 39 (MH); Tirumala (CTR), *GVS* 31917 & 61478 (MH); Microwave station, Tirumala (CTR), *DRC* 1141 (MH); Mundanchala (VSKP), *NPBK* 10958 (MH); Towards Galikonda (VSKP), *GVS* 19615 (MH); Gudem valley (VSKP), *VNS* 481 (CAL); Neelakantapuram (SKLM), *GVS* 62465 (MH); Bandirevu RF (KMM), *RCS* 98920 (BSID); Shevaroy hills (SLM), *Sine coll. s.n.* (MH); Kongampallm, Yercaud (SLM), *DBD* 31423 (MH);

INDIA: Peninsular, E. & NE. India, Tropical Himalaya.

WORLD: Myanmar, Laos, Thailand, Vietnam.

Smilax zeylanica L., Sp. Pl. 1029. 1753; FBI 6: 309. 1892; Fischer 3: 1519. 1928; Matthew, Fl. Tamilnadu Carnatic 3: 1648. 1983. *S. macrophylla* Roxb., Fl. Ind. 3: 793. 1832; FBI 6: 310. 1892.

A large climber; stem smooth, striate, armed with a few small distant prickles or almost unarmed. Leaves alternate, oblong-broadly elliptic or ovate-lanceolate, 8-20 x 4-10 cm, coriaceous, glabrous, base obtuse-rotund, margin entire, apex obtuse, somewhat notched, cuspidate, leaf-sheath narrow, not auricled. Flowers in pedunculate many-flowered umbels. Male flowers: umbels axillary, 1-3, bracts oblong, bracteoles triangular; perianth lobes greenish, linear, obtuse, erect when young, afterwards reflexed, outer ones broad, inner ones narrow, obtuse; stamens 6. Female flowers: ovary 3-lobed, 3-celled, ovule(s) 1 or 2 per cell, staminodes 3-6, linear. Berry globose, smooth, remaining green for a long time, becoming red when ripe.

Occasional in deciduous forests. Fl. & Fr.: June – September. Vern.: *Kondathamara*; Ori.: *Mutri, Mothuri, Ramdatani*; Tam.: *Kaatukodi.*

Ahobilum (KNL), *TP & KNR* 9855; Peccheruvu (KNL), *RVR* & *PVP* 2186; Gundlabramheswaram (KNL), *MVS* & *DV* 38062; Talakona (CTR), *BR* & *BSS* 30256; Maredumilli, neat Tiger park (EG), *KSK* 23446; Tyada RF (VSKP), *BR* & *KNR* 9893; Maredumilli (EG), *GVS* 24206 (MH); Vathangi (EG), *SS* 4916 (AU); Barnakonda RF (VSKP), *KCJ* 17168 (MH); Veduruwada (VZN), *MV* 4079 (AU); on the way to Pubbada, Donubai (SKLM), *NRR* 83620 (MH); Way to Seedhi, near Donubai (SKLM), *NRR* 86133 (MH); Melpat (SA), *CAB* 1087 (MH); Chulliyar shola, Pannaikadu (DGL), *K.Ravikumar* 92522 (MH); Yercaud (SLM), *SKK* 26830 (MH); on the way to Guthirayan (DMP), *TRS* 84192 (MH); Tikarpada, Satkosia Tiger Reserve, *KCM* 7041 (BSID); Purunakote, Satkosia Wild Life Sanctuary, *D. Hazra* & *D.Das s.n.* (BSID); Chahala area, Similipal Reserve (MBJ), *A.R.K.Sastry* & *G.P.Singh* 12203 (BSID).

INDIA: Throughout India on hill tops.

WORLD: Myanmar, Malaysia, Indonesia, Solomon Islands.

PONTEDERIACEAE

1. Plants with swollen petioles..**Eichhornia**
1. Plants without swollen petioles...**Monochoria**

EICHHORNIA Kunth *nom. cons*

Eichhornia crassipes (C. Martius) Solms-Laub. in DC., Monogr. Phan. 4: 527. 1833; Fischer 3: 1530. 1928; Matthew, Fl. Tamilnadu Carnatic 3: 1649. 1983. *Pontederia crassipes* C. Martius, Nov. Gen. Sp. 1: 9. t. 4. 1823.

A free floating or rooted herb, stolons slender, roots elongate, fibrous. Leaves in a rosette, spoon or paddle shaped, 5-10 x 5-10 cm, margin entire, apex rounded or obtuse; petiole elongate, 6-30 cm long, spongy with a fusiform bulbous portion to form floats. Scape erect from the centre of the rosette, 15-25 cm long; spike terminal, *c.* 15-flowered. Flowers violet-blue, perianth funnel-shaped, tube often 2-lipped, upper one lilac, with a blue, bearded, yellow median blotch; lower one gradually smaller; stamens 6, declinate, anthers subequal, dorsifixed; ovary 3-celled, oblong, glabrous, ovules numerous per cell on axile placentae. Capsule ovoid-oblong; seeds ovoid, many-ribbed.

Very common in still or slow-flowing waters, often clogging irrigarion canals. Fl. & Fr.: September – February. Vern.: Tel.: *Gurrapudekka*; Ori.: *Bilatidala, Kajaropati.*

Palakonda hills (KDP), *CS* 9374; Jaladanki (NLR), *PMR* 2236; Kolleru (EG), *BR & KNR* 10622; Tatipudi (VZN), *MCK* 18867; K. Kota (KSN), *PV* 5122 (AU); Rampa (EG), *SS* 9782 (AU); Yercaud (SLM), *SKK* 28280 (MH); Surpanam chamadi (SA), *VNS* 5048 (MH).

INDIA: Throughout India.

WORLD: Native of Brazil, widely introduced and naturalized in tropics and subtropics.

MONOCHORIA Presler

1. Roots stock elongate, creeping; leaves hastate......................................**M. hastata**
1. Roots stock short, suberect; leaves ovate or subreniform................**M. vaginalis**

Monochoria hastata (L.) Solms-Laub. in A. DC., Monog. Phan. 4: 523. 1883. *Pontederia hastata* L., Sp. Pl. Pl. 288 1753. *Monochoria hastaefolia* Presl, Reliq. Haenk. 1: 128. 1827; FBI 6: 362. 1892; Fischer 3: 1529. 1928.

A freshwater floating erect herb; root stout, creeping and rooting below, spongy, clothed with the remains of old sheaths. Leaves 10-20 x 5-15 cm, sagittate, hastate or cordate, obtuse, acute or acuminate, smooth and glossy; many-nerved; petioles of the floral leaves tumid above and embracing spathe-like, short, stout peduncle; petioles of the radical leaves 45-60 cm long, with a broad sheathing base. Inflorescence centrifugal, flowers of a brilliant purplish blue, crowded racemose or subumbellate; perianth-segments twisted round the fruit when withering, large segments obovate, the smaller segments oblong; stamens 6, one usually longer than the rest with its filament horned on one side; ovary 3-celled, ovules many in each cell. Capsule 8 mm long, oblong; seeds many, ovoid, many-ribbed.

Occasional in still or slow flowing waters, sometimes in brackish water, at low elevarions. Fl. & Fr.: August – December. Vern.: Tel.: *Nirtamara*.

Bukkapatnam RF (ATP), *NY* 640; Kona – Tadipatri (ATP), *TP* 929; Krishna Nandi (KNL), *TP & KNR* 9865; Saralanka (EG), *SS* 4896 (AU); Arilova (VSKP), *S.India Flora* 17095 (MH); Near Vadavalasa (SKLM), *GVS* 62405 (MH & CAL); Tikarpada, Satkosia Tiger Reserve, *KCM* & *J.Swamy* 7622 (BSID); Purankote, Satkosia Wild Life Sanctuary, *D.Hazra* & *D.Das* 7260 (BSID); Poondi (CPT), *DN* 874 (MH).

INDIA: Throughout India.

WORLD: Sri Lanka, Bangladesh, Myanmar, China, Indonesia, Malaysia, Philippines, Thailand, Laos, Vietnam, N Australia.

Monochoria vaginalis (Burm. f.) Presl., Reliq. Haenk 1: 128. 1827; FBI 6: 363. 1892; Fischer 3: 1529. 1928; Matthew, Fl. Tamilnadu Carnatic 3: 1650. 1983. *Pontederia vaginalis* Burm. f., Fl. Ind. 80. 1768.

A rooted aquatic herb, 20-60 cm long; root stock short, suberect, spongy. Leaves very variable, 5-10 x 1.5-5 cm, from linear to ovate or ovate-cordate, margin entire, apex gradually acuminate. The peduncle emerging from the channeled sheaths of the upper most leaves. Inflorescence centripetal. Flowers blue, usually spotted with red, regular; perianth campanulate, deeply lobed, lobes 3 + 3, oblong-obovate, inner ones broad, outer ones narrow; stamens 6, inserted at the base of the perianth; ovary 3-celled, globose, ovules many per cell on deeply forked, swollen axile placentae. Capsule oblong, to 1 x 0.8 cm, glabrous; seeds oblong, ribbed.

Common in stagnant ponds, tanks and slow flowing water throughout Eastern Ghats. Fl. & Fr.: August – December. Vern.: Tel.: *Nirkancha*; Ori.: *Mirmira*; Tam.: *Karumkuvalai*.

Aluru kona (ATP), *TP* 929; Near Mudigubba (ATP), *KNR* & *DAM* 9824; Penakacherla (ATP), *NY* 591 (MH); Owk RF (KNL), *RVR* 1477; Pulipadu – Darsi (PKM), *MCK* 22856; Macherla RF (GNT), *VRK* 5887; Diguvametta – Nallamalais (PKM), *JLE* 32384 (MH); Balapalli (KDP), *JLE* 15776 (MH & CAL); Komaticheruvu (CTR), *GVS* 45863 (MH); Amalapuram (EG), *KR* & *RG* 88814; Mudurlanka, Maredumilli (EG), *MM* 105140 (MH); Araku valley (VSKP), *DDSR* 21391 (MH); towards Kinchipada (VSKP), *GVS* 19761 (MH); Munneru river (KMM), *R.Rajan* 113865 (BSID); Khammam to Dhanaiyaigudem (KMM), *R.Rajan* 108516 (BSID); Purunakote, Satkosia Tiger Reserve, *KCM* 5295 (BSID); Pampasor, Satkosia Wild Life Sanctuary, *D.Hazra* & *D. Das* 19020 (BSID); Jamnamarathur, Javadi hills (NA), *EV* 51966 (MH); Kometteri (NA), *KS* 6038 (MH); Palur (SA), *VNS* 4151 (MH); Sonanchvadi (SA), *KRM* 64162 (MH); Chambarambakkam (CPT), *S.India Flora* 14837 (MH); Gandigam lake, Pennagaram (SLM), *EV* 22440 (MH); Yercaud (SLM), *SKK* 23088 (MH).

INDIA: N.E.India, Odisha, Tamil Nadu, Andhra Pradesh.

WORLD: Africa, Sri Lanka, Myanmar, Malaysia, Pakistan, Nepal, China, Korea, Japan, Laos, Indonesia, Philippines, Thailand, Australia.

XYRIDACEAE

XYRIS Linnaeus

1. Leaves 3-8 mm broad; heads mostly 10-12 mm diam.; throat of corolla not bearded .... **X. indica**
1. Leaves up to 2.8 mm broad; heads 6.2-7.5 mm diam.:
   2. Peduncle more than 40 cm long, with a distinct wing; leaves not papillate .... **X. capensis**
   2. Peduncle less than 15 cm long, hardly or obscurely winged; leaves papillate:
      3. Staminodes penicillate .... **X. pauciflora**
      3. Staminodes feathery .... **X. coronata**

Xyris capensis Thunb., Prodr. Pl. Cap. 12. 1794. var. **schoenoides** (C. Mart.) Nilsson in Kongl., Vet. Akad. Forh. 3: 154. 1891; Royen in Steenis, Fl. Males Ser. 1.4: 374. 1953; Matthew, Fl. Tamilnadu Carnatic 3: 1652. 1983. *X. schoenoides* C. Mart. in Wall., Pl. Asiat. Rar. 3: 30. 1832; FBI 6: 365. 1892; Fischer 3: 1532. 1928.

An erect, tufted, rush-like, scapigerous, glabrous herb, apex curved, acute. Scape 40-60 cm long, slightly triquetrous, ribbed, with one distinct wing. Flowers in terminal heads, bisexual; bracts dark brown, yellow at margins, basal one obovate; sepals 3, median one ovate-elliptic, laterals boat-shaped, apex crested; petals 3, yellow, limb obovate, fimbriate, equal to claw; stamens 3, staminodes present; ovary trigonous, 1-celled, ovules on 3 parietal placentae. Capsule 4 mm; seeds ellipsoid, ribbed.

Occasional in marshy localities. Fl. & Fr.: Throughout the year.

Kakkashola, Balmadies estate, Yercaud (SLM), *KS* 6592 (MH), Shevaroy hill top, Yercaud (SLM), *AVNR* 26757 (MH); Kaveri peak, Yercaud (SLM), *DBD* 31292 (MH).

INDIA: Tamil Nadu, Andhra Pradesh, N.E.India.

WORLD: South America, Africa, Bhutan, Nepal, China, Laos, Cambodia, Indonesia, Malaysia, Thailand, Vietnam.

Xyris coronata Hianes, Bot. Bihar & Orissa 3: 1072. 1924; Mishra *et al.*, Indian J. For. 6: 294. 1983.

Slender herb, 60-75 cm long. Leaves 0 or one, sheathing the scape and much shorter than it, blade flat, 2.5 mm broad at base, apex acuminate. Scape obtusely angled and with a single ridge, very minutely scaberulous. Head subglobose, usually broader than long, 6.2-7.5 mm diam.; outer bracts empty, ovate, nerved, obtuse or subcuspidate, flowering 6.2 mm long, broadly cymbiform, keeled near the top of some cuspidate, 3 lateral nerves at some distance each side of midrib; median sepal caduceus, lateral cymbiform, keeled, persistent; corolla *c.* 11.2 mm long including *c.* 6.2 mm long tube; petals yellow, obovate; stamens half as long as petals, throat of corolla between the stamens and in addition to the feathery staminodes densely bearded; style and its branches stout, stigma lobed. Capsule broadly elliptic, to 5 x 3.7 mm; seeds ellipsoid or oblong, 0.5 mm long, longitudinally ridged.

Majhiguda, Koraput (Mishra *et al., loc cit.*).

INDIA: Bihar, Madhya Pradesh, Odisha.

Xyris indica L., Sp. Pl. 42. 1753; FBI 6: 364. 1892; Fischer 3: 1532. 1928; Saxena & Brahmam, Fl. Orissa 3: 1979. 1995.

Robust herb, 2-60 cm high. Leaves 12-30 x 0.3-0.8 cm, apex obtuse or acute. Scape terete, strongly ridged, head ovoid, globose, subglobose or elliptic, 1-2 cm long, 0.7-1.5 cm diam., bracts closely imbricate, dark red-brown, shining, orbicular or cuneate- obovate, glabrous, scarious. Flowers yellow; lateral sepals narrowly boat-shaped with dorsal serrulate wing; throat of corolla not bearded; petals orbicular, erose, claw as long as sepals; filaments short, broad, staminodes penicillate.

Frequent in rice fields, ditces and other marshy land in Eastern Ghats of Odisha. Fl. & Fr.: October – February.

INDIA: Meghalaya, Assam, Sikkim, West Bengal, Bihar, Odisha.

WORLD: Sri Lanka, Cochinchina, China, Malesia, Australia.

Xyris pauciflora Willd., Phytogr. 2. t. 1. F. 1. 1974; FBI 6: 365. 1892; Fischer 3: 1532. 1928; Matthew, Fl. Tamilnadu Carnatic 3: 1653. 1983.

An erect tufted, reed-like scapigerous herb. Leaves narrowly linear, 10 x 0.2 cm, papillose. Scape 15 cm long, slightly compressed, hardly winged, papillate, ribbed. Flower-head globose, bracts chestnut brown, apex dorsally keeled, median bract elliptic, mucronate, basal bract orbicular, margin membranous, apex triangular, papillate, spinulose; sepals 3, median one broad, laterals curved; petals 3, yellow, limbs obovate, emarginated; stamens 3; ovary oblong, 1-celled, ovules on parietal placentae. Fruit capsule.

Common in marshy localities in most of the districts, sea level to 800 m. Fl. & Fr.: January – February. Fruit February onwards.

Talupula (ATP), *KRKS & BSS* 41910; Kailasakona – Puttur (CTR), *GVS* 32035 (MH), Thalakona RF (CTR), *GVS* 46962 (MH & CAL); Thanjavanam (VSKP), *GVS* 42622 (MH); check Veerambakkam (NA), *V.S.Ramachandran* 51530 (MH); Vandalur Zoological park (CPT), DN 966 (MH); Kambakkam hills (NA), *sine coll.*, 8906 (MH);

INDIA: Peninsular, E. & NE. India.

WORLD: Sri Lanka, Bangladesh, Bhutan, Nepal, Myanmar, Malaysia, Vietnam, Australia.

COMMELINACEAE

1. Fruit indehiscent .... **Pollia**
1. Fruit a dehiscent capsule:
    2. Inflorescence leaf-opposed (atleast some of them) .... **Commelina** *p.p.*
    2. Inflorescence terminal or axillary:
        3. Petals connate below into a tube:
            4. Cymes not enclosed in the leaf-sheaths .... **Cyanotis**
            4. Cymes enclosed in leaf sheaths .... **Commelina** *p.p.*
        3. Petals free:
            5. Flowers actinomorphic:
                6. Fertile stamens 6 .... **Amischotolype**
                6. Fertile stamens 1-3:
                    7. Leaves aggregate, enlarged upwards; fertile stamens 2 .... **Dictyopsermum**
                    7. Leaves spreading, not enlarged upwards; fertile stamens 3 .... **Murdannia**
            5. Flowers zygomorphic:
                8. Fertile stamens 5-6 .... **Floscopa**
                8. Fertile stamens 3 .... **Rhopalephora**

AMISCHOTOLYPE Hasskarl

Amischotolype mollissima (Blume) Hassk., Flora 46: 392. 1863. *Campelia mollissima* Blume, Enum. Pl. Jav. 1: 7. 1827. *Amischotolype glabrata* Hassk., Flora 46: 392. 1863. *Campelia marginata* Blume, Enum. Pl. Jav. 1: 7. 1827. *Amischotolype mollissima* var. *marginata* (Blume) R.S. Rao, M.V.M. Patrika 6: 53. 1971. *Forrestia marginata* (Blume) Hassk., Flora 47: 630. 1864. *F. mollis* Hassk., Flora 47: 631. 1864; FBI 6: 383. 1892. *F. glabrata* Hassk., Flora 47: 630. 1864. *F. marginata* var. *rostrata* (Bl.) C.B. Clarke in DC., Monogr. Phan. 3: 237. 1881; FBI 6: 383, 1892. *F. mollissima* (Bl.) Kds., Exk. Fl. Java 1: 282. 1911. *F. mollissima* var. *glabrata* (Kunth) Backer in Backer & Bakh. f., Fl. Java 3: 15. 1968; Subba Rao, Bull Bot. Surv. India 12: 209. 1970.

A robust perennial herb, erect to ascending; rooting at lower nodes, internodes 3.5 cm long. Leaf sheath glabrous with a line of cilia; mouth golden yellow, ciliate. Leaves lanceolate, 10-25 x 3-7 cm, glabrous, base narrowed into petiole, margin with stellate hairs, apex acute. Flowers subsessisle, in dense globose cymes, violet or reddish violet, in crowded axillary panicles; bracts ovate, ciliate, hairy without on the midrib towards apex; calyx lobes 3, free, unequal, linear-oblong; petals 3, free, subequal, linear-oblong, glabrous; stamens 6; ovary 3-angular, style slender. Capsules ellipsoid, to 1.2 x 0.3 cm, apex hairy, violet; seeds one per locule, ovate-triangular, striate-rugose, lustrous bluish-black.

Occasional in the hill areas of Visakhapatnam district at 1000 m. Fl. & Fr.: October – December.

Padalammagudi (VSKP), *GVS* 42584 (MH); Minumuluru (VSKP), *GVS* 29598 (MH).

INDIA: S. & NE. India.

WORLD: Bangladesh, Myanmar, China, Indonesia, Thailand, Vietnam.

COMMELINA Linnaeus

1. Spathes funnel-shaped or cucullate:
    2. Anticous cells of the ovary 2-ovuled, the posticous 1-ovuled:
        3. Leaves glabrous .... **C. petersii**
        3. Leaves pubescent .... **C. benghalensis**
    2. All the cells of the ovary 1-ovuled:
        4. Seeds studded with white granules .... **C. suffruticosa**
        4. Seeds not as above:
            5. Capsules 3-celled:
                6. Capsules subequally 3-valved:
                    7. Stout glabrous herb; capsule trigonous .... **C. paludosa**
                    7. Weak pubescent herb; capsule subglobose .... **C. ramulosa**
                6. Capsules 2-valved .... **C. erecta**
            5. Capsules 2-celled .... **C. ensifolia**
1. Spathes complicate, margins free or connate only at the base:
    8. All the cells of the ovary 1-ovuled .... **C. appendiculata**
    8. Anticous cells of the ovary 2-ovuled, posticous cell 1-ovuled or obsolete:
        9. Capsules 2-celled, rarely posticous cell present usually empty or with an imperfect seed:
            10. Spathes base not auricled .... **C. clavata**
            10. Spathes base distinctly sagittate-auricled .... **C. attenuata**
        9. Capsules 3-celled:
            11. Spathes sessile .... **C. subulata**
            11. Spathes distinctly peduncled:
                12. Seeds tuberculate and reticulate .... **C. diffusa**
                12. Seeds smooth:
                    13. Seeds appendaged at one end .... **C. undulata**
                    13. Seeds not appendaged .... **C. caroliniana**

Commelina appendiculata C.B.Clarke, Commelyn. Cyrtandr. Bengal t 13. 1874; FBI 6: 374. 1892; Saxena & Brahmam, Fl. Orissa 3: 1984. 1995.

Diffusely branched herb, 30-90 cm long. Leaves sessile, linear or narrowly lanceolate, 6-15 x 0.6-1.5 cm, narrowed at both ends, hairy. Spathe complicate, long-peduncled, ovate-lanceolate or lanceolate, 3.7-7.5 cm long and 1.6 cm broad at base when unfolded, caudate, acuminate, glabrous without and hairy within, base cordate. Flowers blue, all the cells of the ovary 1-ovuled. Capsule *c.* 6.2 mm, 2-valved and 2-celled and often with a smaller indehiscent third dorsal cell; seed 1 in each cell, smooth.

Occasional in Eastern Ghats of Odisha. Fl. & Fr.: May.

Joranda, Similipahar (MBJ), *SX* & *MB* 4364 (RRL-B).

INDIA: North Bengal, Sikkim, Himalaya and Odisha.

WORLD: Bangladesh, Sri Lanka.

Commelina attenuata K.D. Koenig ex Vahl., Enum. Pl. Obs. 2: 168. 1806; FBI 6: 372. 1892; Fischer 3: 1539. 1931; Matthew, Fl. Tamilnadu Carnatic 3: 1656. 1983.

A much slender, diffuse, annual herb, stem and branches angled; stem 30-60 cm long, slender, much branched, glabrous or sparsely hairy. Leaves sessile, linear or linear-lanceolate, 4-8 x 0.5-1 cm, chartaceous, sparsely scattered-pubescent above, glabrous below, base obtuse, margin entire, apex tapering, acuminate; sheath 0.4 -1 cm. Flowers blue, axillary with slender peduncle, spathes narrowly ovate, 2-4 cm long, base auriculate; sepals 3, pale white – pale green, paired sepals united basally, odd sepal boat- shaped; petals 3, two petals whitish yellow, one rose-purple and bluish; fertile stamens 3, sterile stamens 3; ovary ovoid, 2-celled, the cells 2-ovulate, style long, white, apex coiled, bends to one side, stigma capitate, trifid. Capsules oblong, to 8 x 5 m, dorsal locule not developed; seeds elliptic, whitish-appendaged at the ends.

Common in grassy places in the plains and on drier slopes of hills throughout Eastern Ghats. Fl. & Fr.: July – December.

Gundumala RF (ATP), *TP & NY* 737; Aluru kona (ATP), *TP* & *NY* 1131; Venganna bavi (KNL), *AA* 2062; Guvvalacheruvu (KDP), *R.V.Reddy* 7993; Near Molakalacheruvu (CTR), *KNR & DAM* 9830; Jonnavalasa (NLR), *PMR* 23091; Mahanandi (KNL), *JLE* 25463 (MH & CAL); Nallamalais (KNL), *Joby* & *Ratheesh* 117 (SJC); Near Cuddapah (KDP), *JSG* 21192 (CAL); Machilipatnam (KSN), *JSG* 12593 (MH); Pydeputta (EG), *SS* 583; Tyda RF (VSKP), *Joby* & *Ratheesh* 1139 (SJC); Banji, Gandhamardan (Bargarh), *SX* & *MB* 6042 (RRL-B); Purunakote, Satkosia Tiger Reserve, *KCM* 5286 (BSID); Kollaimadu, Pelakuppam (SA), *VNS* 4058 (MH); Sithanur (CPT), *S.India Flora* 11184 (MH); Vellore (NA), *S.India Flora s.n.* (MH); Tippukadu RF (NA), *KRM* 17608 (MH); Krishagiri (KGR), *S.India Flora* 13886 (MH);

INDIA: Almost throughout India.

WORLD: Sri Lanka.

Note: *C. attenuata* is allied to *C. wightii* in its densely branching habit, stout stem, attenuate spathe apex and smooth seeds. But *C. attenuata* has glabrous peduncles, blue flowers and appendaged seeds against pubescent peduncles, yellow flowers and non –appendaged seeds as in *C. wightii*. The pits on the appendages of seeds are a peculiar for this species.

Commelina benghalensis L., Sp. Pl. 41. 1753; FBI 6: 370. 1892; Fischer 3: 1539. 1931; Matthew, Fl. Tamilnadu Carnatic 3: 1657. 1983.

An ascending herb; root stock with cleistogamous flowers; stem 30-80 cm long, slender, dichotomously branched from the base upwards; branches diffuse, glabrous or pubescent, creeping and rooting below. Leaves ovate or oblong, 2-5 x 2-3 cm, base subtruncate, margin ciliate, apex acute-obtuse; sessile or shortly with 1 cm long petiole, sheaths short or long, pubescent or villous, the margins ciliate or sometimes bearded with rufous hairs. Spathes clustered, funnel-shaped, pubescent; upper cymes 2-3-flowered; lower one 1 or 2-flowered. Flowers blue; sepals 3, outer one linear, inner ones orbicular; petals 3, two 6 x 5 mm, lanceolate-boat-shaped; fertile stamens 3, antisepalous, sterile stamens 3; ovary 3-celled, 2-ovulate, 1 cell 1-ovulate, glabrous, style 3.5 mm long, stigma capitate. Capsule pyriform, membranous, trilocular, anterior locules 2-seeded, posterior 1-seeded; seeds oblong, closely pitted.

Very common throughout Eastern Ghats, weed in agricultural fields, road sides and waste lands. Fl. & Fr.: July – November. Vern.: Tel: *Vennadevikura*; Ori: *Khet papra, Ranasiri.*

Reddipalli farm (ATP), *NY* 258; Mudigubba (ATP), *KNR* & *DAM* 9822; Kambagirikonda (KNL), *BR* & *SS* 20008; Thummalapalle (KDP), *TP* & *SSR* 33669; Leguntapadu (NLR), *PMR* 16423; Macherla RF (GNT), *VRK* 3914; Devarapalli (EG), *KSK* 23458; Simhachalam (VSKP), *BR & KNR* 9877 & *GVS* 21497 (MH); way to Pathalaganga, Srisailam (KNL), *JLE* 22110 (MH); Siddavatam forest (KDP), *KS* 6354 (MH); Satyavedu (CTR), *MCB* 45266 (MH); Kondapalli RF (KSN), *PV* 5312 (MH); Gaddada, Maredumilli (EG), *MM* 105090 (MH); Karepally (KMM), *R.Rajan* 105959 (BSID); Kangundi (NA), *S.India Flora* 10325 (MH); Komattiyur (NA), *KS* 7477 (MH); Pagoda point, Yercaud (SLM), *SKK* 26829 (MH);

INDIA: Throughout Inda.

WORLD: Tropical and subtropical Africa, Asia.

Commelina caroliniana Walter, Fl. Carol. 68. 1788. *C. hasskarlii* C.B. Clarke, Commelyn. Cyrtandr. Bengal. 13, t. 3. 1874; FBI 6: 370. 1892; Fischer 3: 1538. 1931; B.R.P. Rao in Pullaiah, Fl. Andhra Pradesh 3: 1002, 1997; Saxena & Brahmam, Fl. Orissa, 3: 1987. 1995.

A herb with much branched stem, glabrous or pubescent, sometimes scaberulous. Leaves narrowly lanceolate, 2-6 x 1-1.8 cm, subacute; sheaths 1.2 cm long, base broad, ciliate. Spathes 1-3 cm long, ovate-lanceolate, cordate at the base with rounded lobes, glabrous, scabrid or hispid. Flowers in pubescent cymes, the upper branch 2-4, the lower 1-2-flowered; sepals 3; petals 3, paired petals reniform, blue; fertile stamens 3, sterile stamens 3. Capsules quadrate, subtruncate, membranous; seeds cylindric, truncate at one end, rounded at the other without an appendage.

Common throughout Eastern Ghats. Fl. & Fr.: August – November.

Krishna river (GNT), *VRK* 6741; Tyada RF (VSKP), *BR & KNR* 9896; Chelama (KNL), *JLE* 17999 (MH & CAL).

INDIA: Throughout India.

WORLD: USA, Philippines.

Note: After scrutinizing the *Commelina* specimens at Walter herbarium and specimens of *C. hasskarlii* Clarke at Kew and British Museum, Faden (in Lit) concluded that the specific name of this species was given by Walter in 1788, hundred years before the Clarke's binomial. More over it was an old introduction to South Carolina, possibly with rice seeds from India.

Commelina clavata C.B. Clarke, Commelyn., Cyrtandr. Bengal. t. 5. 1874; FBI 6: 371. 1892; Fischer 3: 1539. 1931; Matthew, Fl. Tamilnadu Carnatic 3: 1657. 1983.

A herb with 30-90 cm long stem, sparingly branched; branches diffuse, rooting at the nodes. Leaves ovate, elliptic-lanceolate or linear, 3-5 x 0.6-1 cm, glabrescent, base tapering, margin entire, apex acute-acuminate; sheath 2.5 cm, apically hirsute, peduncle 3.5 cm. Spathes ovate-lanceolate, 2.5 cm long, base cordate, apex attenuate; sepals with small, brown linear spots; petals lilac. Capsules usually 2, oblong-quadrate, constricted in the middle; seeds 4, 2 per cell, black, two lower often imperfect.

Occasional in southern Eastern Ghats. Fl. & Fr.: June – December.

Kalavabugga (KNL), *RVR* 1875; Thummalapalle (KDP), *SSR & TP* 33620; Bakarapeta (CTR), *KNR* & *PSPB* 9842; Vinukonda RF (GNT), *VRK* 6858; on the waay to Somasila dam (NLR), *P.Venu* 110197 (BSID); Shevaroy hill (SLM), *G.Bidie s.n.* (MH); Vandalur (CPT), *S.India Flora* 11492 (MH); Amirthi (NA), *MBV* 1075(MH).

INDIA: Peninsular India.

WORLD: Sri Lanka, Myanmar, China, Malysia, Laos, Thailand, Vietnam.

Commelina diffusa Burm. f., Fl. Ind. 18. t. 7. f. 2. 1768; Panigr. & Kammathy, J. Indian Bot. Soc. 43: 299. 1964; Matthew, Fl. Tamilnadu Carnatic 3: 1658. 1983. *C. nudiflora* sensu Hook. f., Fl. Brit. India 6: 369. 1892; non L. 1753; Fischer 3: 1538. 1931.

A herb with creeping stem and rooting at the nodes; stems 60-90 cm long, branching from the base. Leaves sessile, ovate or lanceolate, 4-6 x 0.5-0.8 cm, chartaceous, base subcordate, margin entire, apex acuminate. Cymes usually 1-3-flowered. Spathes complicate, oblong-lanceolate, 3-5 cm long, glabrous, base cordate; sepals 3, petals 3 and two interior petals obovate with long claws, dark blue, the exterior sub-sessile, pale blue or nearly white; fertile stamens 3, sterile stamens 2, ovary globose, 3-celled of which two cells are 2-ovulate, the third 1-ovulate, style up

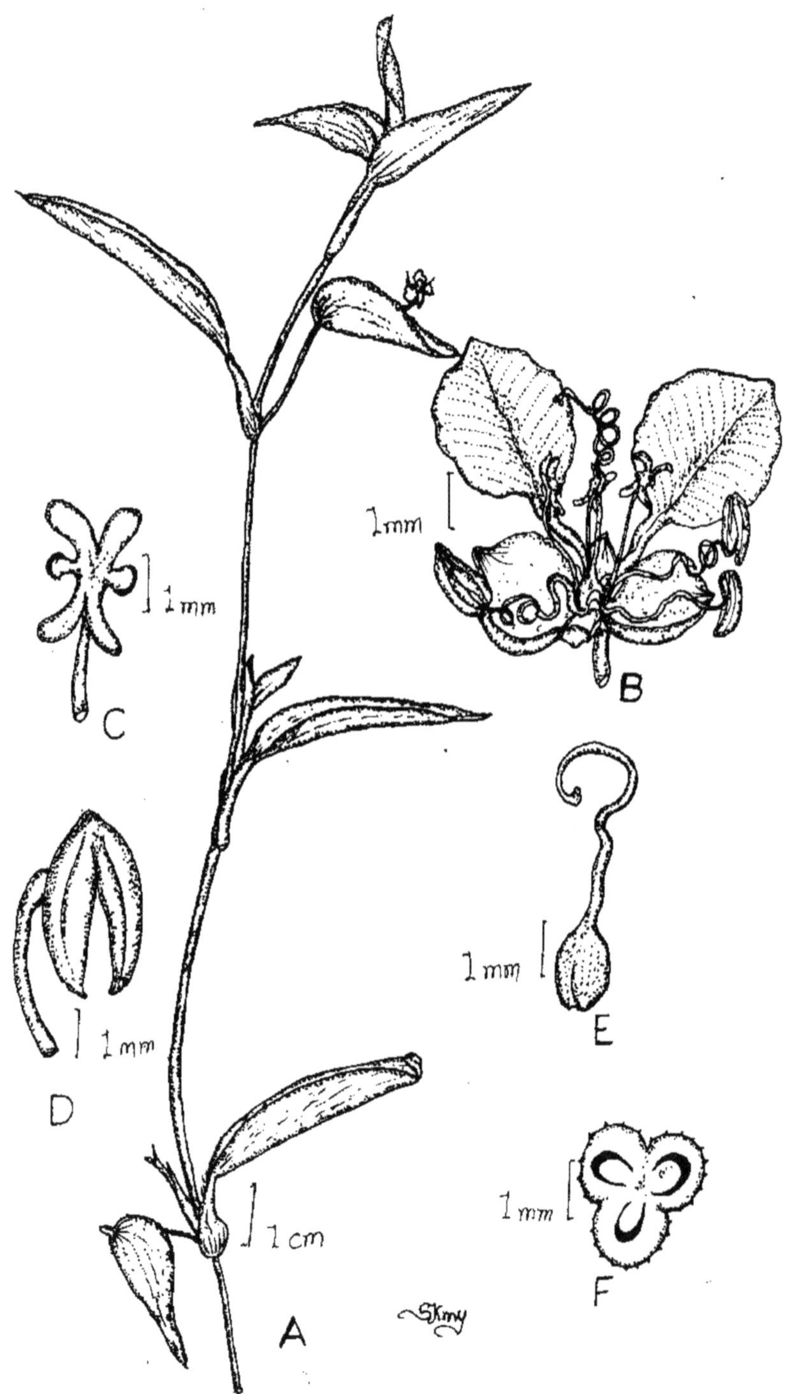

Figure 22. **Commelina clavata** C.B. Clarke
A. Twig; B. Flower; C & D. Stamens; E. Pistil; F. Ovary c.s.

to 5 mm long, coiled at apex, stigma capitate. Capsule broadly oblong, 6 mm long, acuminate, coriaceous, 5-seeded; seeds oblong-cylindric, tuberculate and brown.

Common on riversides and paddy fields, in exposed and partial shady habitats. Fl. & Fr.: July – December.

Rampa hills (EG), *BR & KNR* 10605; Simhachalam (VSKP), *BR & KNR* 9876; Srisailam (KNL), *JLE* 23791 (MH); Horsley hills (CTR), *CECF* 4444 (CAL); Bhupathipalem (EG), *VNS* 240 (CAL); Khatnighada area (VSKP), *GVS* 19518 (MH); Vandalur (CPT), *S.India Flora* 11492 (MH); near Bhuvanagiri (SA), *KRM* 90419 (MH).

INDIA: Peninsular, NE. & NW. India, E.Himalaya, Andaman & Nicobar Islands.

WORLD: Tropics and subtropics worldwide.

Commelina ensifolia R. Br., Prodr. 269. 1810; FBI 6: 374. 1892; Fischer 3: 1540. 1931; R. S. Rao, Blumea 14: 352. 1966; Matthew, Fl. Tamilnadu Carnatic 3: 1658. 1983.

A rhizomatous, diffuse, slender herb; stem 30-45 cm long, slender. Leaves linear or linear-lanceolate, 4-10 x 0.6-0.8 cm, glabrous above, puberulous below, base narrow, margin entire, apex acute-obtuse; sheath 2.5 cm, glabrescent or puberulous. Spathe solitary, cucullate, ovate, 1.5 cm long, truncate on one margin, apex acute, basally elongate pubescent, apically short and appressed-pubescent. Flowers in sheaths, white, blue and violet; sepals 3, paired sepal 4 mm long, connate, glabrous, odd sepal 2 mm long, boat-shaped, glabrous; petals 3, paired petals reniform, blue, odd petal lanceolate, pale white, middle blue; fertile stamens 3, sterile stamens 3; ovary globose, style 9 mm, coiled, white, stigma capitate, blue. Capsule 2-celled; seeds 1-3, ovoid, smooth.

Occasional in fields and wastelands. Fl. & Fr.: July – December.

Thanakallu (ATP), *KNR & DAM* 9828; Dharmapuri RF (ATP), *NY* 1008 (MH); Gulemallebad RF (KNL), *RVR* 1436; Guvvalacheruvu (KDP), *RVR & R.V.Reddy* 7899; Macherla RF (GNT), *VRK* 3564; Balapalle-Mamandur border, Pandigunta (CTR), *JHFB* 117015 (BSID), *GVM* 115196 (BSID); Kailasakona – Puttur (CTR), *GVS* 32033 (MH); Kambakam Range (CTR), *MCB* 45173 (MH); Chitvel (NLR), *P.Venu* 110070 (BSID); Bollapale RF (GNT), *sine coll. s.n.* (MH); Sirumalai (DGL), *S.India Flora* 9108 (MH);

INDIA: Peninsular India.

WORLD: Sri Lanka, Australia.

Commelina erecta L., Sp. Pl. 41. 1753; Saxena & Brahmam, Fl. Orissa 3: 1986. 1995; Subbarao & Kumari, Fl. Visakhapatnam Dist. 291. 2008; Matthew, Fl. Tamilnadu Carnatic 3: 1659. 1983.

A slender herb; stem 30-60 cm long, hairy or glabrate. Leaves lanceolate or oblong-lanceolate, 5-13 x 1.5-3 cm, chartaceous, appressed-pubescent above, glaucous below, base oblique, obtuse, margin entire, apex gradually tapering, acuminate; pubescent, margin ciliate. Spathes 3-5 in a terminal cluster, cucullate, broadly cordate, glabrous, apex acute-acuminate, subsessile. Flowers 3-6, pale violet in racemes; all the cells of the ovary 1-ovuled. Capsule 3-celled, 2-valved; seeds 3, ashy, adnate to capsule wall.

A weed of cultivated fields and wastelands. Fl. & Fr.: August – September.

Bugga-Tadipatri (ATP), *TP & NY* 438; Thanakallu (ATP), *KNR & DAM* 9827; Kolanbarathi (KNL), *BR* 20150; Simhachalam (VSKP), *BR & KNR* 9878 & 9879; Tyada RF (VSKP), *BR & KNR* 9892; Punyagiri hills (VZN), *MCK* 18816; Diguvametta (PKM), *JLE* 42175 (MH); Cuddapah, *JSG* 21280 (CAL); Talakona (CTR), *DRC* 1770 (SVU); Kondapalle (KSN), *PV* 5279 (MH); Rampa hills (EG), *GVS* 24464 (MH & CAL); Araku valley (VSKP), *NPBK* 10798 (MH); Karakakonda (VSKP), *CAB* 1711 (MH); Elwinpeta (SKLM), *GVS* 62468 (MH); Parnasala RF (KMM), *RCS* 104242 (BSID); Bhadrachalam RF (KMM), *RCS* 104304 (BSID); Kwadoli, Raigoda, Satkosia Tiger Reserve, *KCM* 6725 (BSID); Nilavur, Yelagiri hills (NA), *MBV* 1106 (MH).

INDIA: Peninsular & NW. India.

WORLD: USA to Tropical & Subtropical America, Tropical and South Africa, Arabian Peninsula.

Commelina maculata Edgew., Trans. Linn. Soc. London 20: 89. 1851. *Commemlina obliqua* Buch.-Ham. ex D.Don, var. *viscida* C.B.Clarke Mongr. Phan. 3: 178. 1881. *C. paludosa* Blume var. *viscida* (C.B.Clarke) R.S.Rao & Kammathy, Bull. Bot. Surv. India 3: 168. 1962.

Perennial herbs, stems procumbent or creeping. Leafsheaths ciliate at mouth, glabrous elsewhere, leaf blade ovate-lanceolate, 4-10 x 1.5-2.5 cm, villous on both surfaces. Flowers blue in terminal heads; involucral bracts 2 or 3, funnelform, sepals membranous; petals 2 anterior ones to 10 mm, posterior one 4 mm. Capsule globose, 3-valved, seeds 1 per valve, grey-ellipsoid.

Occasional in moist forests, road sides and grasslands. F. & Fr.: September – November.

Kalasamudram (ATP), *KRKS* 36486; Tirumala hills (CTR), *LR* & *M.S.Rao* 2553 (BSID); Putturu (CTR), *MVS* & *V.S.Rao* 30660 (BSID); Kondaveedu (GNT), *J.Swamy* & *S.Nagaraju* 7177 (BSID); R.Udaygiri hills (Gajapathi distr.), *Kalidass & Murugan* 9489 (RPRC).

INDIA: Throughout India.

WORLD: Myanmar, Bhutan, China.

Commelina paludosa Blume, Enum. Pl. Jav. 1: 2. 1827; R.S.Rao & Kammathy, J. Bombay Nat. Hist. Soc. 59: 60. 1962; Fischer 3: 1539. 1931. *C. obliqua* Buch.-Ham ex D. Don, Prodr. Fl. Nep. 45. 1825 (non Vahl 1805); FBI 6: 372. 1892.

A stout subscandent herb; stem 60-90 cm long, branched, glabrous. Leaves lanceolate or elliptic-lanceolate, 10-20 x 2-5 cm, sessile or petiole, acute or caudate-acuminate, membranous, glabrous, scabrous or villous; sheaths reaching 3 cm long, the mouth bearded with long hairs. Spathes sessile, solitary or crowded in terminal heads, funnel-shaped, acute, glabrous, subscabrid, usually filled with a clear glutinous liquid. Flowers blue, in simple raceme, large petals clawed, orbicular, stamens whitish yellow, ovary 3-celled, the cells 1-ovulate. Capsules trigonous-obovoid, subequally 3-valved, 3-celled, 3-seeded; seeds oblong or ellipsoid, smooth, lead-coloured.

Common throughout Eastern Ghats. Fl. & Fr.: June – October.

Bukkapatnam RF (ATP), *NY* 603; Yadiki (ATP), *TP* & *NY* 1112; Ahobilum (KNL), *TP & KNR* 9853; Madhavaram RF (KNL), *RVR* 1639; Palakonda hills (KDP), *CS* 7721; Macherla RF (GNT), *VRK* 3295; Balapalle (KDP), *JLE* 14309 (MH); Kailasakona-Puttur (CTR), *GVS* 32033 A (MH); Kailasakona (CTR), *DRC* 1726 (MH); Maredumilli (EG), *GVS* 68516 (MH); Valamuru, Maredumilli (EG), *MM* 101288 (BSID); Anantagiri (VSKP), *GVS* 32833 (MH); way to Siddharam (KMM), *PVS* 84075 (BSID); Gingee (SA), *S.R.Raju* 17937 (MH).

INDIA: More or less through out India.

WORLD: Myanmar, Nepal, Bhutan, China, Laos, Cambodia, Malaysia, Indonesia, Thailand, Vietnam.

Commelina petersii Hassk., Flora 46: 385. 1863; Faden, in Dassanayake (ed.), Rev. Handb. Fl. Ceylon 14: 191. 2000. *C. persicariifolia* Wight ex C.B.Clarke in DC. Mon. Phan. 3: 171. 1881. *nom. illegit.* non Dwile, 1815; FBI 6: 372. 1892; Fischer 3: 1536. 1931. *C. imberbis* sensu R.S. Rao, Blumea 14: 352. 1966; B.R.P.Rao in Pullaiah, Fl. Andhra Pradesh 3: 1002. 1997.

A herb with diffuse stem; leaves lanceolate, acuminate, base rounded and then narrowed into short petiole, 5-12 x 2-4 cm long, sheaths 1-2.5 cm long, mouth usually minutely pubescent. Spathes ovate, base truncately rounded. Capsule oblong, obtuse, seeds oblong, terete, smooth or obscurely rugose.

Rare in Eastern Ghats of Odisha and Nellore district of Andhra Pradesh. Fl. & Fr.: December – February.

Gudur (NLR), *KCJ* 18513 (MH).

INDIA: Peninsular India.

WORLD: Tropical and South Africa, Sri Lanka, Indo-China.

Commelina ramulosa (C.B.Clarke) H.Perrier, Notul. Syst. (Pari) 5: 185. 1936. *C. forskalaei* var. *ramulosa* C.B.Clarke Monogr. Phan. 3: 168. 1881.

A herb with procumbent stem; rooting at lower nodes. Leaves ovate-lanceolate, 3-6 x 1-2.5 cm, finely pubescent, base cordate, sessile into a sheath, shortly acuminate at apex; sheath 1-1.5 cm long, mouth bristillate. Spathe solitary, sessile, broadly ovate, base trunately rounded, pubescent. Flowers blue, in simple raceme, large petals clawed, orbiculat; stamen yellow, fertile stamens bluish. Ovary 3-celled, uni-ovuled. Capsule 3-valved, 3-seeded; seeds ellipsoid or oblong, grey coloured, smooth.

Rarely distributed in grasses among the ghats. Fl. & Fr.: November – February.

Agasthyeswara kona, Lingapuram (KDP), *J.Swamy & S. Nagaraju* 7984 (BSID).

INDIA: Andhra Pradesh, Karnataka, Madhya Pradesh.

WORLD: South Africa, Madagascar.

Commelina subulata Roth, Nov. Pl. Sp. 23. 1821; FBI 6: 369. 1892; Fischer 3: 1538. 1928; Matthew, Fl. Tamilnadu Carnatic 3: 1660. 1983.

A herb with slender erect stem, 20-40 cm long, simple or branched. Leaves narrow lanceolate or elliptic, 4-9 x 0.2-0.5 cm, glabrescent, base obtuse, margin entire, apex acute; sheath 1.2 cm long, margins glabrous or ciliate. Spathe ovate, 6-8 cm long, base cordate, margin cilate, apex acute. Flowers small, orange-purple, or chrome-yellow, when dry violet, sepals 3, unequal, broadly ovate-orbicular, paired sepals laterally fused, pale white, odd sepal boat-shaped; petals 3, paired petals reniform, golden yellow, free, odd petal ovate, pale white, margin pink; fertile stamens 3, staminodes 3; ovary oblong, 3-lobed, style 1.5 mm, dark purple, stigma capitate, pale purple. Capsules oblong, 3-celled, ventral locules 2-seeded, dorsal locule 1-seeded; seeds ovoid, dark, 4-grooved.

Occasional in Horsley hills. Fl. & Fr.: August – October.

Horsley hills (CTR), *Joby & Ratheeh* 1152 (SJC).

INDIA: Peninsular India.

WORLD: Tropical Africa, South Africa,

Commelina suffruticosa Blume, Enum. Pl. Jav. 1: 3. 1827; FBI 6: 374. 1892; Panigr. & Kamm., J. Indian Bot. Soc. 43: 305. 1964.

A stout herb with slender, tuberous roots, stems stout, branched nearly glabrous. Leaves sessile, lanceolate, 3.5-4 x 2-3.5 cm, scabridly pubescent, base oblique, margins entire, apex acuminate; sheaths auricled, up to 2 cm long. Spathes on axillary branches short-peduncled, broadly ovate-cordate; sepals 3, pale white, paired sepals ovate, base connate, odd sepal ovate; petals 3, paired petals reniform, white, odd petal ovate; fertile stamens 3, sterile stamens 3; ovary trilocular, style 4.5 mm long, stigma trilobed. Capsules 2-celled, locules 1-seeded; seeds ellipsoid, brown, pitted with concave depressions, puberulous.

Rare in Rampa hills. Fl. & Fr.: June – December.

Devammapalli (EG), *VNS* 405 (CAL); Rampa hills (EG), *VNS* 56 & 228 (CAL).

INDIA: C, N, E & NE. India.

WORLD: Bangladesh, China, Indonesia, Thailand.

Commelina undulata R.Br., Prodr. Fl. Nov. Holl. 270. 1810. *C. kurzii* C.B. Clarke, J. Linn. Soc., Bot. 11: 444. 1869; FBI 6: 373. 1892; Fischer 3: 1539. 1931. *C. longifolia* Lam., Tabl. Encyl. 1: 129. 1791; R. S. Rao, Taxon 10: 254. 1961.

A slender, decumbent herb; stem slender, sometimes rooting, glabrous, with long internodes. Leaves linear or linear-lanceolate, 3-8 x 0.5-1 cm, chartaceous, glabrous, base obtuse, margin entire, apex acute-acuminate; sheath 2 cm, margin ciliate. Spathes ovate-lanceolate, axillary, glabrous, base rounded. Flowers small, polygamous; branches of the cyme equal, usually 1-2-flowered; sepals free, ovate, obtuse, the 2 inner connate below, larger than the outer; petals dark blue, the 2 larger ovate, entire, subsessile or with a very short claw; stamens 3, fertile, one anther large, lunate, the other two smaller, ellipsoid, staminodes 3, clavate. Capsules oblong, 3-celled; seeds 2-5, ovoid-subglobose, truncate at one end, black.

Occasional in waste lands and open forests throughout Eatern Ghats. Fl. & Fr.: July – October.

Tadipatri – Bugga (ATP), *TP* 437; Gani RF (KNL), *RVR* 1540; Upper Ahobilam (KNL), *TP* & *RVR* 2709; Kanigiri reservoir (NLR), *MCK* 22893; Vinukonda (GNT), *VRK* 6840; Pulusumamella (KNL), *VBH* 83929 (BSID); Rajapalem (NLR), *P.Venu* 109892 (BSID); Kondapally RF (KSN), *PV* 5941 (AU); Bhupathipalem (EG), *VNS* 327 (CAL); Mudurlanka, Maredumilli (EG), *MM* 102561 (MH, BSID); Barnakonda RF (VSKP), *KCJ* 17129 (MH); Araku valley (VSKP), *NPBK* 595 (CAL); Duggeru (VZN), *MV* 6858 (AU); Elwinpeta block (SKLM), *GVS* 62468 (MH); Ramanagutta (KMM), *R.Rajan* 107971 (BSID); Mahendragiri (GJT), *VNS* 5669 (MH); Kairkchala, Similipal (MBJ), *D.Hazra* & *D.Das* 12255 (BSID); Pampasar RF, Satkosia Tiger Reserve, *KCM* 5112 (BSID); Bhuvanagiri (SA), *KRM* 58198 (MH).

INDIA: Peninsular, E. & NE. India.

WORLD: China, Indonesia, Philippines, Tropical Oceania.

CYANOTIS D. Don *nom. cons.*

1. Cymes enclosed in leaf sheaths:
    2. Filaments bearded .... **C. axillaris**
    2. Filaments naked .... **C. cucullata**
1. Cymes not enclosed in leafsheaths:
    3. Plants cobwebby-woolly throughout:
        4. Cymes densely aggregate on elongate flowering shoots; basal leaves prominently dissimilar to upper ones .... **C. arachnoidea**
        4. Cymes lax, terminal and axillary on stems; leaves similar:
            5. Basal rosette absent, plants creeping; seeds pyramidal, rugose .... **C. fasciculata**
            5. Basal rosette present; plants erect or ascending; seeds ovate-elliptic, striate .... **C. thwaitesii**
    3. Plants not cob-webby-woolly:
        6. Bracteoles prominently cross-nerved; cymes lax, clustered .... **C. arcotensis**
        6. Bracteoles not cross-nerved:
            7. Roots tuberous; plants pubescent; leaves up to 15 cm long .... **C. tuberosa**

7. Roots not tuberous; plants almost glabrous; leaves about 5 cm long:
    8. Cymose solitary; leaf sheath more than 5 mm:
        9. Leaf ovate, pubescent; cyme deeply recurved .... **C. cristata**
        9. Leaf oblong-lanceolate, glabrescent; cyme straight......... **C. vaga**
    8. Cymose clustered; leaf sheath 1-2 mm:
        10. Leaves silky villous below; cymes terminal; seeds rugose .... **C. villosa**
        10. Leaves appressed pilose below; cymes terminal and axillary; seeds pitted........................ **C. pilosa**

Cyanotis arachnoidea C.B. Clarke in DC., Monogr. Phan. 3: 250. 1851. FBI 6: 386. 1892; Fischer 3: 1550. 1931; Matthew, Fl. Tamilnadu Carnatic 3: 1661. 1983.

A herb, prostrate or creeping with cottony or cobwebby woolly stem; stems 15-60 cm long, often decumbent and rooting below. Leaves very variable, basal ones oblong-elliptic or lanceolate, 6-15 x 0.4-0.8 cm, chartaceous, cobwebby, base obtuse, margin entire, apex rounded, sheath 2 cm. Cymes terminal and/or axillary; bracts ovate-lanceolate, 1 cm, exceeding cymes; bracteoles 5-8 pairs, falcate. Flowers blue; sepals 3, green, lanceolate, cobwebby, bearded; petals 3, violet; stamens 6, filaments purple-violet, bearded, hairs blue-violet or purple, with sub apical bulging, anther lobes orange yellow; ovary 3-celled, style pilose. Capsules to 3 mm; seeds terete, pitted.

Common in hills, on exposed slopes by edges of rocks as discrete clumps. Fl. & Fr.: August – October.

Somulavaripalli (ATP), *KRKS* & *KRS* 37830; Tirumala (CTR), *LR* & *M.S.Rao* 2531 (BSID); Papanasanam (CTR), *KS* 6951 (MH); Horsley hills (CTR), *JSG* 21188 (DD); Between Naidupeta & Chitvel (NLR), *PV* 112247 (BSID); Dummakonda (EG), *SS* 10149 (AU); Peddakonda (EG), *SS* 5606 (AU); Sunkarimetta (VSKP), *NPBK* 10945 (MH); Sapparlagedda (VSKP), *GVS* 42768 (MH); Mahendragiri (GJM), *VNS* 4625 & 5575 (MH); Yercaud, near Shervarayan Temple (SLM), *P. Venu s.n.* (BSID).

INDIA: Peninsular India.

WORLD: Sri Lanka, Myanmar, China, Laos, Thailand, Vietnam.

Cyanotis arcotensis R.S. Rao, Blumea 14: 345. 1966; Matthew, Fl. Tamilnadu Carnatic 3: 1661. 1983. *C. papilionacea* C.B. Clarke in DC., Monogr. Phan. 3: 246. 1881. *p.p.* non (L.) Roem. & Schult. 1880; FBI 6: 384. 1892; Fischer 3: 1549. 1931.

A hispid annual herb, 10-25 cm long, sparingly branched from near the base, the branches decumbent, often rooting; stem red, terete, striate, patently pilose. Leaves sessile, linear-oblong or lanceolate, 2-5 x 0.3 cm, pubescent below, glabrous above, the upper leaves larger than the lower, base cordate, margin ciliate, apex obtuse-acute; sheaths very short, pubescent. Cymes terminal, solitary or 2-4 in a cluster, bracts ovate-lanceolate, falcately recurved, often longer than the spike. Capsule valves separating from 3-dentate central column during dehiscence; seeds 6, pitted.

Rare in southern Eastern Ghats. Fl. & Fr.: August – September.

Jamnamarathur, Javadi hills (NA), *MBV* 868 (MH); Andiappanoor (NA), *MBV* 1147 (MH); Tippukadu RF (NA), *KRM* 17655 (MH); Egmore, Madras (CPT), *sine coll., s.n.* (MH); Perungundimedu, Madras(CPT), *sine coll.s.n.* (MH); Gudiyal (CPT), *DN* 731 (MH); Korai RF, Gingee (SA), KRM 52867 (MH); Puluhalli (SLM), *S.India Flora* 9876 (MH); Krishnagiri (KGR), *S.India Flora* 14949 (MH); Shevaroy hills (SLM), *G.Bidie s.n.* (MH);

INDIA: South India. Endemic.

Cyanotis axillaris (L.) D. Don., Prodr. Fl. Nepal 46. 1825; FBI 6: 388. 1892; Fischer 3: 1550. 1931. *Tonningia axillaris* (L.) Kuntze, Revis. Gen. Pl. 2: 721. 1891. *Commelina axillaris* L., Sp. Pl. 42. 1753. *Amischophacelus axillaris* (L.) R.S.Rao & Kamm., J. Linn. Soc. Bot. 59; 306. 19663; Matthew, Fl. Tamilnadu Carnatic 3: 1655. 1983.

An annual glabrous herb, with variable, prostate or erect branches, long and spreading, rooting at nodes, stem 10-40 cm long, glabrous. Leaves narrow-lanceolate, 4-10 x 0.5-1 cm, succulent, plicate or flat, base obtuse, margin entire, apex acuminate; sheaths 6 mm, pilose, pouched. Flowers violet, blue and white clustered in the axils of leaf sheaths; bracteoles linear. Flowers 6 mm across; calyx lobes lanceolate; petals ovate; stamens 6, filaments blue with pilose hairs, anthers oblong, ovary woolly, 3-celled, ovules 2 per cell on axile placentae, pilose, stigma shortly 3-fid. Capsules ellipsoid, apex beaked; seeds oblong, pitted.

Common weed in cultivated lands and moist places throughout Eastern Ghats. Fl. & Fr.: July – November.

Bukkapatnam RF (ATP), *NY* 641; Kammavaripalli (KNL), *TP* & *PVP* 3380; Rachakuntapalle (KDP), *KRKS & KRS* 37626, 37901; Namalagundu (KDP), *KRKS* & *BSS* 39656; Varikuntapadu (NLR), *PMR* 16477; Gundlamotu reservoir (PKM), *BSS* & *SKB* 32780; Jayanthipuram (GNT), *VRK* 6945; Rampa hills (EG), *BR & KNR* 10604; Lakkavarapu reservoir (VSKP), *MHR* & *MCK* 14878; Jonnavalasa (VZN), *MCK* 18847; Nallamalais (KNL), *JLE* 16851 (MH); Bramhagundam (CTR), *GVS* 46930 (MH); Kondavedu forest (GNT), *CAB* 4643 (MH); Addatigala (EG), *GVS* 68573 (MH); S. Kota (VZN), *GVS* 32808 (MH); Karepally (KMM), *R.Rajan* 105961 (BSID); Murugampadu (KMM), *RCS* 104258 (BSID); Koriguttalu (KMM), *R.Rajan* 106046 (BSID); Tikarpada RF, Satkosia Tiger Reserve, *KCM* 5997 (BSID); Gandigam lake, Pennagaram (SLM), *EV* 22455 (MH); Way to Kambukudi (NA), *KS* 7421 (MH); Tippukadu RF (NA), *KRM* 17653 (MH); Peria Ranganar Kuppam (SA), *VNS* 5070 (MH); Chidambarampillai (SA), *K.M.Sebastine* 5208 (MH); Vandalur (CPT), *S.India Flora* 11556 (MH).

INDIA: Almost through out India.

WORLD: Sri Lanka, Myanmar, China, Laos, Cambodia, Indonesia, Malaysia, Philippines, Thailand, Vietnam, Oceania.

Cyanotis cristata (L.) D. Don, Prodr. Fl. Nepal 46. 1825; FBI 6: 385, 1892; Fischer 3: 1549. 1931; Matthew, Fl. Tamilnadu Carnatic 3: 1662. 1983. *Tradescantia cristata* L., Sp. Pl. 42. 1753.

A herb, prostate or creeping; stem branched from the base, the branches 15-45 cm long, slender, creeping and rooting below. Leaves sessile, ovate or oblong-lanceolate, to 2.5 x 1 cm, chartaceous, sparsely pubescent, base cordate, margin ciliate, apex obtuse; sheath loose, pubescent. Flowers in scorpioidly recurved cymes, bracts ovate-lanceolate, exceeding cymes; bracteoles 5-8 pairs; sepals 3, pale white; petals 3, rose to purple, united up to middle; stamens 6, filaments blue to violet, bearded, hairs blue, anther lobes yellow; ovary globose, style glabrous, stigma capitate. Capsules oblong, 2 mm long, trigonous, seeds trigonous, with 2 large pits on 2 sides.

Common throughout Eastern Ghats in exposed to partially shaded areas, moist crevices, shallow road side ditches and banks, near streams. Fl. & Fr.: July – November.

Gundumala RF (ATP), *TP & NY* 745, Upper Ahobilam (KNL), *TP & RVR* 2720; Palakonda hills (KDP), *CS* 7615; Kalikiri (CTR), *KNR* & *PSPB* 9832; Mahanandi (KNL), *JLE* 25461 (MH & CAL); on the way to Jamalkota (CTR), *GVS* 45817 (MH); Durgam (NLR), *PV* 111157 (BSID); Jammalakona (CTR), *GVS* 45817 (MH); Kandapalle (KSN), *P.Venkanna* 5635 (MH); Dummakonda (EG), *GVS* 68593 (MH & CAL); Kottravada, Maredumilli (EG), *MM* 102554 (MH); Sigumari (VSKP), *GVS* 44357 (MH); Araku valley (VSKP), *NPBK* 10772 (MH); Maraigudem RF (KMM), *RCS* 104250 (BSID); Krishnaraja sagar (KMM), *RCS* 102460 (BSID); Parnasala RF (KMM), *RCS* 102465 (BSID); Khalasumi RF (SBP), *Sauris Panda* 961 (BSID); Tikarpada, Satkosia Tiger Reserve, *KCM* 8302 (BSID); Kathavanchimmankoil, Nilavur, Yelagiri hills (NA), *MBV* 1105 (MH); way to Kuttathur (NA), *KS* 7441 (MH); Avadi (CPT), *sine coll., s.n.*(MH); Vepampatti RF, Harur (DMP), *EV* 51833 (MH); Hogainakkal RF (KGR), *EV* 21907 (MH); Yercaud (SLM), *AVNR* 18269 (MH).

INDIA: Throughout Tropical India.

WORLD: Sri Lanka, Myanmar, Bhutan, China, Laos, Cambodia, Indonesia, Malaysia, Thailand, Vietnam.

Cyanotis cucullata (Roth) Kunth, Enum. Pl. 4: 107. 1843; FBI 6: 389. 1892; Fischer 3: 1550. 1931. *Tonningia cucullata* (Roth) Kuntze, Revis. Gen. Pl. 2: 722. 1891. *Amischophacelus cucullata* (Roth) Rao & Kamm., J. Linn. Soc. Bot. 59: 306. 1966. *Tradescantia cucullata* Roth, Nov. Pl. Sp. 189. 1821.

An annual prostrate slender herb, rooting at nodes. Leaves oblong-lanceolate, succulent, 1.5-7 x 0.2-0.8 cm, obtuse, sessile, 2.5-6 x.5-1 cm, distichous, recurved; sheath prominently inflated. Flowers lilac, in axillary fascicles, almost included in the sheath pouch; corolla violet-blue; stamens with filaments densely bearded with long beaded blue hairs; anthers yellow; style blue with colourless 2-lipped sitgma. Capsule depressed at apex with 3 projections; seeds obscurely pitted, brown

A weed of cultivated fields. Fl. & Fr.: July – December.

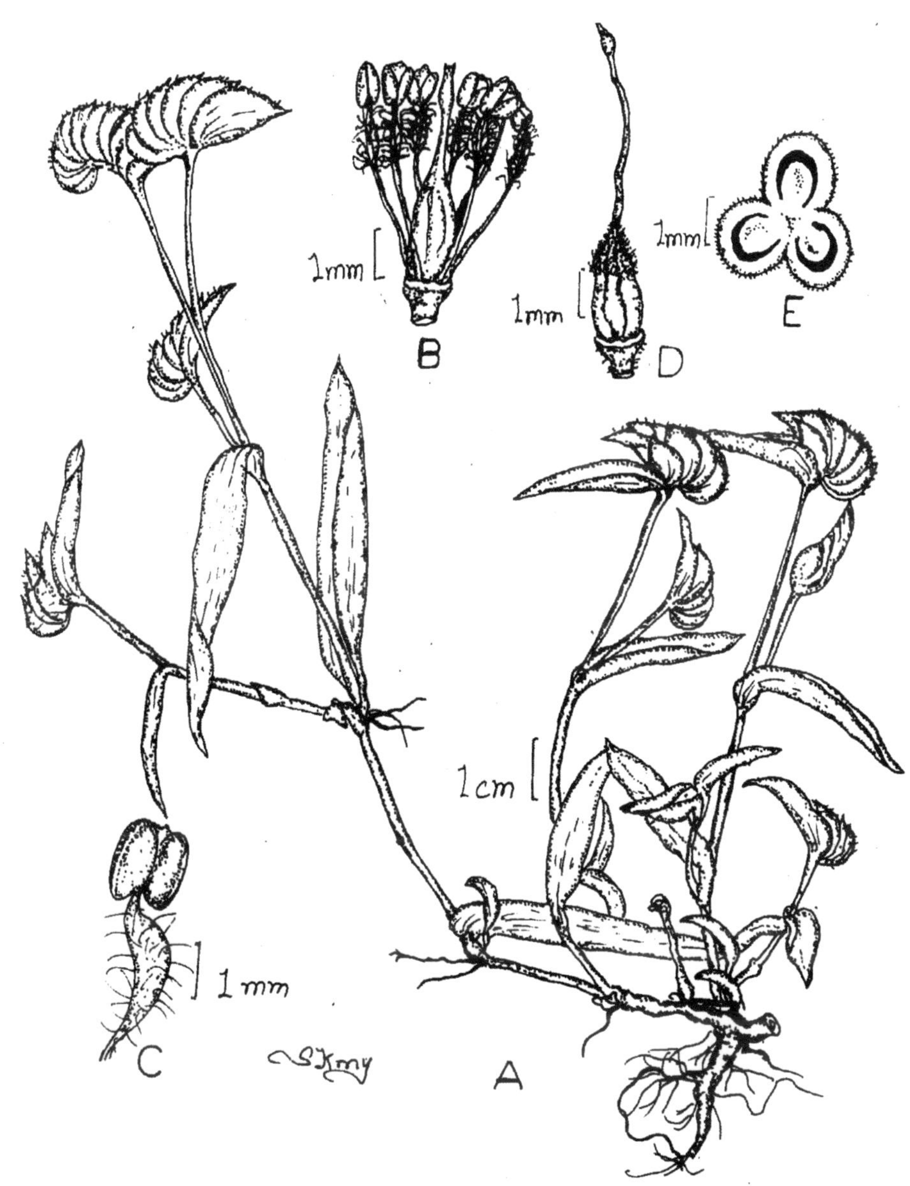

Figure 23. **Cyanotis cristata** (L.) D. Don
A. Habit; B. Pistil & stamens; C. Stamen; D. Pistil; E. Ovary C.s.

Near Kadiri (ATP), *KNR & DAM* 9826; Erramala hills (KNL), *RVR* 1585; Ramalaakota (KNL), *RVR* 1858; Rampa hills (EG), *VNS* 90 (CAL); way to Vaniampadi (NA), *KS* 7491 (MH).

INDIA: Andhra Pradesh, Karnataka, Madhya Pradesh, Maharashtra, Rajasthan.

WORLD: Introduced in Antigua and Barbados.

Cyanotis fasciculata (Heyne ex Roth) Schult. & Schult. f., in L. Syst. Veg. 7: 1152. 1830; FBI 6: 387. 1892; Fischer 3: 1550. 1931; Matthew, Fl. Tamilnadu Carnatic 3: 1662. 1983. *Tradescantia fasciculata* Heyne ex Roth, Nov. Pl. Spec. 189. 1821.

A decumbent herb with slender stems, 5-18 cm long, branchlets cottony or cobwebby. Leaves linear or oblong-lanceolate, 1-3 x 0.4-0.8 cm, chartaceous, silky-cobwebby or floccose, base cordate, margin entire, apex subacute; sheath 5 mm. Cymes axillary and/or terminal, solitary or 2 or 3 in a cluster; pink or green, cobwebby; peduncles slender, bracts narrowly lanceolate, recurved, bracteoles narrowly ovate, 3-5 pairs, apex acuminate. Flowers 5 mm across; calyx cottony, lobes lanceolate; petals 3, rose; stamens 6, filaments blue-purple; ovary ovate-oblong, style blue, bearded, hairs blue-purple. Capsules oblong; seeds 2-6, oblong, faintly rugose.

Common throughout Eastern Ghats. Fl. & Fr.: June – December.

Thalaricheruvu (ATP), *TP* 906; Amagondapalem RF (ATP), *TP & NY* 1026 (MH); Gulemallebad RF (KNL), *RVR* 1691; Kammavaripalli (KNL), *TP & GO* 3395; Guvvalacheruvu (KDP), *RVR & R.V.Reddy* 7888; K. K. Kottala (KDP), *KRKS & KRS* 36429; Kailasakona (CTR), *GVS* 46050 (MH); Near forest (KDP), *KS* 6829 (MH); Udayagiri (NLR), *ASR* 6231 (BSID); Kondavedu fort (GNT), *CAB* 4661 (MH); Pathalaganga (KNL), *JLE* 22112 (MH & CAL); Sigumari (VSKP), *GVS* 44359 (MH); Anantagiri (VSKP), *GVS* 21761 (MH); Pennagaram RF (SLM), *EV* 224110 (MH); Hogainakkal (KGR), *EV* 21915 (MH); Krishnagiri (KGR), *S.India Flora* 14892 (MH); Bettamogililam, Melagiri hills (SLM), *S.India Flora* 9788 (MH); Melpat (SA), *CAB* 949(MH); Mangalam RF, Yelagiri hills (NA), *MBV* 621 (MH); way to Kuttathur (NA), KS 7444 (MH); Mahendragiri (GJM), *VNS* 5755 (MH).

INDIA: Peninsular & NE. India.

WORLD: Sri Lanka, Bhutan, Nepal, Vietnam.

Cyanotis pilosa Schult. & Schult.f., Syst. Veg. 7: 1154. 1830; FBI 6: 387. 1892; Fischer 3: 1549. 1931; Matthew, Fl. Tamilnadu Carnatic 3: 1663. 1983.

Herb, stems 30-90 cm high, erect or ascending. Leaves narrowly lanceolate or elliptic, 5-10 x 1-2.5 cm; sheaths loose, 1.2-3 cm long, villous. Cymes *c.* 3 in a cluster, axillary and terminal, subcorymbose to 1.5 cm; flowers pink to pruple, to 1 cm across, petals ovate-suborbicular, to 6 mm. Capsule ellipsoid or obovoid, to 3 mm; seeds 6, pyramidal, pitted, brown.

Hills above 900 m in Central Tamil Nadu (Matthew, 1991). Fl. & Fr.: October – Decmber.

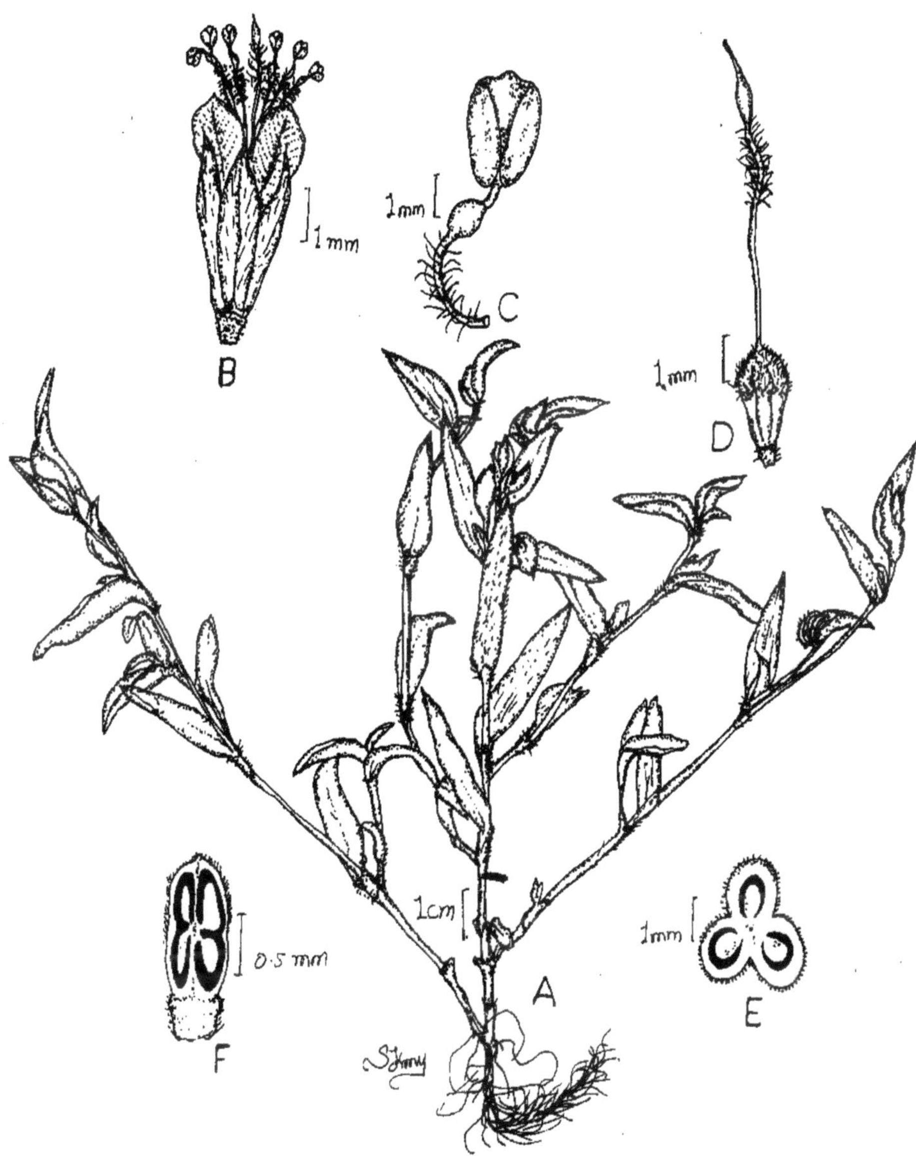

Figure 24. **Cyanotis fasciculata** (Heyne ex Roth) Schult. & Schult.f.
A. Habit; B. Flower; C. Stamen; D. Pistil; E. Ovary c.s; F. Ovary l.s.

Kaveri peak, Yercaud (SLM), *DBD* 31267 (MH); Pulluhalli (SLM), *S.India Flora* 9876 (MH); Javadi hill (NA), *MBV* 1278 (MH).

INDIA: Peninsula.

WORLD: Sri Lanka.

Cyanotis thwaitesii Hassk., Commel. Ind. 136. 1870. *C. fasciculata* (Roth) Schult.f. var. *thwaitesii* (Hassk.) C.B.Clarke in Hook.f., Fl. Brit. India 6: 388. 1892.

Perennial, erect or ascending herb, all parts covered with cobwebby pubescence. Sheath green, cobwebby. Rosette leaves spiral, linear-lanceolate, 1.5-9.5 x 0.3-0.9 cm, base obtuse, margins ciliate; cauline leaves 2.8- 8.5 x 0.3-1.2 cm. Flowers rose-red or purple, terminal or axillary; bracts foliaceous, ovate to lanceolate,; bracteoles usually falcate; sepals 3, basally connate, pale white; petals 3; stamens 6, blue, bearded. Capsule stramineous; seeds brown-reddish brown.

Occasional in rocky hills and slopes, grasslands. Fl. & Fr.: August – October.

Mahanandi (KNL), *BSS* & *SKB* 29634; Talakona (CTR), *DAM* & *KNR* 9846; Papanasanam (CTR), *BSS* & *SKB* 30420; Hanumanbavi thanda (GNT), *BSS* & *SKB* 30675.

INDIA: Peninsular India.

WORLD: Sri Lanka, Indo-China.

Cyanotis tuberosa (Roxb.) Schult. & Schult. f. in L., Syst. Veg. 7: 1153. 1830; FBI 6: 386. 1892; Fischer 3: 1549. 1931; Matthew, Fl. Tamilnadu Carnatic 3: 1663. 1983. *Tradescantia tuberosa* Roxb., Pl. Coromandel t. 108. 1799.

A perennial herb with roots of fusiform tubers; stem 15-75 cm long, swollen, hirsute at the base, suberect or prostrate below, almost glabrous. Basal leaves oblong, 10-30 x 1.5 cm, upper ones linear-ensiform, chartaceous, glabrous or villose above, downy-sericeous below, base obtuse, margin ciliate, apex subacute; sheath 1-2 cm. Cymes terminal and/or axillary, corymbose; bracts ovate, equal or slightly exceeding cyme; bracteoles 10-15 pairs, in 2 series; calyx-tube villous; corolla bluish-purple, lobes ovate, subacute; stamens 6, blue, bearded, hairs blue to violet; style violet-blue, stigma simple. Capsules ellipsoid, the upper half hairy, the lower half glabrous; seeds 4-6, pyramidal, obscurely rugose, brown.

Common throughout Eastern Ghats as forest undergrowth. Fl. & Fr.: August – September.

Thalaricheruvu (ATP), *TP* 914; Gundumala RF (ATP), *TP* & *NY* 726; Kalasamudram RF (ATP), *EC* 2881; North Dhone RF (KNL), *RVR* 2523; Palakonda hills (KDP), *CS* 9394; Horsley hills (CTR), *KNR & DAM* 9837; Thalakona (CTR), *KNR* & *PSPB* 9848; Chelama (KNL), *JLE* 16711 (CAL); Lukke (KNL), *JLE* 42219 (MH); Dugganakonda temple (KDP), *KS* 6357 (MH); Tirumala (CTR), *GVS* 31924 (CAL); Rapur – Chitvel Ghat (NLR), *A.S.Rao* 6543 (BSID); Malkondapenta (PKM), *VBH* 83976 (BSID); Gangireni (KSN), *PV* 5684; Vedurupalli (VSKP), *GVS* 42764 (MH);

Figure 25. **Cyanotis tuberosa** (Roxb.) Schult. & Schult. f.

A.Habit; B. Inflorescence; C. Single spathe with flower; D. Flower-entire; E. Stamen;F. Pistil; G. Capsule; H. Seeds.

Araku valley (VSKP), *NPBK* 508 (CAL); Mannanur RF (MBNR), *VBH* 86624 (BSID); Similipahar (MBJ), *SX* & *MB* 4833, 4945 (RRL-B); Barakutni, Deomali (KPT), *SX* & *MB* 6593 (RRL-B); Krishnagiri (KGR), *S.India Flora* 14934 (MH); Hogainakkal forest (KGR), *EV* 20578 (MH); Bethamogilalam, Melagiri hills (SLM), *S.India Flora* 9781 (MH); Periamalai RF, Javadi hills (NA), *MBV* 1339 (MH); Kangundi (NA), *S.India Flora* 10324 (MH); Karuppumalai (CPT), *DN* 634 (MH).

INDIA: Peninsular India.

WORLD: Sri Lanka.

Cyanotis vaga (Lour.) Schult. & Schult.f., Syst. Veg. 2: 1153. 1830; Saxena & Brahmam, J. Bombay Nat. Hist. Soc. 78: 916. 1980 & Fl. Orissa 1993. 1995. *Tradescantia vaga* Lour., Fl. Cochinch. 1: 193. 1790. *Cyanotis barbata* D.Don, Prodr. Fl. Nepal 46. 1825; FBI 9: 385. 1892.

Slender herb, stems erect or ascending from creeping and rooting base. Leaves sessile, oblong-laceolate, 2.5-5 x 0.4-0.8 cm, apex acute or acuminate, margin entire, glabrescent (or cobwebby beneath). Flowers blue, in axillary and terminal cymes, bract longer than the inflorescence, bracteoles oblong-falcate, acute-acuminate; filaments bearded, staminodes 0; ovary tipped with hairs. Capsule quadrate; seeds brown, reticulate.

Mahendragiri, 1000 m, in moist situations. Fl.: October.

Mahendraagiri (GJT), *SX* & *MB* 3648 (RRL-B).

INDIA: Subtropical Himalayas from Kashmir to Khasia hills, Orissa.

WORLD: Myanmar, China, Java.

Cyanotis villosa (Spreng.) Schult. & Schult.f., Syst. Veg. 7: 1155. 1830; FBI 6: 387. 1892; Fischer 3: 1550. 1931; Matthew, Fl. Tamilnadu Carnatic 3: 1664. 1983. *Tradescantia villosa* Sprengel in L. Syst. Veg. 2L: 116. 1825. *Cyanotis lanceolata* Wight, Ic. Pl. Ind. Orient. t.2085. 1853.

Herb, 1-1.2 m high, silky or villous with spreading hairs. Leaves all cauline, narrowly lanceolate, 3-6 x 0.5-1 cm, silky villous below, purplish, apex acuminate, sheaths moderately loose, pilose, sometimes lanate. Cymes terminal and in the upper axils, often sessile, bracteoles semielliptic, falcate or nearly straight, acute or acuminate; sepals fulvous-pilose; petals purple, to 6 x 5 mm. Capsule to 3 mm; seeds 5, rugose.

Occasional in Shevaroys and Kolli hills. Fl. & Fr.: September - Decmber

Kathevunachiamman koil, Nilavur, Yelagiri hills (NA), *MBV* 1107 (MH); Pulluhalli (SLM), *CAB* 9874 & 9876 (MH); Krishngiri (KGR), *CAB* 14939 (MH); Brooklyn, Shevaroys, Yercaud (SLM), *AVNR* 26943 (MH); Kollimalai (TRP), *KCJ* 11373 (MH);

INDIA: Peninsula.

WORLD: Sri Lanka.

DICTYOSPERMUM Bentham

1. Fertile stamens 3; seeds reticulate .... **D. ovalifolium**
1. Fertile stamens 2; seeds rugose .... **D. montanum**

Dictyospermum montanum Wight, Icon. Pl. Ind. Orient. 6: t. 2069. 1853. *Aneilema montanum* (Wight) Thwaites, Enum. Pl. Zeyl. 322. 1864; C.B. Clarke, Monogr. Phan. 3: 217. 1881; FBI 6: 381. 1892. *Commelina montana* Steud., Nomencl. Bot. ed. 2, 1: 491. 1840. *Tradescantia montana* B. Heyne ex Wall., Numer. List 5203. B. 1831. *T. paniculata* B. Heyne ex Roth, Nov. Pl. Sp. 188. 1821.

Erect herbs to 75 cm tall. Leaves clustered towards apex of stem, 5-13 x 2.5- 5 cm, elliptic-lanceolate, base narrowed, apex acuminate; petiole to 1.5 cm long; sheath to 3 cm long. Panicles terminal, 8-17 cm long, covered with hooked hairs; bracts leafy. Flowers white, 8-10 mm across; pedicels 4-7 mm long; sepals 3, 3-4 x 2-3 mm, obovate, obtuse, concave; petals 3, white, 5-6 x 3-4 mm, ovate, obtuse; stames 3; filaments 2-3 mm long; staminodes 0; ovary 3- celled; *c.* 1.5 mm long; ovule 1 in each cell; style *c.* 2 mm long. Capsule 2-3 mm across, globose, glabrous. Seeds hemispheric, rugose.

Occasional clumps in ravines of Alagar hills. Fl. & Fr.: September – February.

Periya Aruvi valley, Alagar hills (MDU), *SKS* 792 (SGH).

INDIA: South and North-east India.

Dictyospermum ovalifolium Wight, Icon. Pl. Ind. Orient. 6: t.2070.1853. *Aneilema ovalifolium* (Wight) Hook.f. ex C.B. Clarke in DC. Monog. Phan. 3: 218. 1881; FBI 6: 382. 1892; Fischer 3: 1546. 1931; Saxena & Brahmam, Fl. Orissa 3: 1981. 1995.

Erect or decumbent herb, 20-45 cm high; roots long, fibrous, thick and tuberous. Leaves aggregated and enlarged upwards, broadly elliptic, 6-12 x 2.5-6.2 cm, apex acuminate or caudate acuminate, base narrowed into the petiole, glabrous or glabrescent, sheath glabrous or pubescent, mouth usually ciliate. Flowers in terminal panicles, bracts not funnel-shaped, caducous; fertile stamens 2, cells of ovary 1-ovuled. Capsule globose, 3.5-6.2 mm across, glabrous, cells 1-seeded; seeds reticulate.

Occasional in damp places in forests in Eastern Ghats of Odisha. Fl.& Fr.: August – October.

Similipahar (MBJ), *SX* 3773 (RRL-B), *SX* & *MB* 4612, 4873, 5186 (RRL-B); Gonasika forest, Keonjhar, *MB* & *Dhal* 8457 (RRL-B); Tulka RF, Satkosia Tiger Reserve, *KCM* 5277 (BSID).

FLOSCOPA Loureiro

Floscopa scandens Lour., Fl. Cochinch 1: 193. 1790; FBI 6: 390. 1893; Fischer 3: 1552. 1931.

A erect or subscandent herb with fibrous roots; stem rather slender, rooting below; internodes long; branches ascending, leafy above, glabrous or pubescent.

Figure 26. **Dictyospermum montanum** Wight

A. Habit; B. Flower-entire; C. Capsule; D. Stamen; E. Seed.

Leaves elliptic-lanceolate, 5-10 x 1-2 cm, base much narrowed, apex acuminate, shortly petiolate, scaberulous above, glabrous or pubescent beneath; with long hairs. Flowers in terminal, sessile or shortly pedunculate, villous or hirsute panicles; the lower flowers bracteate, the upper ebracteate; bracteoles minute; sepals 3, rotund-ellipsoid, concave, villous; petals 3, slightly longer than the sepals, broadly obovate, white, lilac or pink; fertile stamens 6, filaments longer than the petals, equal, purplish, anthers golden yellow; ovary 2-celled, ovules 1 in each cell, style simple. Capsules 2-celled; seeds pale glaucous, dorsally transversely wrinkled.

Occasional in northern Eastern Ghats, in damp places and near banks of streams. Fl. & Fr.: October – November.

Maredumilli (EG), *GVS* 27295 (MH & CAL); Near Maddhiveedu, Tadepalli (EG), *NRR* 84382 (BSID); Birmisala (VSKP), *GVS* 42665 (MH); on the way to Sapparla peak from Sileru (VSKP), *GVS* 29702 (MH); Araku valley (VSKP), *NPBK* 618 (CAL); Duggeru (VZN), *MV* 6857 (AU); Raigoda, Satkosia Tiger Reserve, *KCM* 6537 (BSID).

INDIA: Throughout Tropical India.

WORLD: Myanmar, Bhutan, China, Laos, Thailand, Vietnam, Oceania.

MURDANNIA Royle *nom. cons.*

1. Roots tuberous .... **M. edulis**
1. Roots fibrous:
    2. Leaves linear, 10 times longer than broad:
        3. Leaves 1-2 mm broad; bracts ochreate; staminal filaments naked .... **M. semiteres**
        3. Leaves 3-8 mm broad; bracts ovate-oblong; staminal filaments bearded:
            4. Erect herbs; roots thick; leaves recurved........................ **M. esculenta**
            4. Decumbent, procumbent or diffuse herbs; root fibrous; leaves flat:
                5. Capsule subglobose; seeds 2 per cell ....**M. nudiflora**
                5. Capsule globose, seeds 1 per cell........................ **M. vaginata**
    2. Leaves ovate-lanceolate, to 3 times longer than broad:
        6. Flowers in branched panicles, many flowered:
            7. Stems erect; leaves cordate or rounded at base; seeds minutely striate ....**M. dimorpha**
            7. Stems decumbent; leaves amplexicaul; seeds reticulate.. **M. japonica**
        6. Flowers in simple racmes, 1-3-flowered:
            8. Flowers blue in terminal racemes .... **M. spirata**
            8. Flowers crimson red to yellow in axillary and terminal racemes.. **M. pauciflora**

Murdannia dimorpha (Dalz.) Brueckner in Engl. & Prantl, Pflanzenfam, ed. 2. 15a: 173. 1930. *Aneilema dimorphum* Dalz., Hook. J. Bot. Kew. Gard. Misc. 3: 138. 1851; FBI 6: 377. 1892; Fischer 3: 1545. 1931.

A herb of 30 cm long; whole plant glabrous, except the mouths of the sheaths, branching from the root. Leaves linear-oblong to ovate-lanceolate, base rounded or cordate, apex acute-obtuse. Panicles sparingly dichotomously branched, usually few-flowered; bracts small, ovate or orbicular, obtuse, cucullate, persistent; sepals 3; petals 3, blue, obovate, subacute; stamens 3 fertile and 3 staminodes, filaments of fertile stamens bearded, those of the staminodes naked; ovary 3-celled, cells 3-5-ovulate. Capsules oblong, 3-gonous, shining; seeds 3-5, uniseriate in each cell, subcubical, cupped at one end, brown-black, minutely striate.

Rare in Eastern Ghats. Fr.: October – November.

Pingerikonda (EG), *SS* 508 (AU); Simalaguda (VZN), *MV* 62722 (AU); Kavalur (NA), *EV* 52013 (MH).

INDIA: Peninsular India.

WORLD: Sri Lanka.

Murdannia edulis (Stokes) Faden, Taxon 29: 77. 1980. *Commelina edulis* Stokes, Bot. Materia Med. 1: 184. 1812. *Aneilema scapiflorum* (Roxb.) Kostel., Allg. Med. Pharm. Fl. 1: 127. 1831; FBI 6: 375. 1892; Fischer 3: 1544. 1928. *Commelina scapiflora* Roxb., Fl. Ind. 1: 178. 1820. *Murdannia scapiflora* (Roxb.) Royle, Ill. 403. t. 95. 1840.

A tufted herb with elongate pisiform tubers. Leaves radical, erect, 10-20 x 1-1.5 cm, linear-ensiform, finely acuminate, glabrous, slightly narrowed at the base. Flowers blue-mauve in panicles on terminal leafless scape; scape together with the panicles 20-45 cm long; upper bracts amplexicaul, ovate, acuminate or truncate, membranous, often spotted with small spots; sepals elliptic-oblong, purple-green; petals blue, obovate; stamens 3 perfect and 3 staminodes, filaments all bearded with blue hairs, anthers of fertile stamens blue and those of staminodes yellow. Capsules obovoid; seeds 5 or 6 in a cell, superposed, sharply 3-gonous.

Rare in open forests and on roadsides. Fl. & Fr.: May – June.

Appapur (MBNR), *BSS* & *BR* 30703 (SKU & BSID); Diguvametta (PKM), *JLE* 23834 (MH); Similipahar (MBJ), *SX* & *MB* 4312 (RRL-B).

INDIA: Throughout India.

WORLD: Myanmar, Nepal, China, Indonesia, Malaysia, New Guinea, Philippines, Thailand, Vietnam.

Murdannia esculenta (Wall. ex C.B.Clarke) R.S.Rao & Kammathy, Bull. Bot. Surv. India 3: 394. 1962; Matthew, Fl. Tamilnadu Carnatic 3: 1665. 1983. *Aneilema esculentum* Wall. ex C.B.Clarke, Monog. Phan. 3: 206. 1881; FBI 6: 377. 1892; Fischer 3: 1545. 1931.

Erect herb, roots tuberous, stem subsolitary. Leaves linear or oblong-elliptic, usually complicate, often recurved, 3-6.5 x 0.4-0.8 cm, apex acute or obtuse, base rounded. Flowers pink to violet, in terminal panicles, bracts oblong or the lower 1 or 2 foliaceous; petals obovate, to 6 mm. Capsule 3-angled, to 7 mm, glossy; seeds 3 or more per cell, sub-cubical, dark brown, obscurely scabrid or pitted.

In Eastern Ghats of Tamil Nadu, in rice fields, marshy areas and among perennial moist rocks. Fl. & Fr.: September – November.

INDIA: Peninsula.

WORLD: Sri Lanka, E. Asia and Australia.

Murdannia japonica (Thunb.) Faden, Taxon 26: 142. 1977; Saxena & Brahmam, Fl. Orissa 3: 1996. 1995. *Commelina japonica* Thunb., Trans. Linn. Soc. London 2: 332. 1794. *Aneilema lineolatum* Kunth, Enum. Pl. 4: 69. 1843; FBI 6: 376. 1892; Fischer 3: 1544. 1931.

Stout herb, 0.9-1.2 m high. Leaves amplexicaul, erecto-patent, lanceolate, lanceolate-oblong or oblong, 11-20 x 3-4.5 cm, apex acute or acuminate, margin white, undulate, base cuneate, rounded or cordate, glabrous. Flowering stem leafy; panicle stout, axillary and terminal, with spreading and ascending branches; bracts narcescent; flowers blue; stamens 2-3, staminodes 2-3, filaments all bearded. Capsule oblong, globose oblong or broadly ellipsoid, to 7 mm long, 6.2 mm diam., trigonous, mucronate; seeds 1 –seriate, 3-4 in each cell, reticulate and glandular-puberulous.

Occasional in moist and shady places in Eastern Ghats of Odisha. Fl. & Fr.: August – October.

Similipahar (MBJ), *SX* 3813 (RRL-B), *SX* & *MB* 4742, 4879 (RRL-B), Sorispada, Dhenkanal, *Dhal* & *Rout* 8166 (RRL-B); Mahadebjharan (SBP), *GP* 30715 (CAL).

INDIA: Tropical India, Sikkim, Khasi hills.

WORLD: Myanmar, Bhutan, Malay Isles.

Murdannia nudiflora (L.) Brenan, Kew Bull. 7: 189. 1952; Matthew, Fl. Tamilnadu Carnatic 3: 1666. 1983. *Commelina nudiflora* L., Mant. 177. 1771. (non Sp. Pl. 41. 1753). *Aneilema nudiflorum* (L.) R. Br., Prodr. 271. 1810; FBI 6: 378. 1892; Fischer 3: 1545. 1931.

An erect or decumbent, slender branching herb, rooting at nodes; stem simple or branched from the base, branches 15-30 cm long. Leaves linear-lanceolate, 3-8 x 0.4 cm, chartaceous, glabrous, base obtuse, margin entire, apex tapering, acuminate; sheath 0.6 – 1 cm, the margins and mouth strongly ciliate. Flowers in subglobose cymes on terminal (rarely axillary) panicles; sepals 3, oblong, obtuse; petals 3, blue or purple suborbicular; stamens 2 perfect, 2-4 sterile, filaments of the fertile only or of all bearded; ovary ovate, style white, stigma capitate. Capsules trigonously subglobose, membranous, 3-celled, with 2 seeds in each cell; seeds cuboid, deeply pitted, dark brown.

Common throughout Eastern Ghats on river banks, near waterlogged paddy ields *etc.* Fl. & Fr.: June – October.

Kalasamudram (ATP), *KRKS* & *KRS* 37690; Batrepalli (ATP), *BSS* & *KRKS* 40339; Sunnipenta (KNL), *BR* & *BSS* 34004; Horsley hills (CTR), *KNR* & *DAM* 9839; Kinthali (SKLM), *MCK* 18869; Chelama-Nallamalais (KNL), *JLE* 22077 (MH); Balapalle (KDP), *JLE* 15000 (MH & CAL); Papanasanam falls (CTR), *KS* 6958 (MH); Satyavedu (CTR), *MCB* 45241 (MH); Gaddada (EG), *MM* 102637 (BSID); Ettukkuru, Maredumilli (EG), *MM* 102614 (MH); on the way to Javapuram konda (VSKP), *GVS* 42515 (MH); Araku valley (VSKP), *GVS* 21544 (MH); Salur (VZN), *NPBK* 932 (CAL); Kurupam (VZN), *MV* 2802 (AU); Nellippaka RF (KMM), *RCS* 99080 (BSID); Aska (GJM), *sine coll. s.n.* (MH); Purunakote, Satkosia Tiger Reserve, *KCM* 5159 (BSID); Vandalur (CPT), *Sine coll., s.n.* (MH); Yelagiri hills (NA), *MBV* 805 (MH); Javadi hills (NA), *MBV* 1242 (MH); Melpat (SA), *CAB* 1038 (MH); Karai RF, Gingee (SA), *KRM* 52865 (MH).

INDIA: Throughout India.

WORLD: Sri Lanka, Myanmar, China, Japan, Cambodia, Laos, Malaysia, New Guinea, Philippines, Indian ocean and Pacific Islands.

Murdannia pauciflora (G.Brückn.) G.Brückn., Nat. Pflanzenfam. ed.2. 15a: 173. 1930; Saxena & Brahmam, Fl. Orissa 3: 1998. 1995. *Phaeneilema pauciflorum* G.Brückn., Notizbl. Bot. Gart. Berlin-Dahlem 10: 56. 1927. *Aneilema pauciflorum* Wight, Icon. T. 2077. 1853 (*illegitimate*); FBI 6: 378. 1892; Fischer 3: 1545. 19311.

Decumbent herb, rooting at nodes, villous on one side; roots fibrous. Leaves sessile, ovate, 1.2-5 x 0.6-1.6 cm, apex acute or subacute, base usually rounded or cordate, more or less hairy on both sides; sheaths short, usually hairy. Cymes 1-3 (rarely 4-5-)-flowered, axillary, pedicels decurved. Capsule oblong-fusiform, 6 mm long, acute, somewhat trigonous, smooth, 3-celled; seeds 4-6 in each cell, 1-seriate, superposed, cubical, black, nearly smooth.

Occasional in shady moist localities. Fl. & Fr.: September – Decmeber.

On the way to Panchalingalakona (NLR), *P.Venu* 109832 (BSID); Hirakud (SBP), *Sauria Panda* 103 (BSID).

INDIA: W. Peninsula

WORLD: Malaysia.

Murdannia semiteres (Dalzell) Santapau, Poona Agric. Coll. Mag. 41(4): 15. 1951; Matthew, Fl. Tamilnadu Carnatic 3: 1666. 1983. *Aneilema semiteres* Dalzell, Hooker's J. Bot. Kew Gard. Misc. 3: 138. 1851. *Murdannia juncoides* (Wight) R.S.Rao & Kamm., Bull. Surv. India 6: 3. 1964. *Dichaespermum juncoides* Wight, Icon. Pl. Ind. Orient. t. 2078. f. 2. 1853. *Aneilema paniculatum* Wall. ex C.B. Clarke in DC., Monogr. Phan. 3: 215. 1881; FBI 6: 381. 1892. *p.p.*; Fischer 3: 1546. 1931. *p.p.*

An annual herb, 5-20 cm long; stem sheathed at the base with yellowish scarious sheaths, roots tuberous. Leaves linear, 5-6 x 1-3 cm, acute, apiculate. Flowers in paniculate cymes; bracts ochreate, cyathiform, truncate, with a tooth at the apex, persistent; sepals 3, elliptic, obtuse, glabrous; petals 3, obovate, cuneate, blue; stamens 3 fertile and 3 sterile, filaments all connate at the very base, naked; ovary

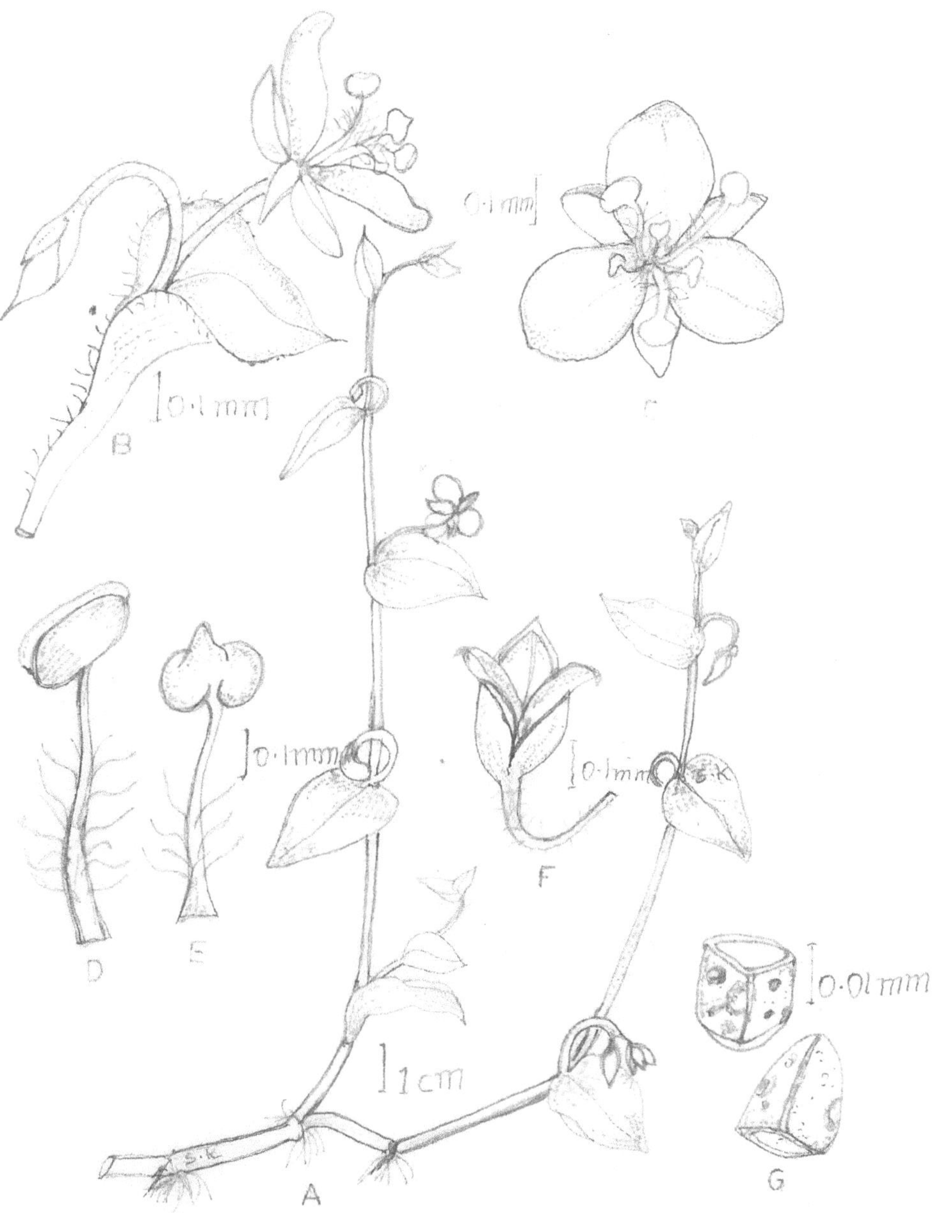

Figure 27. **Murdannia pauciflora** (G.Brückn.) G.Brückn.

A. Habit; B. Inflorescence; C. Flower-entire; D&E. Stamens; F. Capsule; G. Seeds.

ovoid, stigma simple. Capsules ovoid, 3 mm long, smooth, shining; seeds 6-8 in each cell, pale yellow, 2-seriate, irregularly angular, smooth.

Rare in perennially moist exposed rocks and shallow pocket. Fl. & Fr.: August – December.

Hatekeswaram-Srisailam (KNL), *JLE* 32704 (MH); Guvvalacheruvu (KDP), *KS* 6828 (MH).

INDIA: Peninsular India.

WORLD: Kenya, Rwanda, Tanzania, Uganda, Yemen, Iran, Vietnam.

Murdannia spirata (L.) G.Brückn., Nat. Pflanzenfam. (ed. 2) 15 a: 173. 1930; Matthew, Fl. Tamilnadu Carnatic 3: 1667. 1983. *Commelina spirata* L., Mant. Pl. 176. 1767. *Aneilema spiratum* (L.) R. Br., Prodr. 271. 1810; FBI 6: 377. 1892; Fischer 3: 1545. 1931.

A small, erect or diffuse, annual herb, often rooting at nodes; branches 15-25 cm long, decumbent. Leaves sessile, the lower ones broadly linear-lanceolate, the upper ones oblong, cordate, all amplexicaul, glabrous, 1.5-3.5 x 0.6 cm, margin entire, apex acute. Flowers in dichotomously branched panicles; sepals 3, oblong; petals 3, blue, broadly oblong, obtuse; stamens 3 perfect, the anthers apiculate, blue, the filaments bearded, staminodes 3, shorter; ovary 3-celled, cells 3-many-ovulate. Capsules oblong, smooth, shining, trigonous, the cells 3-7-seeded; seeds 1-seriate, minute, cubical, white or straw-coloured.

Common throughout Eastern Ghats. Fl. & Fr.: August – November.

Gundumala RF (ATP), *TP* & *NY* 733; Kalasamudram RF (ATP), *NY* 1168 (MH); Rudrakod (KNL), *BR* & *BSS* 29576; Palakonda hills (KDP), *CS* 7691; Kailasakona (CTR), *MHR* 13999; Jayanthipuram (GNT), *VRK* 6951; On the way to Panchalingalakona (NLR), *PV* 109832 (BSID); Devarapally (EG), *MM* 100789 (BSID); Araku valley (VSKP), *NPBK* 10776 (MH & CAL); Purunakote RF, Satkosia Tiger Reserve, *KCM* 5233 (BSID); Raigoda, Satkosia Tiger Reserve, *KCM* 6539 (BSID); Vazhakadu (NA), *EV* 53439 (MH); Peria Ranganar Kuppam (SA), *VNS* 5072 (MH); Killari RF (SA), *KRM* 58193 (MH); Vandalur (CPT), *S.India Flora* 11465 (MH); Guindy park RF (MDS), *ANH* 47192 (MH); Yercaud (SLM), *SKK* 26879 (MH); Krishnagiri (KGR), *S.India Flora* 13871 (MH).

INDIA: Throughout India.

WORLD: China, Laos, Indonesia, Malaysia, Philippines, Thailand, Vietnam.

Murdannia vaginata (L.) G.Brückn., Nat. Pflanzenfam. (ed. 2.) 15 a: 173. 1903. *Commelina vaginata* L., Mant. Pl 177. 1771. *Aneilema vaginatum* (L.) R. Br., Prodr. 271. 1810; FBI 6: 381. 1892; Fischer 3: 1546. 1931.

A slender grass like herb with erect and prostrate branches from near the root; roots fleshy; stem 15-40 cm long, very slender, flaccid, rooting at the lower nodes; branches sub erect, ending in filiform peduncles with distant flowering bracts. Leaves linear, 5-13 x 0.6-1 cm, apex acuminate, glabrous or sparsely hairy; sheaths short, open ciliate. Flowers small, 1-3 together, fascicled in distant bracts; bracts narrowly oblong, obtuse, strongly ribbed, slender, usually twice jointed; sepals glabrous or pubescent; petals orbicular, blue; stamens 2 perfect, 3-4 sterile, filaments of the perfect stamens bearded, ovary 2-celled, cells 1-ovulate. Capsules globose, cuspidate, shining, 3-celled; seeds trigonously conical, obtuse, slightly rugose, black.

Common in moist places in northern Eastern Ghats. Fl. & Fr.: July – October.

Forest near Anigeri village, near Addatigala (EG), *GVS* 68553 (MH & CAL); Araku valley (VSKP), *NPBK* 10764 (MH); Rathamhutta hills (KMM), *RCS* 104231 (BSID); Chatrapur (GJM), *JSG* 21526 (MH); Aska (GJM), *sine coll., s.n.* (MH); Purunakote RF, Satkosia Tiger Reserve, *KCM* 5168 (BSID); Avadi (CPT), *S.India Flora* 15468 (MH).

INDIA: Throughout India.

WORLD: Sri Lanka, Bangladesh, Myanmar, Nepal, China, Papua New Guinea, Philippines, Thailand, Vietnam.

POLLIA Thunberg

Pollia secundiflora (Blume) Becker, Ann. Jard. Bot. Buitenz. 32: 166. 1922. var. **indica** (Wight) Sanjappa in Fl. Ind. Enum. Monocot. 30. 1989. *Commelina secundiflora* Blume, Enum. Pl. Jav. 1: 5. 1827. *Pollia sorzogonensis* (E. Mey) Endl., Gen. 1029. 1840; FBI 6: 367. 1892. *P. sorzogonensis* var. *indica* (Wight) Hook. f., Fl. Brit. India 6: 367. 1892; Fischer 3: 1536. 1931. *Aclisia sorzogonensis* E. Mey, in Presl., Rel. Haenk. 1: 138t. 25. 1827. *A. indica* Wight, Icon. Pl. Ind. Orient. t. 2068. 1853.

Perennial ascending herb, rooting at base. Leaves oblong-lanceolate, 9-32 x 2-9 cm, scabrous above, puberulous beneath; base cuneate, apex acuminate; sheaths tubular, viscid. Flowers in 2-13 cm long panicles; sepals glabrous; bracts prominent; 3 anthers large and white, 3 yellow. Capsules ellipsoid or oblong, shining blue, many-seeded.

Rare in open forests in northern Eastern Ghats of Andhra Pradesh. Fl. & Fr.: June – August.

Minumuluru (VSKP), *KR & N. Balachandran* 110885 (FRLHT).

INDIA: Peninsular India.

WORLD: Myanmar, China, Indonesia, Laos, Malaysia, Thailand, Vietnam.

RHOPALEPHORA Hasskarl

Rhopalephora scaberrima (Blume) Faden, Phytologia 37: 480. 1977. *Commelina scaberrima* Blume, Enum. Pl. Jav. 1: 4. 1847. *Aneilema scaberrimum* (Blume) Kunth, Enum. Pl. 4: 89. 1843; FBI 6: 382. 1892; Fischer 3: 1546. 1931.

A herb with stout, 1 m long stem, puberulous, sometimes decumbent and rooting at the lower nodes. Leaves scattered, usually distant, not enlarging upwards, lanceolate or oblong-lanceolate, 8-16 x 4-9 cm, base rounded and then narrowed into a petiole, apex acuminate, hispid above, glabrous below, sheaths loose, 1-3 cm long, viscid or hispid, mouth long-ciliate. Panicles with long, slender peduncles and branches; bracts persistent, funnel-shaped. Capsules globose, pubescent, 1-3-seeded; seeds plano-convex, glaucous, black, rugose.

Occasional in Rampa hills, Galikonda hills and Chintapalli hills. Fl. & Fr.: July – December.

Sesharayi (EG), *SS* 279 (AU); Sunkarimetta (VSKP), *NPBK* 10925 (MH), *GVS* 21673 & 21676 (MH); Galikonda (VSKP), *GVS* 19622 (MH); Chintapalli (VSKP), *GVS* 29749 (MH), *JLE* 37168 (MH).

INDIA: S. & NE. India, E. to C. Himalaya.

WORLD: Sri Lanka, Myanmar, Bhutan, China, Malaysia, Laos, Indonesia, Philippines, Thailand, Vietnam.

Zebrina pendula Schnizl. and **Tradescantia spathacea** Sw. are cultivated frequently in gardens for its ornamental foliage.

FLAGELLARIACEAE

FLAGELLARIA Linnaeus

Flagellaria indica L., Sp. Pl. 333. 1753; FBI 6: 391. 1892; Henry & Swamin., Indian J. Forest. 3: 140. 1980. *F. indica* var. *minor* Hook.f., Fl. Brit. India 6: 391. 1892.

A reed-like climber, quite glabrous, climbing over lofty trees by the leaf-tendrils; stem nearly 2.5 cm thick towards the base, terete, smooth; branches clothed with cylindric smooth striate, closed, truncate sheaths, branchlets as thick as a crow-quill. Leaves sessile, 15-25 cm long, variable in breadth, lanceolate from a rounded base, shortly narrowed into the sheath, drawn out at the apex into a slender spiral tendril, many-nerved. Flowers white, in shortly pedunculate, irregularly, laxly branched panicles, 15-30 cm long; outer perianth-segments broadly ovate or suborbicular, obtuse; inner-segments similar, more or less unequal. Drupe pisiform, red smooth.

Near Sesharayi stream in East Godavari district (R.S. Rao, 1964). Fl.: March.

INDIA: Assam, Odisha, Andhra Pradesh, Karnataka.

WORLD: Sri Lanka, Tropical Africa, Tropical SE Asia and N. Australia.

JUNCACEAE

JUNCUS Linnaeus

Juncus prismatocarpus R.Br., Prodr. 259. 1810; FBI 6: 395. 1892 *p.p.*; Fischer 3: 1553. 1931; Matthew, Fl. Tamilnadu Carnatic 3: 1668. 1983.

An erect, usually perennial herb; stem terete or compressed, smooth, 8-50 cm high. Leaves basal and cauline, linear, 3.5-10 x 0.4-0.5 cm, plicate, distantly septate; sheaths 2-4 cm; auricles 2, obtuse. Scape 20-60 cm. Cymes terminal, corymbose; bract ovate, acuminate; flowers 6-10 per cyme, perianth 2-seriate; segments purplish-green, ovate-lanceolate, apex acute, stamens 3; ovary superior, sessile, ovules on 3 parietal placenate, style short. Capsule prismatic, 1-celled; seeds numerous, light brown, ellipsoid.

Common in marshy places at higher elevations. Fl. & Fr.: Throughout the year.

Near Padmapuram gardens (VSKP), *MHR* & *MCK* 14829; Galikonda (VSKP), *MHR* & *MCK* 14858; Araku valley (VSKP), *NPBK* 10864 (MH & CAL), *GVS* 19700 (MH); Minumuluru (VSKP), *GVS* 29746 (MH); Towards Birmisala (VSKP), *GVS* 47332 (MH); Raigoda, Satkosia Tiger Reserve, *KCM* 6541 (BSID); Kaveri peak, Yercaud (SLM), *DBD* 31298 (MH); Graegmore estate,Yercaud (SLM), *SKK* 26993 (MH).

INDIA: Assam, West Bengal, Bihar, Odisha, Andhra Pradesh.

WORLD: Sri Lanka, Myanmar, SE to E. Asia, Malesia, Australia, New Zealand.

ARECACEAE (*nom. alter.* PALMAE)

1. Climbers; leaves scattered; fruit covered with closely imbricating scales .... **Calamus**
1. Non-climbers; leaves in a terminal crown; fruit never scaly:
    2. Leaves simple, palmately divided or flabelliform:
        3. Flowers unisexual, dioecious .... **Borassus**
        3. Flowers bisexual .... **Licuala**
    2. Leaves pinnately compound:
        4. Leaves pinnate:
            5. Leaflets modified into spines; plant dioecious; ovary of 3 free carpels .... **Phoenix**
            5. Leaflets non-spinous; plant monoecious; ovary of unicarpel:
                6. Leaf sheath tubular; stem less than 15 cm across; drupes 3.5 cm long .... **Areca**
                6. Leaf sheath absent; stem more than 40 cm across; drupes 10-30 cm long .... **Cocos**
        4. Leaves bipinnate .... **Caryota**

ARECA Linnaeus

Areca catechu L., Sp. Pl. 1189. 1753; FBI 6: 405. 1892; Fischer 3: 1555. 1931; Matthew Fl. Tamilnadu Carnatic 3: 1670. 1983.

A tall tree, to 50 m high; cylindrical stem 10-15 cm across, unarmed. Leaves in terminal crown, to 1.5 m long, prominently arching; leaflets many, linear to lineat lanceolate, close; lower one plicate, upper ones coherent, base narrow, apex praemorse, nerves 2, 5 or more, prominent. Spadices several on the axis of falled leaves, to 50 cm, branched, shortly peduncled; rachis angular, flattened, apex tapering; spikes flexuous or straight, to 20 cm long; spathe boat like, coriaceous. Flowers monoecious; lower ones female,1-3 at spike-base; upper ones male, uniseriate, many; sepals 3, triangular, 1 mm, basally connate; petals 3, ovate, 2.5 mm; stamens 6; filaments to 1 mm long; anthers 2-5 mm long; pistillode to 3 mm, bifid. Female flowers 1.5 cm long; ovule solitary, basal, erect. Fruit 3.5 x 2 cm, with fibrous mesocarp.

Occasionally cultivated for the nuts in the foot hills of Shervarayans, Kolli hills and Sirumalai hills.

INDIA: Coastal and southern states.

WORLD: Africa and Tropical America.

BORASSUS Linnaeus

Borassus flabellifer L., Sp. Pl. 1187. 1753; FBI 6: 482. 1893; Fischer 3: 1562. 1931. *B. flabelliformis* Murr. in L., Syst. Veg. (ed. 13). 827. 1774.

A tree with tall stem, stout, to 40 m high and 0.5 m diameter near the ground. Leaves simple, palmately fan-shaped, plicate, the margin multifid; leaflets 60-80, induplicate, apex acuminate; petiole stout, spinous, base broad, split. Spadices interfoliar, large, dioecious, branched; peduncles sheathed with open spathes. Male flowers small, clustered, mixed with scaly bracts, second in 2 series in a small spikelet; sepals 3, oblong; petals 3, obovate-spathulate; staments 6; pistillode of 3 bristles. Female flowers larger, globose, perianth fleshy, accrescent; sepals reniform, imbricate; petals smaller, convolute; staminodes 6-9; ovary globose, subtrigonous, entire or 3-4-partite, 3-4-celled, ovules basal, erect, stigmas 3, sessile, recurved. Fruit a large subglobose drupe with 1-3, fibrous pyrenes, yellow when ripe; pericarp thinly fleshy; seed oblong, top 3-lobed, testa adhering to the pyrene.

Common in waste lands, field bunds, along streams all over Eastern Ghats. Fl. & Fr.: February – April. Vern.: Tel.: *Thadi, Thatichettu*; Or.: *Talo, Tala*; Tam.: *Panaimaram*.

Gulemallebad RF (KNL), *RVR* 1421; Gangineni (KSN), *PV* 5572 (MH); Puttagondhilanka, near Tadepalli (EG), *DN* 85549 (BSID).

INDIA: Through out Tropical India.

WORLD: Africa, Madagaascar, Sri Lanka, Indo-China, Indonesia, Australia.

CALAMUS Linnaeus

1. Leaf not ending in a cirrus; sheaths usually flagelliferous; leaflets usually narrowly-linear or –lanceolate:
   2. Leaflets fascicled at least near base .... **C. viminalis**
   2. Leaflets never fascicled, more or less regularly equidistant .... **C. rotang**
1. Leaf rachis produced into a cirrus up to 90 cm long armed with many strong, irregularly aggregated claws; sheaths not flagelliferous:
   3. Leaflets broadly lanceolate or elliptic-lanceolate, sheath spines up to 2.5 cm long.. **C. latifolius**
   3. Leaflets linear-lanceolate, sheath spines up to 1.8 cm long.............. **C. guruba**

Calamus guruba Buch.-Ham. in Mart. Hist. Nat. Palm. 3: 206. 1850; FBI 6: 449. 1893; Haines, Bot. Bihar Orissa 3: 887. 1924; Mooney, Suppl. Bot. Bihar & Orissa 143. 1950; Saxena & Brahmam, Fl. Orissa 4: 2014. 1995.

Large climber; stems covered, with the leaf-sheaths, often glabrous, densely armed with flattened spines of various lengths, longer ones 1.8 cm, shortest conical. Leaves 0.9-1.5 m long, shorter with 30-40 close, equidistant leaflets each side; leafets dark-green, linear, 15-30 x 1.2-1.8 cm, smaller upwards, setose on the 3 or 2 lateral costae beneath, on the central one only above, margins setose, smooth or bristly-spinulose; petiole and rachis armed with long recurved and short, conical spines beneath and on the margins, not flagelliferous, but many of the leaf-sheaths flagelliferous, spines sometimes geminate. Inflorescence with numerous distant spathes and much panicled branches on a very slender ranchis ending in a long flagellum. Spathes at first tubular throughout and enclosing the panicles, splitting open above and 1.8 cm broad, contracted to the tubular base and persistent, brown and shining. Panicles (from the axils of each spathe), pyramidal or thyrsiform, 15-22.5 cm long. Lower branches 5-7.5 cm or longer and again branched, very slender with closely appressed tubular spathes with truncate mouth, shortly acuminate on one side, ultimate male spikes (spikelets) very numerous, erect, 1.2-7.5 mm, with minutely cuspidate spathes and very shallow or flat minutely 2-(laterally)lobed spathellules; calyx tubular, 1.2 mm long with short ovate teeth; corolla 2.5 mm, petals connate one-third up, acute; filaments adnate below; bracts on the female spikes with more prominent cusps; spathellules flat. Fruit globose, 7.5 mm, scales yellowish-green with brown and scarious margins.

Along streams in the hills of Odisha (Haines *loc. cit.*); abundanrt in the hills of Ranpur, less common in the forests of Nayagarh and rare elsewhere. Fl. & Fr.: March-April.

INDIA: North, Central and North East India.

WORLD: Bangladesh, Myanmar, Thailand and Malay Peninsula.

Calamus latifolius Roxb., Fl. Ind. 3: 775. 1832; Fischer 3: 1568. 1931. var. **marmoratus** Becc., Ann. Roy. Bot. Gard. Calcutta. 12: 107, 409. 1908; Haines, Bot. Bihar & Orissa 3: 886. 1924.

A large scandent shrub, armed with numerous, often subverticillate, close, reflexed or young patent, horizontally flattened, lanceolate prickles over 2.5 cm long and numerous very short triangular ones, young parts rusty tomentose. Leaves very long with the rachis produced into a long flagellum armed with recuved prickles; sheaths with marbled appearance, not flagelliferous, petiole pulvinate below, on the sheath with sharp narrow spines; leaflets very inequidistant, mostly geminate with members of the pair on the same side of the rachis, oblong-lanceolate or oblanceolate-oblong, *c.* 10 x 3.7 cm or sometimes 5 cm, x 11 ending in a bristly acumen, margin with distant setiform erect prickles not setose on the nerves. Spadix decompound; branches and spathes usually armed with broad spines; male flowers 5 mm long; calyx 2.5 mm long. Fruit subglobose, pale yellow fruiting calyx pedicelliferous.

Rare along streams in hills in Northern Eastern Ghats. Fl. & Fr.: September – March.

Near Dharawada (WG), *DN* 85515 (BSID); Madgol hills (VSKP), *A.W.Lushington s.n.* (MH); Madugula hills (VSKP), *NRR* 85515 (MH); Simlipahar (MBJ), *SX* & *MB* 636 (RRL-B); Malayagiri (DK), *MB* & *Dhal* 8399 (RRL-B).

INDIA: Arunachal Pradesh, Meghalaya, Nagaland, Assam, Sikkim, West Bengal, Odisha.

WORLD: Bangladesh.

Calamus rotang L., Sp. Pl. 325. 1753; FBI 6: 447. 1892; Fischer 3: 1568. 1931; Matthew, Fl. Tamilnadu Carnatic 3: 1671. 1983.

An armed, scandent shrub; stem ringed at nodes. Leaves pinnatisect, 30-60 x 0.6-1 cm, base cuneate, margins bristly ciliate, apex gradually acuminate, setose, nerves prominent, midnerve with scattered, straight spines below; petiole very short, with straight or recurved spines; rachis apically 3-gonous, with recurved spines; leaf-sheath glabrous; ochrea short, truncate. Spadices axillary, elongate, branched, 60 cm long, apical flagellum with recurved spines; spathes tubular, sparingly armed. Flowers small, polygamodioecious. Male flowers in 3 or 4 series; calyx cupular, 3-lobed; petals 3, yellow, free; stamens 6, shortly connate at base. Female flower: ovary imcompletely 3-celled, closed with retrorse scales, ovules 3, basal erect, style terminal. Fruit subglobose, mucronate, with scales in vertical series; seed solitary.

In all the drier tracts; from sea-level to 500 m. Fl. & Fr.: September – February.

Mahanandi (KNL), *JSG* 10930 (MH); Krishna Nandi (KNL), *JLE* 25527 (MH); Kambakkam hills (CTR), *DRC* 1374 (MH), *MVS, VSR* & *BR* 31086 (BSID); Neelakantavaram (VZN), *MV* 19829 (AU); Adyar (MDS), *JSG s.n.* (MH).

INDIA: Peninsular India.

WORLD: Sri Lanka.

Calamus viminalis Willd., Sp. Pl. 2: 203. 1799 var. **fasciculatus** (Roxb.) Becc. in Hook. f., Fl. Brit. India 4: 444. 1892; Fischer 3:1568. 1931. subvar. **pinangianus** Becc., Ann. Roy. Bot. Gard. Calcutta 15: 209. 1856.

A stout scrambling and climbing plant; stem rather stout. Leaves to 1 m, leaflets numerous in several planes pointing in different directions, distinctly grouped in fascicles of 2-4 on each side, those near the apex more regular, nearly equidistant and in one plane, narrowly lanceolate, up to 25 cm long and 2 cm wide, densely bristly on the margins and on the midrib above; rachis nearly terete, armed below with a solitary or ternate, needle-like, pale, spreading or deflexed spines, up to 3 cm long; primary spathes tubular, clawed mainly on the back. Fruit globose or slightly turbinate, distinctly beaked, scales broad, not channeled, uniformly pale-straw-coloured.

Occasional in Northern Eastern Ghats. Fl. & Fr.: October – March.

Rampa hills (EG), *JSG* 14029 (MH); Valamuru, Ramapa (EG), *JSG* 15933 (MH); Vedurupalli (EG), *GVS* 42756 (MH); near Madhiveedu, Tadeplli (EG), *NRR* & *DN* 84370 (MH); Near Kakur teak plantation (EG), *GVS* 68507 (MH); Palakonda (VSKP), *JSG* 14029 (MH); near Vedurupalli (VSKP), *GVS* 42756 (MH); Isakagadda-Seethampeta (SKLM), *GVS* 62464 (MH); Ajaigad forest, Serango (GJM), *MB* 3096 (RRL-B); Daringabadi to Mahagudi (GJM), *CAB* 1430 (MH).

INDIA: West Bengal, Odisha, Andhra Pradesh, Andamans.

WORLD: Bangladesh, Myanmar.

CARYOTA Linnaeus

Caryota urens L., Sp. Pl. 1189. 1753; FBI 6: 422. 1892; Fischer 3: 1560. 1931; Matthew, Fl. Tamilnadu Carnatic 3: 1872. 1983.

A tree up to 20 m high; stem annulate, cylindric, smooth, grey, shining, covered with long shallow cracks with corky edges. Leaves bipinnate, 5 m; pinnae ca 10 pairs; subopposite, curved. Leaflets induplicate, clustered or alternate, 15-20 x 7-10 cm, base cuneate, oblique, margin praemorse (ragged), irregularly serrate, upper serration caudate, nerves flabellate; petiole stout. Spadices interfoliar, 3 m long; branches simple, forming a dense tassel drooping from the stout short peduncle, all reaching the same level. Spathes 0.5 m long, closely embracing the peduncle of the spadix. Flowers monoecious, solitary and/or in threes with the central one pistillate between the 2 staminate. Male flowers: sepals 3, rounded, imbricate; petals linear-oblong, valvate; stamens *c.* 40, filaments short; pistillode 0. Female flowers subglobose, sepals broader; petals rounded, valvate; staminodes 3; ovary trigonous, 3-celled, ovules 1 per cell, erect, stigma sessile, 3-lobed. Fruit globose; seed(s) 1 or 2.

Common in Eastern Ghats. Grown also as ornamental in plains. Fl. & Fr.: February – August. Tam.: *Koonthalpanai, Thalipanai.*

Velgode to Gundlabrahmeswaram (KNL), *BSS* & *KP* 34099; Talakona to Rudragal (CTR), *BR* & *SKB* 32759; Srungavarapukota (VZN), *BR & KNR* 9883; Bakarimanchala, way to Ramanapenta, Nallamalais (KNL), *JLE* 32645 (MH); Thumbura theertham (CTR), *J.Swamy* & *S.Nagaraju* 7281 (BSID); Near Geddapalli (WG), *DN* 85512 (BSID); Vokurth (EG), *SS* 7692 (AU); Neelakantapuram (VZN), *MV* 19827 (AU); Erukkampattu, Melpat, Javadi hills (NA), *MBV* 1038 (MH); Kolli hills (NMK), *A.Mohan* 13270 (MH); Boragharo (GJM), *VNS* 5992 (MH).

INDIA: Through out the hotter parts of India.

WORLD: Sri Lanka, Myanmar, Malaysia.

COCOS Linnaeus

Cocos nucifera L., Sp. Pl. 1189. 1753; FBI 6: 422. 1892; Fischer 3: 1560. 1931; Matthew, Fl. Tamilnadu Carnatic 3: 1672. 1983.

Tall tree, to 30-40 m tall; stem annulate, shining. Leaves pinnatisect, 4-6 m; leaflets reduplicate, 50-100 x 1.5-5 cm, base narrow, apex tapering, acute; petiole elongate, 1-2 m, stout. Spadices interfoliar, 60-90 cm long, panicled; branches to 40 cm, flexuous; lower spathes 50 cm to 1.5 m, oblong, woody. Flowers monoecious, shortly pedicellate. Male flower often paired; sepals ovate to 3 mm; petals narrowly ovate, to 8 mm; stamens 6, filaments to 4 mm long, anthers to 6 mm long, pistillode short. Female flower globose, 1 per branch; perianth lobes broadly ovate, 2.5 cm across, inner lobe orbicular, to 1 cm across; ovary 3-celled, uni-ovuled, style short. Drupe trigonous, ovoid or globose, to 30 cm, greenish-yellow; pericarp fibrous; endocarp stony with 3 basal pores; seeds coherent with the endocarp; endosperm lining with endocarp, with a hollow centre filled wth coconut milk.

Common in lower hills of all the areas cultivated for its fruits. Fl. & Fr.: Round the year. Tel.: *Tenkaya*; Tam.: *Thengai*.

INDIA: Mostly southern India.

WORLD: Cultivated in all the tropical countries.

LICUALA Thunberg

Licuala peltata Roxb., Fl. Ind. 2: 179. 1832; Becc. & Hook.f. in Hook.f., Fl. Brit. India 6: 430. 1892.

Erect palm, 0.6-1.5 m high. Leaves erect, palmate, orbicular, 0.9-1.5 m diam., segments narrowly obcuneate, *c.* 20, 10-25 cm wide at the top, 3-5-lobed with the lobes again 2-lobed, apex retuse or emarginated; petiole flat above, keeled below, with 2 rows of short, recurved, black prickles; young leaves with deciduous rufous tomentum. Spadix erect, 1.2-2.4 m long with 5-8 simple, drooping racemes, 20-45 cm long, woolly or tomentose; flowers subsessile, white and green, turning brown with age; spathe to 30 cm long; calyx obconic, 7.5-12.5 mm long, toothed or shallowly lobed; corolla-tube with spreading or reflexed, acutely triangular downy lobes, 6.2-7.5 mm long; carpels 3-gonous, apprised, united above in the style. Drupe orange, ellipsoid, to 1.2 cm long.

Along muddy streams or swamps with moving water in the hill forests of Eastern Ghats of Odisha. Fl.: November – April; Fr.: March – May.

Malayagiri, Dhenkenal, *MB* & *Dhal* 8366 (RRL-B).

INDIA: Assam, Sikkim, Andamans.

WORLD: Bangladesh, Myanmar.

PHOENIX Linnaeus

1. Stem 8-30 m high .... **P. sylvestris**
1. Stem 0.3 – 7 m high:
    2. Leaflets stiff:
        3. Fruit ellipsoid, up to 1.5 cm long .... **P. acaulis**
        3. Fruit sub-globose, up to 2 cm long .... **P. pusilla**
    2. Leaflets pliable:
        4. Stems less than 2 m high .... **P. loureiroi** var. **lourerioi**
        4. Stems 5-7 m high .... **P. loureroi var. pedunculata**

Phoenix acaulis Roxb., Pl. Coromandel 3: 69. 1820; FBI 6: 427. 1892; Haines, Bot. Bihar & Orissa, 3: 882. 1924.

Short palm, stem not more than 30 cm high and about as broad as high, covered with the persistent petiole bases or stem O. Leaves 0.6-1.8 cm long; leaflets stiff, more or less fascicled and in different planes, 25-50 x 0.7-1.8 cm, finely acuminate, lowest reduced to strong spines, 5-15 cm long, base of leaflets thickened and decurrent on the rachis; petiole to 30 cm long. Spadix 15-25 cm long in flower elongating to 30-90 cm in fruit, suberect, fruiting peduncles short. Drupe oblong-ellipsoid, 1.2-1.5 cm long, orange or orange-red, finally black.

Common in drier forests and open grassy forest. Fl.: April; Fr.: May –June.

Anantagiri (VSKP), *GVS* 19567 (MH); Murhi RF (Dhenkenal Distr.) *A.R.K.Sastry* 11011 (BSID).

INDIA: Throughout India.

WORLD: Myanmar.

Phoenix loureiroi Kunth., Enum. Pl. 3: 257. 1841. var. **loureiroi.**

A plant with usually well developed stem, up to 2 m, densely covered with the bases of the fallen petioles, more or less spirally arranged; root suckers developing when the primary stem has been burnt or injured. Leaves 2.5 m long; leaflets 15-30 x 0.8-1 cm, pliable, base narrow, apex acuminate, spinous. Spadices interfoliar. Male: spadix 40 cm; spikes 10 cm, in clusters; spathes 15-20 cm; calyx lobes triangular; petals oblong; stamens 6. Female: Spadix 1 m; spikes 8 cm. Drupe oblong, 1.5 cm, at first orange, black when ripe.

Occasional in open forests in all the hilly districts. Fl. & Fr.: March – May.

Peccheruvu RF (KNL), *RVR* 2169; Gapakathirtham (CTR), *KS* 7857 (MH); Balapalli (KDP), *JSG* 10835 (DD); Banigocchaa rest house, Satkosia Tiger Reserve, *KCM* 8372 (BSID).

INDIA: Kerala, Andhra Pradesh.

WORLD: India to SE Asia.

Phoenix loureiroi Kunth var. **pedunculata** (Griff.) Govaerts, World Checklist of palms 171. 2005; Mattew, Fl. Tamilnadu Carnatic 3: 1674. 1983 *"loureirii'*. *P. pedunculata* Griff., Palms Brit. E. India 139. 1850. *P. humilis* Royle var. *pedunculata* Becc., Malesia 3: 379 & 387. 1890; FBI 6: 427. 1892; Fischer 3: 1560. 1931. *P. robusta* (Becc.) Becc. & Hook.f. in Hook. f., Fl. Brit. India 6: 427. 1892; Fischer 3: 1559. 1931. *P. humilis* var. *robusta* Becc., Malesia 3: 384. 1890.

A plant with trunk 5-7 m high and 40 cm diameter, clothed and appearing tessellated from the spirally arranged small leaf sheaths. Leaves 1-2 m long, glabrous, shining, shorter, broader, thinner and smoother than those of *Phoenix sylvestris*, leaflets fascicled, quadri-farious, shining, 30 cm long and 1.5 cm wide, strongly conduplicate. Spathes narrow, coriaceous, up to 20 cm long, fringed with brown wool on the keel; Male flowers 3-5 mm long, dense; Female flowers distant. Fruiting spadix about 0.5 m long. Fruit oblong-ellipsoid, 1.5 cm long, brown.

Common throughout Eastern Ghats. Fl. & Fr.: January – April. Vern.: Tam.: *Malai eecham, Chiru eecham.*

Peddakonda RF, Nallamada (ATP), *NY* 1220; Nallamalais (KNL), *JLE* 23817 (MH); Pullampeta (KDP), *KCJ* 83225 (MH); way to Japalithirtham (CTR), *KS* 7857 (MH); Vojili, near Naidupeta (NLR), *KCJ* 83355(MH); Talur (GNT), *CAB* 4302 (MH); Bhimadole (WG), *KCJ* 459 (MH); Cheedipalem (EG), *GVS* 27428 (MH); Karaka (VSKP), *CAB* 1650 (MH); Anantagiri (VSKP), *GVS* 19567 (MH); Cherukonda (VSKP), *GVS* 29735 (MH); Bodikonda, Laxmipuram konda (SKLM), *GVS* 62366 (MH); Ettapur (SLM), *R. Ansari* 49982 (MH); Athanur- Yelagiri hills (NA), EV 53472 (MH); Kambakkam hills (NA), *sine coll.* 8975 (MH); Damudua to Odaba (Audaba) (GJM), *CAB* 1463 (MH); Khaiguda (GJM), *JSG* 13750 (MH).

INDIA: Assam, Manipur, Bihar, Odisha, South India.

WORLD: Myanmar, Cochinchina.

Phoenix pusilla Gaertn., Fruct. Sem. Pl. 1: 24. 1788; FBI 6: 426. 1892. *P. farinifera* Roxb., Pl. Coromandel 1: 55, t. 74. 1796 & Fl. Ind. 3: 783. 1832; FBI 6: 426. 1892; Fischer 3: 1559. 1931.

A small shrub, 15 cm – 3 m high, stem thickly clothed and hidden by the old leaf-sheaths. Leaves 1-1.5 m long, leaflets fascicled more or less 4-farious, rigid, shining, usually with an orange-red pulvinus at the junction with the rachis, 7-20 x 1-1.5 cm, several basal forming stout flat spines of 5-7.5 cm long. Fruit 1.2-2 cm long, first green, black when ripe.

Coromandel, at low elevations. Vern.: Tel.: *Chitti-ita, Churuta-ita.*

Balapalle (KDP), *JLE* 15789 & 17590 (MH); Nilagiri view Point (MBNR), *S.R.Srinivasan* 109720 (BSID); Kangampallam, Yercaud (SLM), *DBD* 31430 (MH); Guindy park (MDS), *ANH* 47181 (MH); Chingleput, *JSG s.n.* (MH); Takarai RF (SA), *KRM* 52846 (MH); Komaratchi to Kattumannargudi (SA), *KRM* 60188 (MH).

INDIA: South India.

WORLD: Sri Lanka.

Phoenix sylvestris (L.) Roxb., Fl. Ind. 3: 787. 1832; FBI 6: 425. 1892; Fischer 3: 1559. 1931; Matthew, Fl. Tamilnadu Carnatic 3: 1674. 1983. *Elate sylvestris* L., Sp. Pl. 1189. 1753.

A tall graceful palm, up to 15 m high with a large thick hemispherical crown; trunk clothed with the persistent bases of the petioles. Leaves greyish green, 3-4 m long; leaflets 10-30 x 1-2.5 cm, glaucous, rigid, base narrow, apex acuminate, spinous; lower leaflets spiny. Spadices 1 m; spathes as long as spadices, coriaceous;

spikes numerous, in clusters, 10-30 cm long. Male flowers white, scented, angular, oblique; calyx cup-shaped, with 3 short rounded teeth; petals 3 or 4 times as long as the calyx, concave, warted on the outside; stamens 6, anthers linear, filaments very short, free. Female flowers distant, roundish; calyx cup-shaped, obsoletely 3-toothed; petals 3, very broad, convolutedly imbricate; staminodes 3 or 4. Fruit oblong – ellipsoid, orange-yellow, edible; seeds rounded at the ends, grooved on the face, pale brown.

Common along streams and waste lands in plains. Fl. & Fr.: January – April. Vern.: Tam.: *Eecham, Periya eecham*.

Racherla RF (KNL), *RVR* 1361; Pulivendula (KDP), *KRKS* 42054; Talakona (CTR), *SKB* & *BSS* 32753; Nallapadu (GNT), *VRK* 6775; Tanuku (WG), *KCJ* 457 (MH); Karakamokasa (VSKP), *KCJ* 449 (MH).

INDIA: Through out the plains of India.

WORLD: Myanmar, Sri Lanka.

PANDANACEAE

PANDANUS Linnaeus f.

Pandanus odorifer (Forssk.) Kuntze, Revis, Gen. Pl. 2: 737. 1891. *Keura odorifera* Forssk., Fl. Aegypt. Arab. 172. 1775. *Pandanus fascicularis* Lam., Encycl. 1: 372. 1785; FBI 6: 485. 1893. *P. odoratissimus* L. f., Suppl. Pl. 424. 1781. *nom. illeg. P. tectorius* sensu Fischer 3: 1570. 1931.

A much branched shrub or tree, 6 m high; rarely erect, stem supported by aerial roots. Leaves linear-ensiform, stoutly armed with white prickles on margins and dorsal midrib, under-surface of leaf glaucous; leaves usually less than 2 m long and 8 cm wide. Male inflorescence of several spikes with white or cream, axillant bracts. Spikes 4-12 cm long, composed of numerous phalanges, these racemose, of numerous stamens; anthers elongate, briefly apiculate, white. Female inflorescence a solitary terminal cephalium, pendulous in fruit. Phalanges mostly 5-15-celled, the carpels concentrically arranged. Fruit an oblong syncarpium, 15-25 cm long, yellow or red.

Occasional in marshy localities. Fl. & Fr.: July – December.

Narpala (ATP), *KNR* & *DAM* 10638; Gutturkona (ATP), *NY* 994 (MH); Owk tank (KNL), *RVR* 3055; Mogalikona (CTR), *MVS* & *V.S.Rao* 31105; Srungavarapukota (VZN), *BR* & *KNR* 9884; Kailasakona – Puttur (CTR), *GVS* 32016; on the way from Mandasa to Jalantrakota (SKLM), *GVS* 62325 (MH); Hogainakkal (KGR), *EV* 24158 (MH).

INDIA: Through out the coasts of India.

WORLD: Myanmar, Indonesia, Malaysia, China.

TYPHACEAE

TYPHA Linnaeus

1. Leaf sheath angular, keeled .... **T. elephantina**
1. Leaf sheath not angular, not keeled:
    2. Leaf sheaths taper into lamina; colour of female flower bracts translucent; staminate bracts laciniate; anthers not twisted after dehiscence .... **T. domingensis**
    2. Leaf sheaths auriculate at tip; colour of female flower bracts dark brown; staminate bracts forked at apex; anthers twisted after dehiscence .... **T. angustifolia**

Typha angustifolia L., Sp. Pl. 2: 971. 1753; Matthew, Fl. Tamilnadu Carnatic 3: '684. 1983; Halder *et al.*, Rheedea 24: 17. 2014.

Perennial, rhizomatous, erect herbs, 1.5–3 m high. Stems unbranched, green, glaucous, terminating in an inflorescence. Leaves linear, flat or slightly convex on abaxial face; blades 4–12 mm wide when fresh, 3–8 mm when dry, base sheathing, margin entire, apex acute; sheath membranous, auriculate at tip, often deciduous at maturity; mucilage glands present on adaxial face, usually from middle to tip of sheath, linear-oblong, brown; distal blades usually exceeding inflorescence. Flowering shoots 5–12 mm thick in middle and 2–3 mm thick near inflorescence. Inflorescence a long, cylindrical, compact spike with staminate and pistillate flowers in separate aggregations, separated by 3 (0.6) – 8 (12) cm long interval of naked axis. Staminate spike not subtended by any bracts, 21–23 x 0.6–0.8 cm, relatively less dense, yellow; staminal filament half as long as anthers or smaller; anthers 2–5, slender, oblong, 2–2.5 x 0.2–0.3 mm, yellow, turn dark brown at apex when dried, twisted when old; bracts 3–5, filiform, 2–2.5 mm long, forked at apex, white when young, turn brown when dried; remnants of staminate bracts dry and fall off the spike leaving rachis bare after pollen release. Pistillate spike not subtended by any bracts, 21.5–23 x 0.5–0.6 cm; female flowers originate on peg-like 0.5–0.7 mm long, compound pedicels; gynoecium filiform, 0.8–1.5 mm long, borne on a carpopodium; ovary fusiform, pale green, unilocular, small, with a single pendulous ovule; style slender, 0.3–0.4 x *c.* 0.1 mm long; stigma simple, as broad as style, erect, elongating, bending to form surface mat in flower; dark brown, deciduous in fruit; perigonial hairs linear, arising from the carpopodium, remain below the stigma; pistillate bracts arise from carpopodium, slender, spathulate at tip, dark brown. Fruits minute follicles, fusiform, *c.* 1.2 mm long; seed 1, pendulous, oblanceolate, 0.8-0.9 mm, orange brown, a thin-walled pericarp surround the seed; surface vertically ribbed with papillae.

Common in ditches, ponds and stagnant water. Fl. & Fr.: April – December. Vern.: Tam.: *Sampai.*

Nellore district, *JSG* 12231 (CAL).

INDIA: Himachal Pradesh, Madhya Pradesh, Jharkhand, West Bengal. Andhra Pradesh, Tamil Nadu.

WORLD: Pantropical.

Typha domingensis Pers., Syn. Pl. 2: 532. 1807; Napper in Redhead & Polhill, Fl. Trop. E. Afr. 1. 1971; Cook in Tutin & *al.*, Fl. Eur. 276. 1980. *T. angustata* Bory & Chaub., Exp. Sci. Morec. Bot. 1: 338. 1833; FBI 6: 489. 1893; Fischer 3: 1571. 1931.

A robust herb, 2 m high, stem terete. Leaves distichous, linear, erect, 2 m long and 3 cm broad, exceeding the flowering stem, apex tapering, acute, base sheathing. Spikes cylindric, the male and female spikes often separated by a considerable interval. Male spike 10-20 cm long; rachis (separating male from female) 3 cm long; stamens 2 or 3; anthers linear, connections shortly produced, apex rounded, hairs around stamens linear, spathulate. Female spike 20 cm long, 1-1.5 cm broad, cushion-like; bracteoles numerous, gynophores white, ovary linear, stigma flattened. Fruit fusiform, 2 mm.

Common in stagnant water and slow flowing streams. Fl. & Fr.: May – July.

Peddapalli (ATP), *JSG* 20881 (MH); Gani RF (KNL), *RVR* 1556; Pulipadu tank – Darsi (PKM), *MCK* 22862; Macherla RF (GNT), *VRK* 3299; Maduravada tank (VSKP), *MHR* & *MCS* 14956; Guvvalacheruvu (KDP), *KS* 7786 (MH & CAL); Kalyani dam (CTR), *DRC* 1399 (MH); Somasila dam (NLR), *PV* 110201 (BSID); Pedyedlagdi (KSN), *PV* 116537 (BSID); Kolleru (WG), *KS* 5076 (MH); Thatipudi dam (VSKP), *GVS* 44250 (MH); Majhipada, Satkosia Tiger Reserve, *KCM* 6152 (BSID); Marappur (SLM), *S.India Flora* 11650 (MH); Hogainakkal (KGR), *KCJ* 17996 (MH); Chinnakalarayan (SLM), *R.Ansari* 49950 (MH); Mattukkannuru Kolla (NA), *TRS* 95567 (MH, BSID); Inapudle – Puliyur (NA), *KS* 6523 (MH); Sivapuri (SA), *KRM* 90422 (MH); Pakamalai RF – Gingee (SA), *KRM* 53502 (MH); Poondi (CPT), *DN* 875 (MH); Chinnar dam (DMP), *EV* 57886 (MH).

INDIA: Through out India.

WORLD: Tropical and subtropical regions of Europe, Asia and North America.

Typha elephantina Roxb., Fl. Ind. 3: 566. 1832; FBI 4: 489. 1894; Sharma & Gopal, Aquatic Bot. 9: 381. 1980.

Perennial herb, 1.5-4 m high. Leaves linear or broadly linear; trigonous above the sheath, angularly keeled, dorsally, 3-angled; lamina 25-40 mm broad. Upper part flat, about as long as inflorescence. Male and female parts separate; axis of male spike coverd with hairs; female spike cylindric, blackish brown or brown; pistillodes present; female flowers bracteate; stigma lanceolate; bracts spathulate and longer than the hairs.

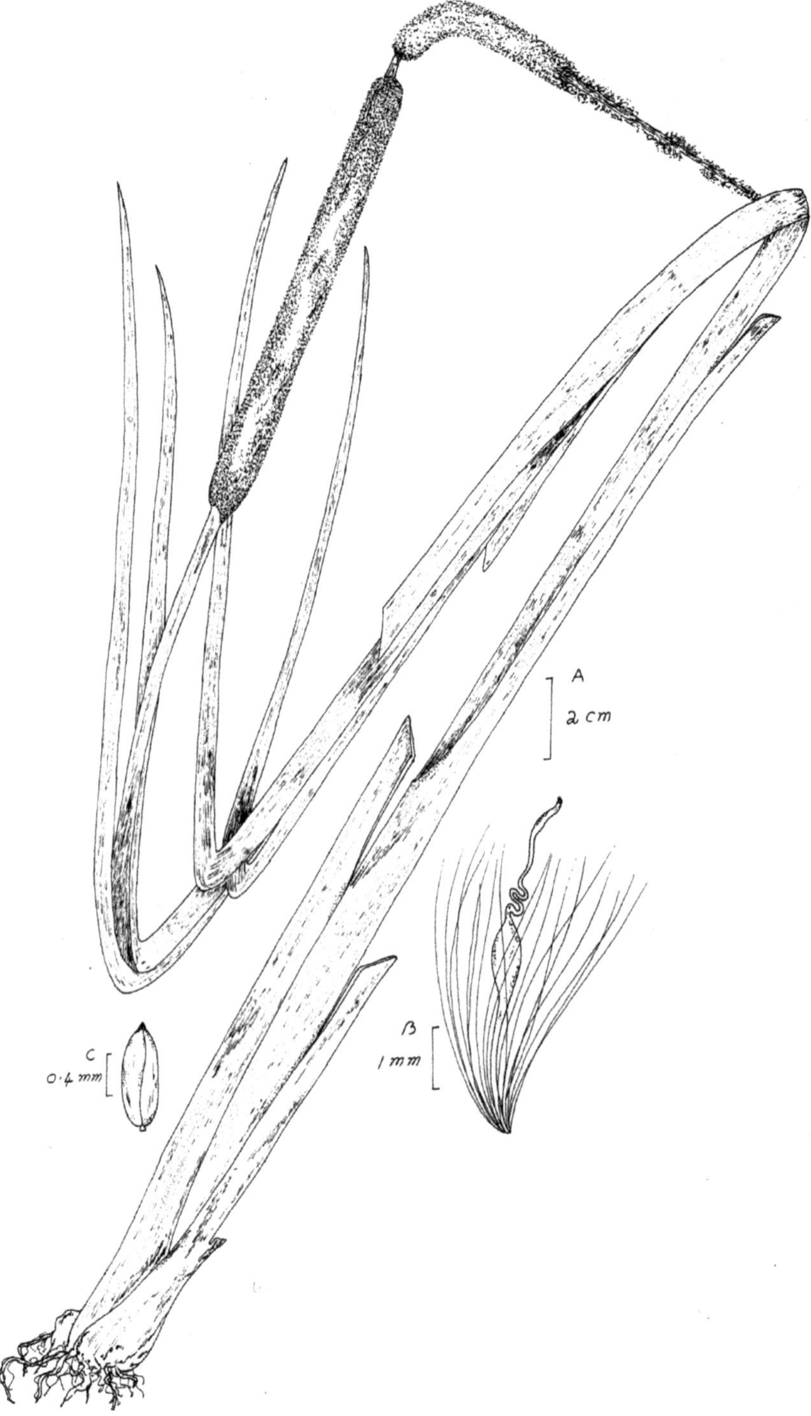

Figure 28. **Typha dominegensis** Pers. (Syn.: *T. angustata* Bory & Chaub.)
A. Habit, B. Female flower, C. Seed.

Occasional in ponds, ditches and slow flowing water in most districts. Fl. & Fr.: May – June.

INDIA: Andhra Pradesh, West Bengal, Rajasthan, Assam.

WORLD: Northern Africa, West Asia, Pakistan, Nepal, China, Turkmenistan, Tajikistan, Uzbekistan.

ARACEAE

1. Floating herbs; leaves in close spirals .... **Pistia**
1. Terrestial or marsh or epiphytic; leaves not in close spirals:
    2. Climbers:
        3. Petioles winged; leaves entire:
            4. Spathe linear-lanceolate, 3 cm across, apex beaked .... **Scindapsus**
            4. Spathe obovoid-globose, 5 mm across, apex deflexed .... **Pothos**
        3. Petioles unwinged; leaves perforate or pinniatifid .... **Rhaphidophora**
    2. Non-climbers:
        5. Marshy plants, armed .... **Lasia**
        5. Teresstrial, unarmed:
            6. Leaves simple:
                7. Epiphytic or in pockets of bare rock .... **Remusatia**
                7. Terrestrial:
                    8. Spadix covered with sterile flowers to apex and the sterile appendix is clearly roughened:
                        9. Ovules many, parietal .... **Colocasia**
                        9. Ovules few, basal .... **Alocasia**
                    8. Spadix with smooth and naked terminal appendix which is green or purplish:
                        10. Spadix included .... **Theriophonum**
                        10. Spadix exserted .... **Typhonium**
            6. Leaves compound:
                11. Inflorescence appearing with the leaves; leaf blade pedatisect or radiatisect.. **Arisaema**
                11. Inflorescence appearing without leaves; leaf blade trisect, each division again variously cut .... **Amorphophallus**

ALOCASIA (Schott) G. Don

1. Leaves sagittate; posterior lobes subtriangular:
   2. Petiole to 30 cm in length; leaves 10-12 cm across........................ **A. fornicata**
   2. Petiole exceeding 50 cm in length; leaves 10-20 cm across ....**A. decipiens**
1. Leaves deeply cordate; posterior lobes ovate............................ **A. macrorrhizos**

Alocasia decipiens Schott, Bonplandia 7: 28. 1859; FBI 6: 526. 1893. *Arum fornicatum* Wight, Icon. Pl. Ind. Orient. t. 789. 1844, non Roxb. 1832. nec Wight, t. 792. 1844.

A stout herb or undershrub, stems usually rhizomatous, but often forming a distinct above ground caudex. Leaves oblong, sagittate, 20-40 x 10-20 cm, lobes 15 cm, sinus broad open. Peduncle 30-60 cm long, tube of spathe shorter than the oblong cymbiform acuminate limb. Male and female flowers separated by neuter flowers; stamens 3-6; ovary globose, style short, stigma capitate, entire. Fruiting tube of spathe ellipsoid or pyriform, berries red when ripe; seed subglobose.

Rare along streams under shade in northern Eastern Ghats. Fl. & Fr.: July – September.

Rampa hill (EG), *VNS* 209 (CAL).

INDIA: Andhra Pradesh, Andaman and Nicobar Islands.

WORLD: Bangladesh, Myanmar,

Note: Karthikeyan *et al.* (1989) have given the distribution of this taxon in Andaman and Nicobar Islands in India. However R.S. Rao (1964) reported it as new record for Peninsular India. His report was based on the collection of V. Narayanaswamy which was wrongly identified by Fischer (1931) as *A. macrorrhiza* and the correct identify of that specimen is *A. decipiens*.

Alocasia fornicata (Roxb.) Schott, Oestr. Bot. Wochenbl. 4: 410. 1854; FBI 6: 526. 1893. *Arum fornicatum* Roxb., Fl. Ind. 3: 501. 1832 non Wight (excl. t. 792) 1844.

A herb with inclined or prostrate stem, to 1 m long. Leaves ovate-lanceolate, 20-30 x 10-12 cm, base cordate, acute or obtuse, margin slightly wavy, apex curved into a short cuspidate tip; nerves 6-8 pairs; petiole 20-30 cm. Spadix nearly as long as the spathe; spathe 6-10 cm, greenish yellow, tube of the spathe about half as long as the oblong cymbiform acuminate limb; female portion about 1 cm long, male portion up to 2 cm long, neuter portion as long as the male portion, ovary narrowed into a distinct style, stigma 3-4-lobed. Berries and seeds globose.

Occasional along streams under shade. Fl. & Fr.: July – October. Vern.: Tel.: *Bulla sema.*

Maredumilli RF (EG), *BR & KNR* 10615; Near Borra caves (VSKP), *KSK* 23439; on way to Kakur (EG), *GVS* 67954 (CAL); Near Tadepalli (EG), *NRR* & *DN* 84345 (BSID); way to Keesariguda, Kedaripuram (VZN), *NRR* 83640 (BSID); Rudrupura, Rairakhol (SBP), *HFM* 3963 (DD).

INDIA: NE and South India.

WORLD: Sri Lanka, Bangladesh.

Alocasia macrorrhizos (L.) G. Don in Sweet, Hort. ed. 3. 631. 1839; FBI 6: 526. 1893; Fischer 3: 1582. 1931. *Arum macrorrhizon* L., Sp. Pl. 965. 1753. *Alocasia indica* Spach., Hist. Nat. Veg. Phan. 12: 47. 1846; FBI 6: 526. 1893; Fischer 3: 1582. 1931. *A. montana* (Roxb.) Schott, Oestr. Bot. Wochenbl. 4: 410. 1854; FBI 6: 525. 1893; Fischer 3: 1582. 1931. *Arum montanum* Roxb., Fl. Ind. 3: 497.1832.

A stout undershrub with subcylindric rootstock with long suckers from the crown. Leaves broadly ovate-cordate, obtuse and shortly apiculate, 30-100 x 15-60 cm, base cordate, margins undulate, apex obtuse. Spathe 10-15 cm long, limb cucullate; spadix nearly as long as the spathe; appendix thickened at the base, subulate, subacute. Male and female flowers separate; stamens 3-8; ovary ovoid or oblong, style short, stigmas 3-4-lobed. Berries enclosed in the ellipsoid accrescent tube of the spathe; seeds subglobose.

Occasional in moist places close to water courses in Northern Eastern Ghats; also cultivated for its tubers. Fl. & Fr.: September-February.

Atthi (SKLM), *PVS* & *NRR* 76884 (MH).

INDIA: Wild and cultivated all over India.

WORLD: SE Asia.

AMORPHOPHALLUS Blume ex Decne *nom. cons*

1. Male and female inflorescences separated by a row of neuter flowers:
   2. Spathe covered only basal part of stipe .... **A. sylvaticus**
   2. Spathe covered almost all the part of stipe............................ **A. margaritifer**
1. Male and female inflorescences continuous without and separation, neuter flowers absent:
   3. Leaves bulbiliferous at the forks .... **A. bulbifer**
   3. Leaves not bulbiliferous:
      4. Spathe campanulate; stipe apex conical and variously crumbled .... **A. paeoniifolius**
      4. Spathe tubular; stipe apex obtusely attenuate .... **A. konkanensis**

Amorphophallus bulbifer (Roxb.) Blume, Rumphia 1: 148. 1837; FBI 6: 515. 1893; Fischer 3: 1598. 1931. *Arum bulbiferum* Roxb., Fl. Ind. 3: 510. 1832. *Amorphophallus aculatum* Hook.f., Fl. Brit. India 6: 515. 1893.

A cormous herb; corm globose, 5-10 cm across. Leaves 30-45 cm across, bulbiliferous at the base, forkings and on the nerves; leaflets lanceolate-oblovate, 6-20 cm long, margins purple; petiole spotted. Spathe 12-20 cm long, erect, pale pink, rose-pink inside; peduncle 12-25 cm long, green or pink; spadix sessile, appendage 7-10 cm long, pale flesh-coloured or white. Berries red, 2-3-seeded.

Rare in the hills of Northern Eastern Ghats. Fl. & Fr.: April-July. Vern.: Ori.: *Dhara oal.*

Devarakonda (EG), *VNS* 483 (CAL): Papadahandi, Jeypore (KPT), *T.Das* 40 (DD).

INDIA: Assam, West Bengal, Sikkim, Bihar, Odisha, Andhra Pradesh.

WORLD: Myanmar.

Amorphophallus konkanensis Hett *et al.*, Blumea 39: 289. 1994; Chorghe *et al.*, Nelumbo 59: 66. 2017.

Tuberous herbs, up to 40 cm high, tuber globose or depressed globose, 3-6 x 3-4 cm. Leaf dissected, solitary; petiole smooth, 20-50 cm long, brown or greenish brown; leaflets lanceolate, acuminate, 5-15 x 1-4 cm. Inflorescence long-peduncled, 25-30 cm long; spathe erect, ovate, acute, 3.5 – 8 x 2.3-7 cm, pinkish-brown, veins dark purplish brown, inside maroon, base within dark maroon, longitudinally ridged. Spadix stipitate, up to twice as long as spathe, 10-15 cm long, female zone cylindric, 1-2 cm, flowers congested; male zone cylindric, 1.5-3 cm long, flowers slightly distant, staminodal zone between female and zone 0.5-1.5 cm long, staminodes congested; appendix elongate, conic, apex blunt, 5-10 cm long; ovaries depressed globose, diamond-shaped in cross section, 2-3 x 0.8-1 mm, pale green, near the top becoming purplish; stigma large, sub-circular or slightly irregular in cross section. Male flowers consisting of 4-6 stamens; stamens *c.* 1 mm long, filaments 0.2-0.3 mm long, connate. Staminodes ovate or rhomboid in cross section, slightly convex, 3-6 x 2.5-4 mm, pinkish-brown.

Common in open rocky grasslands of dry deciduous forests. Fl. & Fr.: May-July.

Sidhout (KDP), *L. Rasingam* & *M. Sankara Rao* 2867 (BSID).

INDIA: Andhra Pradesh, Goa, Madhya Pradesh and Maharashtra. Endemic.

Amorphophallus margaritifer (Roxb.) Kunth, Enum. Pl. 2: 34. 1837. *Plesmonium margaritiferum* (Roxb.) Schott, Sys. Aroid. 34. 1856; FBI 6: 518. 1893; Fischer 3: 1588. 1931. *Arum margaritiferum* Roxb., Fl. Ind. 3: 512. 1832.

A tuberous herb; tuber 10 cm across across. Leaf solitary, rarely 2, 30-45 cm across; leaflets narrowely lanceolate, acuminate, sometimes forked, 10-20 cm long; petioles 30-75 cm long. Peduncle as long as petiole, stout, pale green streaked with darker green. Spathe broadly ovate, 6-12 cm long, leathery, green without, deep purple at the base within, sometimes flushed with purple upwards. Spadix very stout, stipitate, obtuse, as long as the spathe. Female inflorescence 2-3 cm long. Neuters large, clavate, pure white, occupying a space of about 1.5 cm. Male inflorescence 3-5 cm long; ovary slightly sunk in the spadix, 2-3-celled; ovule solitary, axile. Berries ovoid, 2-3-celled; seeds ellipsoid.

Rare in deciduous forests. Fl. & Fr.: June – September. Vern.: Ori.: *Jhumpagudi.*

Nallamalais (KNL), *JLE* 16722 (MH); Rampa hills (EG), *VNS* 652 (CAL); Lakshmipuram (SKLM), *CAB* 1805 (MH).

INDIA: Uttar Pradesh, Madhya Pradesh, West Bengal, Odisha, Andhra Pradesh, Tamil Nadu.

WORLD: Bangladesh, Myanmar.

Amorphophallus paeoniifolius (Dennst.) Nicolson, Taxon 26: 338. 1977; Saxena & Brahmam, Fl. Orissa 4: 2038. 1996. *Dracontium paeoniifolium* Dennst., Schussel Hort. Malab. 13: 38. 1818.

Key to Varieties

1. Petiole usually purplish brown with light pinkish blotches, strongly muricate especially at basal half; leaflet-bases strongly decurrent on primary rachises to the main junction; spadix-appendix elongate-conoid, height more than breadth; style length about double the ovary height; stigma usually 2-lobed .... var. **paeoniifolius**

1. Petiole usually greenish with white blotches, smooth, rarely slightly rough at basal half; leaflet-bases not decurrent to the junction of the primary rachises; spadix-appendix subglobose, breadth more than height; style length 3-4 times the ovary height; stigma usually 3-lobed... var. **campanulatus**

Amorphophallus paeoniifolius (Dennst.) Nicolson var **paeoniifolius**

Large cormous herbs; corms up to 10 cm across. Petiole stout, fleshy, muricate, 3-partite at apex; leaves tripartitely decompound, leaflets to 10 x 4 cm, elliptic, base decurrent, apex acute or acuminate. Flowers foetid, stigma usually bi- or trilobed with a distinct style. Berries orange red.

Rare in shady moist localities of forests. Fl.: Appear immediately after the first rain. way to Peddalova, Tadepalli, Maredumilli agency (EG), *NRR* & *DN* 84336 (BSID).

INDIA: Through out plains of India, N.E.India, Andaman & Nicobar Islands.

WORLD: Sri Lanka, Bangladesh, Myanmar, Bhutan, China, Laos, Indonesia, Philippines, Thailand, Vietnam, N. Australia, Pacific Islands.

Amorphophallus paeoniifolius (Dennst.) Nicolson, Taxon 26: 338. 1977. var **campanulatus** (Decne.) Sivad., Taxon 32: 130. 1983; Saxena & Brahmam, Fl. Orissa 4: 2038. 1996. *Arum campanulatum* Decne., Nouv. Ann. Mus. Hist. Paris 3: 366. 1834. *Amorphophallus campanulatus* Decne. *loc. cit. nom. superfl.*; FBI 6: 513. 1893; Fischer 3: 1587. 1931.

A stout cormous herb; corm depressed globose, 25-30 cm across. Leaves 30-90 cm across, segments pinnate or bipinnatisect, ultimate segments obliquely oblong, acuminate, 7.5 cm long; petiole dark green with paler blotches. Spathes broadly campanulate, 20-35 cm across with broad, undulate, revolute margins; peduncles 6-10 cm long; elongate, to 100 cm in fruiting. Spadix globose, as long as spathe; appendix 7-18 cm long. Flowers monoecious, crowded in cylindric masses. Male inflorescence to 10 x 6 cm; ovaries densely crowded, sessile; ovules solitary, styles twice the length of ovaries, dark purple; stigma 2-lobed, yellow. Berries obovoid, red; 2-3-seeded.

Occasional in the hills, cultivated and wild. Fl. & Fr.: May – August. Vern.: Tel.: *Manchi kanda*; Ori.: *Ol.*

Upper Ahobilam (KNL), *J.Swamy* & *S.Nagaraju* 9071 (BSID); Rampa hills (EG), *VNS* 890 (CAL); Way to Pedhalova, Tadepalli (EG), NRR & *DN* 84336 (BSID); Dagseru (VZN), *MV* 4142 (AU).

INDIA: South and E. India, cultivated and wild.

WORLD: Sri Lanka.

Amorphophallus sylvaticus (Roxb.) Kunth, Enum. Pl. 3:34. 1841; Fischer 3: 1587. 1931. *Arum sylvaticum* Roxb., Fl. Ind. 3: 511. 1832. *Synantherias sylvaticus* (Roxb.) Schott., Gen. Aroid. t. 28. 1858; FBI 6: 518. 1893.

A cormous herb; corms to 5 x 3 cm. Leaves 20-100 x 10-60 cm, segments lanceolate; petioles purplish with dark greenish-black spots; leaf segments base strongly decurrent on the main petiolules to the junction with the petiole. Spathes convolute with open top, 3-5 cm long, cream-coloured outside, maroon within, 10-50 cm long; spadix 20-30 cm long, very shortly stipitate, appendix linear-subulate, about 20 cm long. Male inflorescence 9-13 cm long, female inflorescence 3-4 cm long.

Occasional in moist localities. Fl. & Fr.: June – September.

Upper Ahobilam (KNL), *BSS* & *BR* 29206; Chelama (KNL), *JLE* 16772 (MH); Somasila (NLR), *A.S.Rao* 6009 (BSID); Kankanapally (VZN), *MV* 2580 (AU).

INDIA: Peninsular & E. India.

WORLD: Sri Lanka.

ARISAEMA Martius

1. Leaves 2, rarely 3, pedatisect; spadix sigmoid, long exserted .... **A. tortuosum**
1. Leaf solitary, radiatisect; spadix cylindrical, barely exserted .... **A. leschenaultii**

Arisaema leschenaultii Blume, Rumphia 1: 93. 1836; FBI 6: 504. 1893; Fischer 3: 1585. 1931; Malhotra & Balodi, J. Econ. Taxon. Bot. 7: 585. 1985. *A. pulchrum* N.E. Br., J. Linn. Soc. 18: 252, t. 6. 1880.

A cormous herb; corm subglobose. Leaf solitary, radiatisect; petiole to 1 m; sheaths to 75 cm, mottled; leaflets 5-13, obovate-lanceolate, to 30 x 10 cm, glossy above and light green below. Spathe to 15 cm long, with longitudinal white stripes, pale green but sometimes purplish; tube to 9 cm, cylindric, shortly outrolled at top. Spadix usually dioecious. Berries with 1-few globose seeds.

Occassional in Kollimalais, Servarayans in Tamil Nadu and Nellore district in Andhra Pradesh (Suryanarayana, 1979). Fl. & Fr.: August – December. Vern.: Tam.: *Kaatukarunai*.

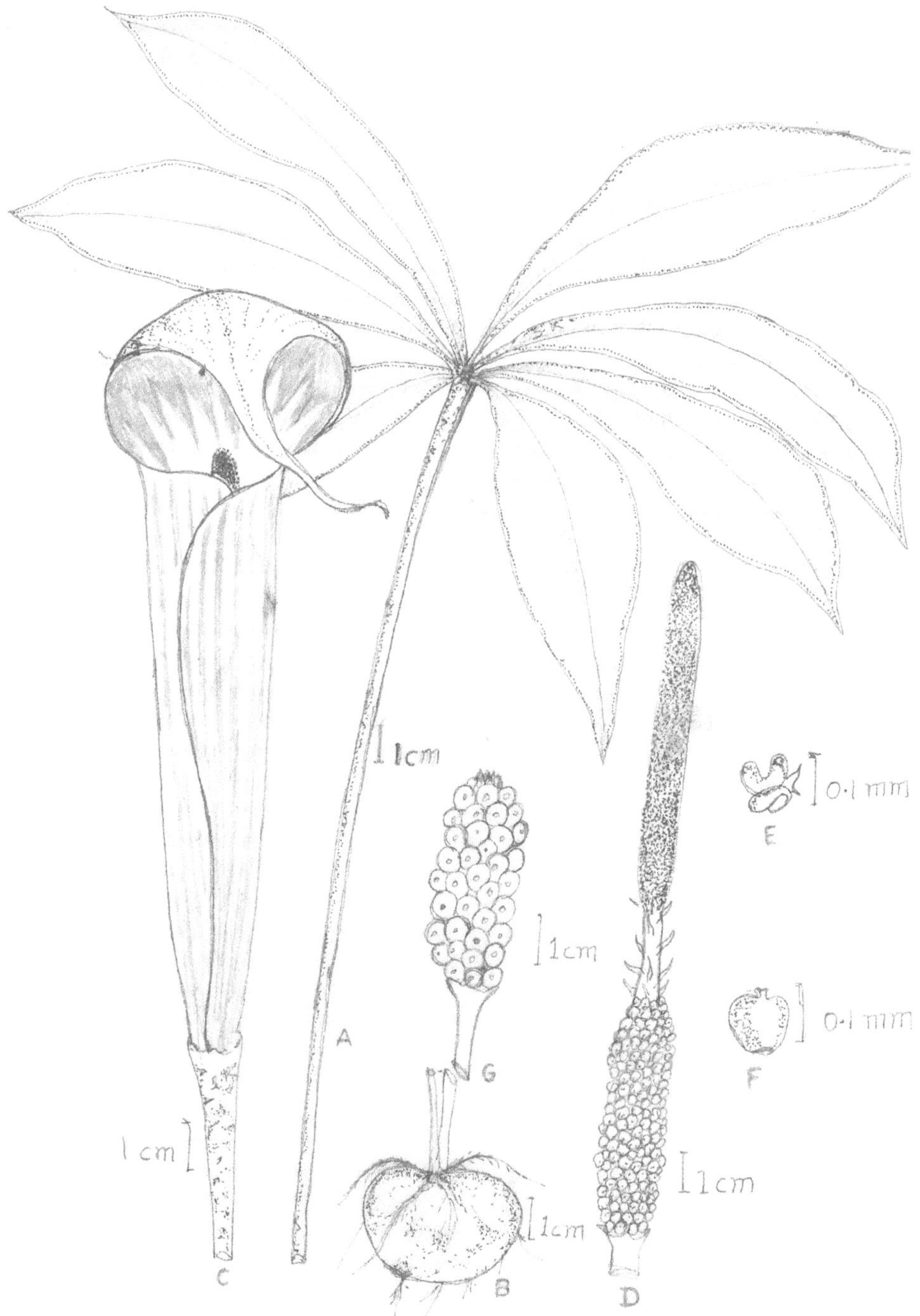

Figure 29. **Arisaema leschenaultii** Blume

A. Leaf; B. Tuber; C. Spathe; D. Spike; E. Stamen; F. Pistil; G. Fruits.

INDIA: South India.

WORLD: Sri Lanka.

Arisaema tortuosum (Wall.) Schott & Endl., Melet. Bot. 17: 1832; FBI 6: 502.1893; Fischer 3: 1584. 1931. *Arum tortuosum* Wall., Pl. Asiat. Rar. 2: 10. t. 114. 1830.

A cormous herb, to 1 m long; corms globose, to 10 cm across. Leaves 2, rarely 3, pedatisect; petioles 30-90 cm long; sheaths often mottled with purple; leaflets ovate-lanceolate or linear-lanceolate, subcaudately acuminate, distant or crowded; spathe 10-15 cm long, green outside, gradualy dilated into ovate or oblong, acuminate, incurved limb; peduncle 60-120 cm long. Spadix monoecious, rarely androecious, to 17 cm long, with a basal female inflorescence portion about 1 cm long followed by the male inflorescence about 2.5 cm and the apical, naked appendix with a few or no neuters at the base. Male flowers stalked; appendix very long, sigmoid, much exserted, tapering; ovary ovoid, attenuate into a short style. Berries orange-red, 4-5-seeded.

Occasional in the forests. Fl. & Fr.: June – September.

Papanasanam (CTR), *BR* & *BSS* 30401; Maredumilli (EG), *BR* & *KNR* 10620; Dammaku – Borra caves (VSKP), *KSK* 23437; Tirumala (CTR), *DRC* 1149 (BSID); Dummakonda (EG), *VNS* 659 (CAL); Tiger camp to valamuru (EG), *MM* 105046 (BSID); Chintapalli (VSKP), *GVS* 28240 (MH); Sunkarimetta (VSKP), *NPBK* 10916 (MH); Near Dabuguda, Lakshmipuram (VSKP), *NRR* & *DN* 84224 (BSID); Kankarapalli (VZN), *MV* 2580 (AU); Purunakote, Satkosia Tiger Reserve, *KCM* 6565 (BSID).

INDIA: Assam, West Bengal, Bihar, South India.

WORLD: Pakistan, Sri Lanka, Bangladesh, Myanmar, Nepal.

COLOCASIA Schott *nom. cons.*

Colocasia esculenta (L.) Schott & Endl., Melet. Bot. 18. 1832. *Arum esculentum* L., Sp. Pl. 965. 1753. *Colocasia antiquorum* Schott in Schott & Endl., Melet. Bot. 18. 1832; FBI 6: 523. 1893; Fischer 3: 1580. 1931.

A robust tuberous herb; tubers to 15 cm across; leaves ovate or suborbicular, 15-50 x 8-30 cm, base cordate, margin undulate, apex rounded and usually apiculate, basal sinus triangular, dark-green, petiole 75-130 cm long, green or violet. Spathe 35 cm long, tube elliptic, 3-5 cm long, green, basally convolute; limb lanceolate, c. 25 cm long, yellow, caducous. Spadix with female portion about 3 cm long, male portion about 6 cm and the apical sterile appendix about 5 cm long. Female flowers many, arranged in close spirals; with few white tipped sterile flowers, stigma sessile, inconspicuously 3-5-lobed, white. Neuter flowers occur at the constriction of the spathe along c. 2 cm. Male flowers and appendix cream-coloured, appendix tapering towards the tip. Berries ovoid, many-seeded.

Wild along canals, irrigation channels, also cultivated. Fl. & Fr.: August – November. Vern.: Tel.: *Chamadumpa, Chamagadda.* Ori.: *Saru, Bansaru.* Eng.: Taro.

Mahanandi (KNL), *TP* & *KNR* 9864; Maddiletiswamy (KNL), *SS* & *AMR* 22546; Thumburatheertham (CTR), *J.Swamy* & *S.Nagaraju* 7423 (BSID); Nukarayi (EG), *VNS* 293 (CAL), Araku valley (VSKP), *NPBK* 10848 (MH); Punyagiri hills (VZN), *GVS* 44239 (MH); Kurupam (VZN), *MV* 2081 (AU); Raigoda, Satkosia Tiger Reserve, *KCM* 6723 (BSID).

INDIA: Through out hotter parts of India, also cultivated.

WORLD: Native of SE Asia, now Pantropically cultivated.

CRYPTOCORYNE Fischer ex Wydler

Cryptocoryne retrospiralis (Roxb.) Kunth, Enum. Pl. 3: 12. 1841; FBI 6: 493. 1893. *Ambrosina retrospiralis* Roxb., Fl. Ind. 3: 492. 1832. *Cryptocoryne unilocularis* (Roxb.) Kunth, Enum. Pl. 3: 13. 1841; Fischer 3: 1575. 1931. *Ambrosina unilocularis* Roxb., Fl. Ind. 3: 493. 1832.

Herbs with creeping root stock and stout fleshy roots. Leaves linear or linear-lanceolate, grass-like, 7.5-40 x 0.5-1.6 cm, base narrowed into a slender petiole, margin often crisped or wavy, apex acute or acuminate. Peduncle up to 3 cm; spathe 7.5-10 cm, subsessile, lower tube rather wider than the upper tube; spadix with a female portion *c.* 6 mm, male portion *c.* 3 mm, ovaries 4-6, 3-more-ovuled. Fruit shortly peduncled, ovoid, ovoid-oblong or ellipsoid-oblong, ca 1.5 cm long

Usually found near streams of forests in Northern Eastern Ghats. Fl. & Fr.: November – April.

INDIA: S, C. & NE. India, Bihar, West Bengal,

WORLD: Myanmar.

LASIA Loureiro

Lasia spinosa (L.) Thwaites, Enum. Pl. Zeyl. 336. 1864; FBI 6: 550. 1893; Fischer 3: 1589. 1931. *Dracontium spinosum* L., Sp. Pl. 967. 1753.

A stout, prickly, marshy plant; rhizome thick, branched, petiole, peduncle and leafy nerves beneath all prickly; stem 4 cm across, creeping and upturning. Leaves hastate when young or sagittate, older ones often broader than long and deeply pedately pinnatifid, 15-45 x 10-30 cm, lobes linear-elliptic or oblong-lanceolate, acuminate, 1-ribbed, spinous on the nerves beneath; petiole terete, spinous. Spathe coriaceous, greenish brown to purplish, 40-55 cm long and slightly twisted, flexing open in lower 7 cm to expose spadix. Spadix 5 x 1 cm, pinkish. Flowers bisexual, perianth of 4, rarely 6, obovate, truncate, segments incurved at the tip; stamens 4-6, filaments short; ovary ovoid. Berries dense, minutely muricate at the apex; seeds compressed, rugose.

Rare in marshy localities in interior forests of Northern Eastern Ghats. Fl. & Fr.: June – December.

Maredumilli (EG), *BR* & *KNR* 10613; Dharawada (WG), *DN* 85519 (BSID); Rampa hills (EG), *JSG* 15936 (MH); Kuntravada (EG), *SS* 3878 (AU); Sileru (VSKP), *GVS* 47413 (MH); Kankanapalli (VZN), *MV* 2564 (AU); Similipahar (MBJ), *SX* 436 (RRL-B), *SX* & *MB* 4626 (RRL-B); Deogarh (DGR), *MB* & *Dhal* 8292 (RRL-B).

INDIA: Tropical Sikkim Himalaya, Assam, West Bengal, Bihar, Odisha, Andhra Pradesh.

WORLD: Sri Lanka, Myanmar, Singapore, Malay Islands, China, New Guinea.

PISTIA Linnaeus

Pistia stratiotes L., Sp. Pl. 963. 1753; FBI 6: 497. 1893; Fischer 3: 1573. 1931.

A floating stemless stoloniferous aquatic herb with a peculiar muriatic odour; roots of tufted simple white fibres clothed with fibrillae. Leaves sessile, obovate-cuneate, 3-10 x 1-5 cm, tufted, rounded or retuse at the apex, densely and closely pubescent on both surfaces; nerves few or many flabellately arranged, converging within the margin. Spathe about 1.2 cm long, obliquely campanulate, white, gibbous and closed below, contracted about the middle, dilated and nearly orbicular above. Spadix adnate to the back of the spathe tube, free above. Male flowers in a whorl of a few connate stamens beneath the apex of the spadix. Neuters few, minute, confluent in a ring below the male. Female flowers solitary, ovary 1-celled, obliquely adnate to the spadix, the apex free and forming a conical style. Berries ovoid; seeds few.

Common in stagnant waters. Fl. & Fr.: December – February. Vrn.: Tel.: *Anthara thamara.*

Owk (KNL), *RVR* 1458; Jadadanki (NLR), *PMR* 22367; Repalle (GNT), *VRK* 5776; Srisailam (KNL), *JLE* 16855 (MH & CAL); Chandragiri Fort (CTR), *KS* 6967 (MH); Kolletikota (KSN), *PV* 512 (MH); Chettarparu (WG), *KS* 5117 (MH); Kurupam (VZN), *MV* 2888 (AU).

INDIA: Throughout India.

WORLD: Tropics and Subtropics.

POTHOS Linnaeus

Pothos scandens L., Sp. Pl. 968. 1753; FBI 6: 551. 1893; Fischer 3: 1592. 1931; Saxena & Brahmam, Fl. Orissa 4: 2047. 1996.

Much branched scandent liane, to 6 m long, root climbing. Leaves obovate-oblong, ovate or oblong, 2.5-10 x 0.4-4 cm, apex attenuate-mucronate, base rounded to acute; petiole 2-14 x 0.5-2 cm, broadly winged. Spathe peduncled, cymbiform, 4.3-6.2 cm broad, cuspidate; spadix ellipsoid, obovoid or globose, 4.2- 5mm in diam., with a pedicel above the spathe which is usully deflexed. Flowers *c.* 1-2 mm, tepals 6, yellow-green to dirty white, stamens 6; ovary 1.6 mm long. Infructescence with 1-5 berries, fruit obclavate; seeds ellipsoid to compressed-globose.

Occasional in shady places in Eastern Ghats of Odisha. Fl.: September.

Similipahar (MBJ), *SX* 374 (RRL-B), *SX* & *MB* 4203 (RRL-B); Upper Barakamara, Similipal (MBJ), *A.R.K.Sastry* & *G.P.Singh* 12403 (BSID).

INDIA: Assam, Bihar, Goa, Karnataka, Kerala, Maharashtra, Meghalaya, Mahe, Odisha, Tamil Nadu, Tripura, West Bengal.

WORLD: Madagascar, Sri Lanka, Bangladesh, Malaysia, Indonesia, Cambodia, China, Philippines, Thailand, Vietnam.

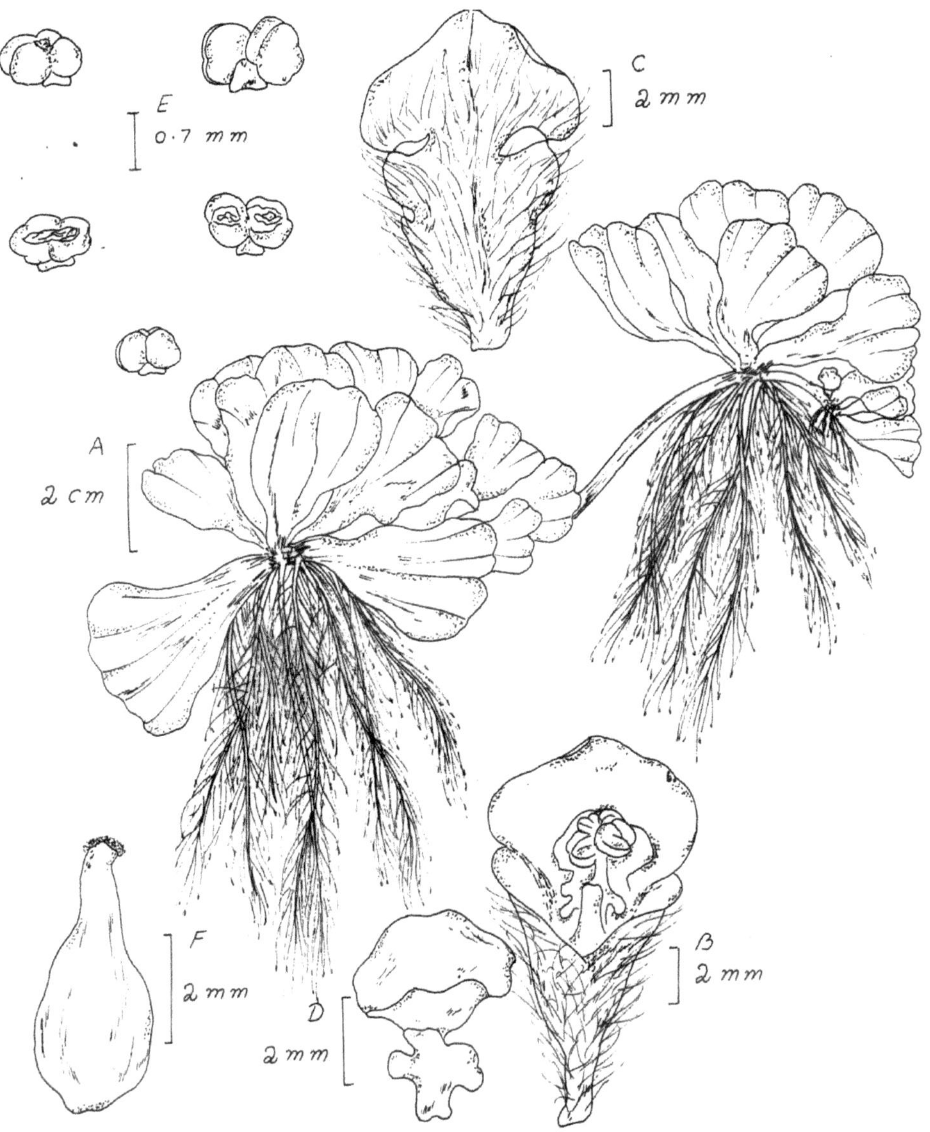

Figure 30. **Pistia stratiotes** L.
A. Habit, B. Spadix with spathe, C. Spathe (outer view), D. Perianth, E. Stamen, F. Pistil.

REMUSATIA Schott

Remusatia vivipara (Roxb.) Schott in Schott & Endl., Melet. Bot. 1: 18. 1832; FBI 6: 521. 1893; Fischer 3: 1583. 1931. *Arum viviparum* Roxb., Fl. Ind. 3: 496. 1832 (*viviparium*). *Caladium viviparum* (Roxb.) Lodd., Bot. Cab. 3: t. 281. 1818.

A tuberous herb, emitting 15-40 cm long leafless bulbiliferous shoots from the crown of the tuber 2.5-4 cm across; bulbils with recurved tipped scales. Leaf usually solitary, sometimes 2-3, broadly ovate, 15-35 x 10-25 cm, base subcordate, acuminate, peltate, sometimes purplish tinge between the primary veins on the lower surface of the lamina; petiole 20-40 cm long. Spathe about 17 cm long, basal portion oblong, green, convolute and upper portion deflexed, yellow, broadly-obovate limb, 5-13 x 2.5 - 9 cm. Spadix sessile, 5-7 cm long. Male and female flowers separated by neuters, stamens 2-3; anthers sessile, connate into a 4-6-angled and subclavate, flat-topped synandrium; synandria densely packed with neuters, ovaries crowded, ovoid, 1-celled or 2-4-celled upwards; stigma sessile, disciform, ovules many in 2 series on 4-6 parietal placentae. Berries small, obovoid, seeds small, ovoid.

Occasional in Northern Eastern Ghats and common in Kolli hills of Tamil Nadu. Fl. & Fr.: February – April.

Punyagiri (VZN), *BR* & *KNR* 9885; Tadepalli (EG), *NRR* 84372 (MH); Nukarayi (EG), *SS* 10044 (AU); Sunkarimetta (VSKP), *NPBK* 10924 (MH) & *GVS* 21694 (MH); Kwadoli, Raigoda, Satkosia Tiger Reserve, *KCM* 6721 (BSID); Kolli hills, Kulivalavu (NMK), *SKS* 1120 (SGH).

INDIA: Subtropical Himalaya, Khasi hills, Mahdy Pradesh, Odisha.

WORLD: Indo-Malesia, Africa, Madagascar, Australia.

RHAPHIDOPHORA Hasskarl

1. Leaves entire .... **R. hookeri**
1. Leaves pinnatifid or pinnatisect, sometimes entire, if so perforate with elliptic holes:
   2. Leaves glaucous beneath; petiole channeled .... **R. glauca**
   2. Leaves not glaucous; petiole terete:
      3. Peduncle 3-5 cm long; spathe ovate-oblong; stigma sessile...... **R. pertusa**
      3. Peduncle 6-10 cm long; spathe subcylindrically cymbiform, stigma raised on the conical top of the ovary .... **R. decursiva**

Rhaphidophora decursiva (Roxb.) Schott, Bonplandia 5: 45. 1857; FBI 6: 547. 1893. *Pothos decursivus* Roxb., Fl. Ind. 1: 456. 1820. *Scindapsus decursivus* (Roxb.) Schott in Schott & Endl., Melet, Bot. 21. 1832. *Rhaphidophora eximia* Hook. f., Fl. Brit. India 6: 547. 1893.

A lofty climber, climbing with the help of roots. Leaf blade oblong, subcordate, 50-60 x 30-40 cm, pinnately divided to segments of 5 or more; sheath 10-15 cm long, deciduous. Peduncle 15-17 cm, erect. Spathe 13-20 cm, yellow, deciduous. Spadix

Figure 31. **Remusatia vivipara** (Roxb.) Schott

A. Habit; B. Spathe; C. Spike; D. Stamen; E. Pistil; F. Bulbil.

sessile, cylindrical, to 14 x 3 cm; at maturity the upper 3-4 mm of the pistil falls away, exposing a fragrant orange pulp full of seeds and trichosclereids.

Rare in Northern Eastern Ghats. Fl. & Fr.: July – November.

Valamuru (EG), *MM* 100867 (BSID), *SS* 3492 (AU); Galikonda (VSKP), *GVS* 19646 (MH); Rella (VZN), *MV* 6841 (AU).

INDIA: E. Himalaya.

WORLD: Sri Lanka.

Rhaphdophora glauca (Wall.) Schott. Bonplandia 5: 45. 1857; FBI 6: 547. 1893; Saxena & Brahmam, Fl. Orissa 4: 2050. 1996. *Pothos glaucus* Wall., Pl. As. Rar. 2: 45. t. 156. 1831.

Large climber; stem *c.* 2.5 cm thick. Leaves obliquely ovate, or ovate-oblong, 15-35 x 15-30 cm, pinnatisect, usually glaucous beneath; segments 2-6, falcate, linear-oblong, 10-19 x 2.8-5 cm, apex acuminate to caudate, base narrowed, petiole as long as the blade, channeled. Peduncle 7.5-18 cm long. Spathe oblong or lanceolate, 6-15 cm long, variable in size, cuspidate, yellow. Spadix shorter than the spathe, pale yellow, cyindric, 5-7 cm long. Fruiting spadix erect.

Occasional in shady forest. Fl.: October – November.

Badomukhabadi, Similipahar (MBJ), *SX* & *MB* 5587 (RRL-B); Gorasika (KJR), *MB* & *Dhal* (RRL-B).

INDIA: Tropical and Subtropical Himalaya, Manipur hills, Odisha.

WORLD: Nepal.

Rhaphidophora hookeri Schott., Bonplandia 5: 45. 1857; FBI 6: 546. 1893; Mooney, Suppl. Bot. Bihar & Orissa 142. 1950; Saxena & Brahmam, Fl. Orissa, 2051. 1996.

Climber, stem 1.8-2.5 cm diam. Leaves oblong, 25-60 x 7.5-30 cm, appex caudate-acuminate, base cuneate, rounded or cordate, dark green, very thinly membranaceous, nerves many, spreading and arched. Spathe green, yellow within and along the margins, thick, base rounded. Spadix *c.* 7.5 cm long, stout, ovoid-oblong, stigma very broad, ovaries 5 mm diam.

Karlapat hills, Kalahandi, scandent on tree close to waterfall in shady places (Mooney *loc. cit.*). Fl.: January.

INDIA: Sikkim, Manipur, Assam, Odisha. Endemic.

Rhaphidophora pertusa (Roxb.) Schott, Bonplandia 5: 45. 1857; FBI 6: 545. 1893; Fischer 3: 1590. 1931; Matthew, Fl. Tamilnadu Carnatic 3: 1698. 1983. *Pothos pertusus* Roxb., Fl. Ind. 1: 455. 1820; *'pertusa'*. *Monstera pertusa* (Roxb.) Schott, Wiener Z. Kunst. 1830: 781. 1830. *Scindapsus pertusus* (Roxb.) Schott, in Schott & Endl., Melet. Bot. India 21. 1832.

A lofty epiphytic climber; stem cylindric. Leaves distichous, ovate to suborbicular, acute, oblique, simple, perforate to pinnately lobed, 20-45 x 15-26 cm; petioles 15-40 cm long with a withering sheath, geniculate at apex. Spathe oblong, shortly pedunculate, yellow, 8-12 x 5-6 cm, acuminate; peduncle 3-5 cm long. Spadix shortly stipitate, cylindric, 6 – 9 x 1.5 cm. Flowers naked, bisexual, stamens 4 with strap-shaped filament; ovary unilocular, ovules many, on 2 parietal placentae at base. Berries many-seeded; seeds oblong or reniform.

Occasional in coastal districts in moist deciduous forests. Fl. & Fr.: January – March.

Vijayapuri South (GNT), *VRK* 6727; Maredumilli (EG), *BR* & *KNR* 10616, *GVS* 68522 (MH & CAL); Rampa hills (EG), *JSG* 15946 (MH); Near Maddhiveedu, Tadepalli (EG), *NRR* & *DN* 84379 (BSID); Near Natchiamman temple, Kolli hills (NMK), *SKS* 673 (SGH).

INDIA: South India.

WORLD: Sri Lanka.

SCINDAPSUS Schott

Scindapsus officinalis (Roxb.) Schott in Schott & Endl., Melet. Bot. 1: 21. 1832; FBI 6: 541. 1893; Fischer 3: 1590. 1931. *Pothos officinalis* Roxb., Fl. Ind. ed. Carey 3: 507. 1832.

A climbing shrub; branches wrinkled when dry. Leaves ovate, elliptic or nearly orbicular, 7-25 x 5-20 cm, caudate-acuminate, base rounded or slightly cordate, primary nerves distinct, petiole 6-15 cm, broadly winged up to the knee. Peduncle solitary, terminal, much shorter than the petiole. Spathe about 10-15 cm long, green without, yellow within, beak slender. Spadix equaling the spathe, elongating in fruit, greenish-yellow. Fruiting spike hemispheric; berries few only ripening, fleshy, seeds ovate-cordate.

Frequently climbing on trees and rocks in damper valleys of northern Eastern Ghats. Fl. & Fr.: November – January. Vern.: Tel.: *Gajapippali, Enugu pippallu;* Ori.: *Gaja pipali, Girudhuni.*

Barmakonda (VSKP), *KCJ* 17166 (MH); Deogam (Dhenkenal Distr.), *sine coll.* 11276 (BSID).

INDIA: Tropical Himalaya, Sikkim, Bihar, West Bengal, Odisha, Andhra Pradessh, Andamans.

WORLD: Bangladesh, Myanmar.

THERIOPHONUM Blume

Theriophonum minutum (Willd.) Baill., Hist. Pl. 13: 457. 1895; Fischer 3: 1579. 1931. *Arum minutum* Willd., Sp. Pl. 4: 484. 1805. *A. crenatum* Wight, Hooker's J. Bot. Kew Gard. Misc. 2: 100 (Suppl.) t. 3. 1831. *Theriophonum crenatum* (Wight) Blume, Rumphia 1: 128. 1837; FBI 6: 512. 1893. *T. wightii* Shcott., Oesterr. Bot. 8: 3. 1858; FBI 6: 512. 1893; Fischer 3: 1579. 1931.

A cormous acaulescent herb; corm 1.5-2 cm across. Leaves with slender petioles, 8-15 cm long, basally sheathing, vaginate 4 cm from base, lamina of various shapes, from broadly ovate to hastate, typically sub-3-lobed; the median lobe 4-7 x 3.5-6 cm, ovate, acuminate, mucronate; posterior lobes 1.5-4 x 1-2 cm. Inflorescence with peduncle 3-4.5 cm long and about 2 mm diameter. Spathe 5-12 cm long with the basal convolute portion 1-2.5 cm long and 0.6-1 cm in diameter and the upper erect, ovate-oblong, acuminate, greenish to purplish green limb portion 4-10 x 2-3 cm, margins purplish and wavy or sometimes entire, the colour later fades to light greenish to white cream; spadix shorter than the spathe, 4-6 cm long. Neuter flowers below and above the male portion, small, subulate. Female flowers: few, in a single whorl, ovate. Male ones scattered and few, sessile, ovary unilocular; ovules 3-6, basal and apical. Berries 7-10 mm long, subconical, 3-seeded; seeds broadly ovoid.

Occasional in forests. Fl. & Fr.: June – December.

Bairlooty (KNL), *BR* & *BSS* 33348; Upper Ahobilam (KNL), *SS* 24407; Thumburatheertham (CTR), *J.Swamy* & *S.Nagaraju* 7440 (BSID); Kondapuram (NLR), *PMR* 9556; Chelama (KNL), *JLE* 16766 (MH); Tikarpada, Satkosia Tiger Reserve, *KCM* 8311 (BSID); Raigoda, Satkosia Tiger Reserve, *KCM* 7092 (BSID).

INDIA: Deccan, Central & SE. India, Konkan, Carnatic.

WORLD: Sri Lanka.

TYPHONIUM Schott

1. Lamina ovate to hastate:
    2. Spathe lanceolate, yellowish green .... **T. inopinatum**
    2. Spathe broadly ovate, deep purplish red .... **T. roxburghii**
1. Lamina distinctly 3-lobed .... **T. trilobatum**

Typhonium inopinatum Prain in King & Prain, J. Asiat. Soc. Bengal Pt. 2. Nat. Hist. 67: 301. 1898 & Bengal Pl. 1107. 1903; Anand Kumar *et al.*, Rheedea 24 (2): 120. 2014; Jagannath & Chaturvedi, Biosci. Disc. 6 (2): 89. 2015; Rasingam & Swamy, Indian J. Forest. 40 (4): 401. 2017; Prameela *et al.*, Annals Plant Sci. 7(4): 2147. 2018.

Tuberous, perennial herbs to 10-45 cm high; tubers sub-cylindric, globose, 1-3 x 0.8-1.5 cm. Leaves ovate to triangular or hastate, 5-14 x 4-10 cm; basal lobes orbicular, margin entire, acute to acuminate at apex; secondary veins 6-10 per side; petioles 25-30 cm long, green. Inflorescence solitary, monoecious; peduncles much shorter than petioles, *c.* 2.5 cm long. Spathes globose to ovoid, convolute tube and an apical limb with a constriction between the two, greenish with light purple externally; tube 0.8-1.5 cm long; limb narrowly ovate to lanceolate, 5.5-7 x 1.2-2 cm, margins entire, acuminate, recurved, coiled apically, glabrous, green with dark purple internally. Spadix 4.3-9 cm long; basal pistillate zone followed by a zone of sterile flowers, a naked zone or interstice, a staminate zone and a terminal barren appendix. Both pistillate and sterile flower zones are enclosed by basal tube of spathe. Pistillate zone conical, 3-3.5 mm long, greenish; flowers sessile, 11.5 mm long; ovary ellipsoid, 1-1.3 mm long, glabrous; style very short; stigma disc-shaped, glabrous. Sterile flower zone yellow, 2-4.5 mm long; sterile flowers filiform, entire, decurved, each 2.5-4 mm long, partially covering pistillate flower zone. Naked zone 6-9 mm long. Staminate zone cylindric, 5-9 x 2-3 mm, pale yellow; flowers sessile, 0.5-1 mm long with 2 thecae; dehiscence by apical short slits or pores. Appendix 4-6 cm long, yellow, yellowish-brown.

Rarely found in shady habitats and open wet sandy soils.

Fl. & Fr.: May – November.

Vizianagaram town, Boggula dibba (VZN), *R.Prameela & J.Swamy* 009256 (BSID); Vizianagaram town, Collectorate area (VZN), *R. Prameela & J.Swamy* 009257 (BSID).

INDIA: Andhra Pradesh, Bihar, Madhya Pradesh, Maharashtra, Telangana, Uttar Pradesh and West Bengal.

WORLD: Myanmar, Thailand.

Typhonium roxburghii Schott, Aroid. 1: 12. 1853; Nicolson & Sivadasan, Blumea 27: 492.1981; Prameela *et al.*, Biosci. Discovery 9: 104. 2018. *T. mottleyanum* Schott, Prodr. 106.1860; FBI 6: 510.1893. *T. schottii* Prain, J. Asiat. Soc. Bengal Pt. 2. Nat. Hist. 67: 303. 1898. *T. divaricatum* Bl. in Dene., Nouv. Ann. Mus. Hist. Nat. 3: 367.1834; Gamble 3: 1578. 1931, nom. illegit. *T. amboinense* Blatt. & McCann, J. Bombay Nat. Hist. Soc. 35: 23.1932

Tuberous perennial herbs, up to 35 cm high; tubers sub-globose, 22.2 x 1.8-2 cm, rooting at the top. Petiole 25-30 cm long, purplish in lower third, sheathing up to 8 cm long; blade ovate to hastate, shallowly 3-lobed, 8-9 x 6-8 cm, entire along margin, cuspidate-acuminate at apex; secondary veins 7-10 pairs. Inflorescence solitary, monoecious; peduncle much shorter than petioles, *c.* 7.7 cm long. Spathe with basal ellipsoid to ovoid, convolute tube and an apical limb with a constriction between the two; tube *c.* 2.1 cm long, dark red-purple inside, purple outside; limb narrowly ovate, much longer than the broad, *c.* 16 x 2.5 cm, abruptly tapering from below the middle, entire along margin, long cuspidate at apex, glabrous, reddish brown. Spadix basal pistillate zone followed by a zone of sterile flowers, a naked zone or interstice, a staminate zone and a terminal barren appendix. Both pistillate and sterile flower zones are enclosed by basal tube. Pistillate zone conical, *c.* 5 mm long, purple-pink; flowers sessile, 11.2 mm long; ovary ellipsoid, *c.* 1 mm long, glabrous; style very short; stigma disc-shaped, glabrous. Sterile flower zone yellow, *c.* 3 mm long; sterile flowers filiform, entire, decurved, each 2.5-3 mm long. Naked zone *c.* 1.3 cm long, creamy. Staminate zone cylindric, 6-6.5 x 11.5 mm, coral pink; flowers sessile, 0.5-1 mm long with 2 thecae; dehiscence by apical short slits or pores. A somewhat contracted stipe *c.* 4.3 mm long subtends the basally swollen appendix. Appendix 12.5-15 cm long, dark red/brown.

Rare in shady places. Fl. & Fr.: July September

Gunupuru (VZN), *R.Prameela* and *J. Swamy* 004341 (BSID).

INDIA: South to Eastern India.

WORLD: Sri Lanka, Malaysia, Indonesia, Philippines, Papua-New Guinea, Brazil, Bangladesh, China, Japan (Bonin Islands), Thailand; introduced in E Africa, W. Australia, and South America.

Typhonium trilobatum (L.) Schott in Weiner Z. Kunst. 3: 72. 1829; FBI 6: 509. 1893; Fischer 3: 1578. 1931. *Arum trilobatum* L., *Sp. Pl. 965. 1753.*

A tuberous herb. Leaves hastately 3-lobed or sub-3-partite, 5-30 cm in diameter, lobes ovate, acute or acuminate. Peduncle 2.5-10 cm long; spathe 7-25 x 4-10 cm, nearly flat, apex of spathe acute or acuminate, not twisted; appendix bright red, stipitate, mucrocate, base expanded and intruded male and female flowers well separated, with neuters above. Female flowers filiform, curved, stamens 1-3; ovary 1-celled, stigma sessile. Berry ovoid, 1 or 2-seeded; seed globose.

Occasional in shady places in Northern Eastern Ghats. Fl. & Fr.: April – May.

Dalayavalasa (VZN), *MV* 5224 (AU).

INDIA: West Bengal, Bihar, Eastern and Western Peninsula,

WORLD: Sri Lanka, Indo-China to Malaysia, introduced elsewhere.

Zantedeschia aethiopica (L.) Sprengel is cultivated in gardens at Yercaud. It escaped and naturalized on banks of streams and water bodies. **Monstera deliciosa** Liebm. is introduced in the gardens at Shervarayans. **Caladium bicolor** Vent. and **Anthurium** cultivars are common ornamental potted plants in gardens.

LEMNACEAE

1. Fronds with 1 or more roots:
    2. Fronds with single root .... **Lemna**
    2. Fronds with tuft of roots .... **Spirodela**
1. Fronds without any roots .... **Wolffia**

LEMNA Linnaeus

1. Root sheath not winged at base; root tips rounded .... **L. minor**
1. Root sheath winged at base; root tips sharply pointed:
    2. Fronds often gibbous .... **L. gibba**
    2. Fronds not gibbous:
        3. Fronds bear 2–3 larger papillae above the node and one smaller near the tip .... **L. perpusilla**
        3. Fronds bear one small papillae above the node and one larger near the tip .... **L. aequinoctialis**

Lemna aequinoctialis Welw., Apont, 578. 1859; Halder & Venu, Curr. Sci. 102: 1632. 2012. *L. paucicostata* Hegelm., Lemnac. 139. t. 8. 1868; FBI 6: 556. 1893; Fischer 3: 1593. 1931.

Fronds light green, usually 1–3 together, oblong or ovate or orbicular, 2–5 x 0.09–0.13 mm, asymmetrical; two distinct papillae on the dorsal surface; nodal (where the veins converge) papillae smaller than apical one. Roots one; root sheath winged, *c.* 0.01 x 0.02 mm. Inflorescence on two lateral pouches. Male flowers two, *c.* 0.1 mm in length; anthers divaricate, bilocular, dehisce by transverse slit. Female flower composed of gynoecium; *c.* 0.2 mm long; ovary globose, hyaline; style one, terminal. Fruit utricle.

In still waters throughout Eastern Ghats. Fl. & Fr.: September – January.

INDIA: Widely distributed in India.

WORLD: All temperate and tropical regions.

Lemna gibba L., Sp. Pl. 70. 1753; FBI 6: 556. 1893; Fischer 3: 1593. 1931.

A small or minute, green, floating, stemless, rootless or with capillary rootlets herb. Fronds floating, suborbicular or obovate, entire, not tailed, 0.8-1.2 cm in diameter, opaque, thick, flat above, at length very convex beneath the young fronds, sessile, each giving rise to a single root-fibre, the under surface at length spongy and greatly swollen; root-sheath elongate, cylindric, root cap acute. Stamens 2. Utricle opening circumssilely.

In still waters in all districts (Fischer). But Halder and Venu (2012) have not reported this species from Eastern Ghats. Fl. & Fr.: September - January.

INDIA: Throughout India.

WORLD: Through out the world.

Lemna minor L., Sp. Pl. 970. 1753; Halder & Venu, Curr. Sci. 102. 1630. 2012.

Fronds dark green, 5–10 together, ovate, 1–8 x 0.5–5 mm, flattened. Upper surface with distinct papillae near apical region; nerves 3–4, not reaching tip, distinct. Stipe small, 0.1–0.2 mm hyaline. Roots one, *c.* 12 mm; root cap rounded; wing absent. Inflorescence on lateral pouches, male and female flowers together. Male flowers two, *c.* 0.3 mm; anther bilocular with transverse dehiscence line, *c.* 0.2 mm dia. Female flowers one, *c.* 0.15–0.2 mm. Fruiting not observed.

In still waters in all districts (Halder and Venu, 2012). Fl. & Fr.: September – January.

INDIA: Jammu & Kashmir, Haryana, Rajasthan, Uttar Pradesh, Sikkim, Gujarat, Andhra Pradesh.

WORLD: Europe, Central Asia, West Coast and Central North America, Eastern half of Africa,

Lemna perpusilla J. Torrey, Fl. New York 2: 245. 1843; van der Plas in Steenis, Fl. Males I. 7: 231. 1971; Matthew, Fl. Tamilnadu Carnatic 3: 1705. 1983.

A small or minute, floating herb. Fronds in groups of 2 or 3, sometimes solitary, ovate-obovate or oblong, 1-4.5 x 1.25 mm; dorsal side flat, with a hook-shaped papilla at the tip; ventral side somewhat flat or convex, base obtuse, margin entire, apex obtuse-rounded, 3-nerved. Root 1, root-sheath with 2 lateral wings, root cap acute. Budding pouches 2, basal, lateral, one on either side of the axis, triangular, opening by a transverse slit, slit in line with the margin of the frond. Inflorescence 1 in the budding pouch, with 1 male and 2 female flowers, anthers transversely dehiscent. Fruit ellipsoid; seed with prominent longitudinal ribs.

Common in still waters. Fl. & Fr.: September – January.

Narpala (ATP), *KNR* & *DAM* 10627; Anantagiri to Borra (VSKP), *NPBK* 814 (CAL); Kolleru (WG), *KS* 5126 (MH).

INDIA: Widely distributed in India.

WORLD: All temperate and tropical regions.

SPIRODELA Schleiden

Spirodela polyrrhiza (L.) Schleiden, Linnaea 13: 392. 1839; Mattew, Fl. Tamilnadu Carnatic 3: 1706. 1983. *Lemna polyrrhiza* L., Sp. Pl. 970. 1753; FBI 6: 557. 1893; Fischer 3: 1593. 1931.

A small or minute floating herb. Fronds herbaceous, not tailed, opaque, thick, flat above, slightly convex below, 6 mm diameter, dark green above, usually purplish beneath, 7-veined; young fronds sessile, each frond giving rise to a tuft of root-fibres. Spathe 2-lipped, stamens 2, ovules 1-2.

Occasional in still waters. Fl. & Fr.: September – January.

Ethipothala falls (GNT), *VRK* 3887; Palem (KSN), *PV* 5567 (MH); Kolleru (WG), *KS* 5125 (MH).

INDIA: Throughout India.

WORLD: Cosmopolitan but absent from S. America.

WOLFFIA Horkel *nom. cons.*

Wolffia globosa (Roxb.) Hartog & Plas, Blumea 18: 367. 1970; Matthew, Fl. Tamilnadu Carnatic 3: 1706. 1983. *Lemna globosa* Roxb., Fl. Ind. 3: 565. 1832. *Wolffia arrhiza* auct non Horkel ex Wimmer 1857; FBI 6: 557. 1893 *p.p.*; Fischer 3: 1593. 1931.

A small or minute rootless herb with very minute fronds, round or ovate, 0.2 – 0.8 x 0.1 – 0.3 mm, base and apex obtuse, margin with a few papillose cells or entire. Flowers in a groove on the upper surface of the frond, naked; male solitary, stamen solitary, anther 1-celled, sessile; ovay solitary, style short, stigma depressed, ovule solitary. Utricle spherical.

Occasional in still waters. Fl. & Fr.: June – October.

SNG Palem (KSN), *PV* 5566 (AU).

INDIA: Throughout India.

WORLD: E. Asia, SE. Asia, Africa, Australia, Sri Lanka.

BUTOMACEAE

BUTOMOPSIS Linnaeus

Butomopsis latifolia (D.Don) Kunth, Enum. Pl. 3: 165. 1841; FBI 6: 562. 1893. *Butomus latifolius* D.Don, Prodr. Fl. Nepal. 22. 1825. *Tenagocharis latifolia* (D.Don) Buchenau, Nachr. Königl. Ges. Wiss. Georg-Augusts-Univ. 1869: 238 1869; Saxena & Brahmam, Fl. Orissa 4: 2061. 1996.

Erect aquatic herb, 20-60 cm high. Leaves erect, long petioled, lamina lanceolate, elliptic-lanceolate or oblanceolate, 5-15 x 1.6-5.5 cm, apex acute, subacute or obtuse, tipped by a hard, blunt mucro at the underside of which is a large hydathode, 3-7-nerved, base cuneate; petiole 8-27 cm long, dilated at the base. Peducnle stout, far exceeding the leaves with a single whorl of 4-15 flowers; bracts ovate, scarious, up to 1.6 cm long, pedicels 2.5-15 cm long; sepals oblong-ovate, wrapping round the young fruit. Follicles 8.7 mm long, shortly beaked, slightly connate below; seeds elliptic, shining brown.

Fairly common in wet places, rice fields *etc.* Fl. & Fr.: November – January.

Raika to Digi (GJM), *CAB* 1313 (MH); Majhipada, Satkosia Tiger Reserve, *KCM* 6148 (BSID).

INDIA: Assam, Kashmir, Madhya Pradesh, Manipur, Odisha, Punjab, Rajasthan, West Bengal.

WORLD: Europe, North Asia, North America

ALISMATACEAE

1. Pistils inserted on a large, globose receptacle; stamens many; achenes laterally flattened .... **Sagittaria**
1. Pistils on a small receptacle; stamens 6; achenes round in section:
    2. Higher flowers male, lower ones bisexual or female .... **Limnophyton**
    2. All flowers bisexual........ **Caldesia**

CALDESIA Parlatore

Caldesia parnassifolia (L.) Parl., Fl. Ital., 3: 599. 1858; Saxena & Brahmam, Fl. Orissa 4: 2064. 1996; Matthew, Fl. Tamilnadu Carnatic 3: 1707. 1983. *Alisma parnassifolia* Bassi ex L., Syst. Nat. (ed. 12) 3 (App.): 230 1768. *A. reniforme* D.Don, Prodr., Fl. Nepal 22. 1825; FBI 6: 560. 1893; Fischer 3: 1595. 1931.

Large, glabrous herb, with an annual stem. Leaves ovate-orbicular, 5-10 x 5-15 cm, apex rounded or retuse, margin entire, base cordate; petiole to 70 cm long, Flowers white or pink-purple, in 30-60 cm long panicles; sepals 3, elliptic, petals 3, orbicular-ovate; stamens 6, ovaries *c.* 15, obovoid, compressed, style subterminal, stigma punctiform. Achenes ellipsoid, to 4.5 x 2.5 mm, laterally compressed, 3-5-ribbed; seed 2 x 1 mm.

Occasional in ponds and tanks in Eastern Ghats of Odisha and Tamil Nadu. Fl. & Fr.: May – September.

INDIA: Throughout the plains of India.

WORLD: From N and C. Africa and Madagascar to S and C. Europe through SE Asia to China, Japan and N. Australia.

LIMNOPHYTON Miquel

Limnophyton obtusifolium (L.) Miq., Fl. Ned. Ind. 3: 242. 1856; FBI 6: 560. 1893; Fischer 2: 1595. 1931; Matthew, Fl. Tamilnadu Carnatic 3: 1708. 1983. *Sagittaria obtusifolia* L., Sp.Pl. 993.1753.

A scapigerous, stemless, palustrine, perennial herb; root stock short, stout, with numerous long root-fibres. Leaves broadly sagittate, 4-20 x 3-15 cm, chartaceous, 7-20-nerved, stellate-pubescent, base broadly cordate; basal lobes lanceolate, margin entire, apex obtuse-rounded; petiole 6-60 cm. Panicles to 1.5 m long, branches in whorls of 2-6, 10-15-flowered per whorl; bracts 3, lanceolate, pedicel 4 cm. Flowers white, the upper whorls mostly male, the lower hermaphrodite. Male flowers: sepals 3, oblong-elliptic, hyaline along margin; petals 3, white, obovate; stamens 6; ovaries 12-20, free, obovoid, ovules on basal placentae. Achene cluster globose, attenuate at base, reticulately ribbed, shortly beaked, turgid with air-chambers between exocarp and endocarp; seed oblong, 4 mm.

Common in stagnant waters. Fl. & Fr.: January – March and July – September.

Kalasamudram RF (ATP), *NY* 679 & 1204; Guvvalacheruvu RF (KDP), *R.V.Reddy* 8121; Near Cuddapah-Dinnela (KDP), *JSG* 21273 (MH & CAL); Ramapuram (KDP), *KS* 7796 (MH & CAL); Kuppam (CTR), *DRC* 1326 (MH); Gudur (NLR), *KCJ* 18511 (MH); Pedamantanala tank (PKM), *RKM* 0451 (CAL); Rollapadu (KMM), *R.Rajan* 108038 (BSID); Majhipada RF, Satkosia Tiger Reserve, *KCM* 6149 (BSID); Sinarangakuppam (SA), *VNS* 5093 (MH); Madurantakam (CPT), *S.India Flora* 11162 (MH); Andiyappanur- Alangayam (NA), *EV* 52051 (MH); Gingee RF (SA), *KMS* 12377 (MH); Gandigam lake- Pennagaraam (SLM), *EV* 22441 (MH); Manali lake behind Madhavaram Milk colony (CPT), *DN* 767 (MH).

INDIA: Throughout India.

WORLD: From Tropical Africa and Madagascar through SE. Asia to Malesia.

SAGITTARIA Linnaeus

1. Mature leaves hastate or sagittate, apex acute or acuminate, basal lobes usually diverging, narrower and often longer than the rest of the blade; male flowers with *c.* 24 stamens; achenes with an entire or subcrenate blade .... **S. trifolia**

1. Mature leaves broadly ovate, deeply cordate, apex rounded, basal lobes rounded; male flowers with 6-12 stamens; achenes with a broad toothed wing all round .... **S. guayanensis**

Sagittaria guayanensis Kunth, Nov. Gen.Sp. 1: 250. 1816 subsp. **lappula** (D.Don) Bogin, Mem. New York Bot. Gard. 9a: 192. 1955. *S. lappula* D.Don, Prodr. Fl. Nepal 22. 1825. *S. guayanensis* non Kunth 1850; FBI 6: 561. 1893.

Scapigerous aquatic herb; roots fibrous, densely tufted. Leaves floating, broadly ovate, 2.5-10 x 2.5-8.7 cm, apex rounded or obtuse, base deeply cordate, auricles rounded, nerves obscure, *c*.9, radiating, distinct when dry, sheaths very broad, suddenly contracted into long, often hairy petiole. Flowers white, 1.75 cm across, scapes and pedicels hairy, upper male flowers more numerous with 6-12 stamens; petals obovate. Achenes flat, surrounded by a broad, prominently toothed wing.

Frequent in ponds, ditches and tanks in Eastern Ghats of Odisha and occasional in Southern Eastern Ghats in Tamil Nadu. Fl. & Fr.: August – November.

Gingee RF (SA), KMS 12325 (MH).

INDIA: Throughout India.

WORLD: Tropics of Africa, SE to E. Asia.

Sagittaria trifolia L., Sp. Pl. 993. 1753. *S. sagittifolia* auct non L.; FBI 6: 561. 1893; Fischer 3: 1596. 1928.

An aquatic, perennial, stoloniferous, deep rooted herb, 30-100 cm high. Leaves radical, dimorphic, submerged leaves ribbon-shaped, 5-20 cm long, emergent leaves lanceolate or elliptic, sagittate, apex acute or acuminate; petiole very spongy, trigonous. Flowers white, often with a purple claw, scape with 3-5 flowers in each whorl, the lower whorls female, upper whorl male; petals 3. Achenes obliquely obovate, flattened.

Rare in Visakhapatnam district (Ramachandran *et al.*, 1976). Fl. & Fr.: October – December.

INDIA: Throughout India.

WORLD: From SE. corner of the Black sea and of Arabia eastwards and southwards to Japan and Malaysia.

NAJADACEAE

NAJAS Linnaeus

1. Internodes with spinous teeth .... **N. marina**
1. Internodes without spinous teeth:
    2. Leaf-sheaths apically produced into a linear-oblong auricle; leaf margin with more than 50 spiny teeth........ **N. graminea**
    2. Leaf-sheaths apically obtuse, rarely with obscure auricle; leaf margins with less than 20 spiny teeth:
        3. Spathe present in male flower; anthers 4-celled .... **N. indica**
        3. Spathe absent in male flower; anthers 1-celled........ **N. minor**

Najas graminea Del., Descr. Egypt. Hist. Nat. 282. t. 50. f. 3. 1813; FBI 6: 569. 1893; Fischer 3: 1604. 1931; Matthew, Fl. Tamilnadu Carnatic 3: 1710. 1983.

A submerged, aquatic herb, rooting from the base and lower nodes; stems or shoots terete, up to 50 cm long. Leaves in pseudowhorls, linear-subulate, obtuse or acute with a spinulosely dentate margin, 1-2.5 x 0.005-0.1 cm, sheaths spinulosely denticulate with triangular, acute or obtuse, entire or rarely shallowly notched, 0.2 – 0.5 cm long, auricles 0.4 – 0.8 cm long. Flowers solitary or 2 – 4 together. Male: Spathe absent, perianth segments rounded, anther ellipsoid-ovoid or oblong. Female: Spathe absent; flowers solitary, axillary, subsessile, perianth lobes closely appressed to ovary, ovary elliptic. Fruit ellipsoid-oblong, attenuate at the apex, 0.15-0.2 cm long, areoles minute, subquadrate or polygonal.

Common in ponds. Fl. & Fr.: August – November.

Garugudukona (ATP), *TP* 947; Kalasamudram (ATP), *NY* 12025; Rangapuram RF (KNL), *RVR* 1478; Palakonda hills (KDP), *CS* 8292; Krishna River (GNT), *VRK* 6744; Srisailam (KNL), *JLE* 16856 (MH & CAL); Guvvalacheruvu (KDP), *KS* 6432 (MH); Gandigam lake to Pennagaram (SLM), *EV* 22460 (MH).

INDIA: Throughout India.

WORLD: Cosmopolitan.

Najas indica (Willd.) Cham., Linnaea 4: 501.1829; Fishcer 3: 1604. 1931; Matthew, Fl. Tamil Nadu Carnatic 3:1711. 1983. *Caulinia indica* Willd., Mem. Acad. Roy. Sci. Hist. (Berlin) 89. 1801. *Najas falciculata* A.Braun, J.Bot. 2: 278. 1864; FBI 6: 569. 1893; Fischer 3: 1604. 1931. *N. lacerata* Rendle, Trans. Linn. Soc. London 5: 416. 1899; Fischer 3: 1604. 1931.

Slender, submerged fresh water herb. Leaves 0.6-2 x 0.1-0.2 cm, base subterete, margins with 6-15 spines on each side, apex flat, sheath oblong, to 35 mm, margin entire. Male flowers: Spathe to 2 mm, beaked, apex spiny; flower solitary; perianth lobes distinct, anthers 4-celled, ellipsoid. Female flowers: Spathe to 2.2 mm; flower solitary; perianth lobes distinct, ovary to 1.8 mm, ellipsoid, style to 0.5 mm, stigma to 1 mm. Fruit subterete, trigonous, 2 x 0.8 mm, with prominent aereoles, longitudinally disposed.

Plains; abundant in Veeranam lake (Matthew, 1983).

INDIA: All the soutnern states to foot hill of Himalaya.

WORLD: Bangladesh, Africa, tropical continental Asia northward to Japan, throughout Malesia.

Najas marina L., Sp. Pl. 1015. 1753; Wilde in Steenis, Fl. Males I. 6: 162. 1962. *N. major* All., Fl. Pedem. 2: 221. 1785; FBI 6: 569. 1893; Matthew, Fl. Tamilnadu Carnatic 3: 1712. 1983.

A submerged aquatic herb; stem often with spinous teeth. Leaves linear, 0.6-1.5 x 0.1-0.2 cm, flat, margins with 5-9 prominent spiny teeth, thick, dorsal surface smooth, septate; sheath oblique, 4 mm, apex obtuse, with 1 or 2 spines; auricles 0. Male flowers: Spathe membranous, apex slightly beaked or sometimes lacerate flower subsessile, perianth-lobe distinct, anthers 2-celled, elliptic. Female flowers: solitary, sessile, spathe closely appressed to ovary, margined at the periphery, ovary obovoid, trigonous, base attenuate, apex rotund, style 3-fid. Fruit ovoid-subglobose.

Occasional in Yerramalais. It is a coastal plant but occurs occasionally inland also. Fl. & Fr.: August – November.

Maddileti stream (KNL), *RVR* 3078.

INDIA: Throughout India.

WORLD: Throughout the World.

Najas minor All., Fl. Pedem. 2: 221. 1785; FBI 6: 569. 1893; Fischer 3: 1604. 1931; Matthew, Fl. Tamilnadu Carnatic 3: 1712. 1983. (incl. var. *spinosa* Rendle).

A submerged aquatic herb; shoots 4-25 cm long. Leaves linear, 0.5-1.5 x 0.1-0.2 cm, base triquetrous, apex flat, dorsal surface smooth, septate, margins with 10-15 spinous teeth oblong. Male: spathe 2-3 mm long, 1 mm broad, beak 1.5 mm, with linear spinous teeth. Flowers solitary, sessile, perianth lobes obscure, anthers 1-celled. Female: spathe 0, flower solitary, ovary terete. Fruit oblong, trigonous, 2 x 0.8 mm, areoles subquadrate, in longitudinal rows.

Rare in ponds. Fl. & Fr.: July – November.

Gani RF (KNL), *RVR* 1550; Chintaparthi river (CTR), *MHR* 13376; Horsley hills (CTR), *JSG* 15719 (MH); Nagari (NA), *S.India Flora* 14749 (MH); Ibrahimpatnam (KSN), *P.Venakanna* 6054 (MH).

INDIA: Throughout India

WORLD: Europe, Tropical and N. Africa, Tropical and Temperate Asia to Japan and Pegu, Malesia.

APONOGETONACEAE

APONOGETON Linnaeus f. *nom. cons.*

1. Tepals attenuate at the base; ovules 8 per ovary .... **A. natans**
1. Tepals broad at base; ovules 2 per ovary .... **A. crispus**

Aponogeton crispus Thunb., Nov. Gen. 4: 73. 1781 (*'crispum'*); FBI 6: 564. 1893 *p.p*; Fischer 3: 1597. 1931; H. Bruggen, Blumea 18: 481. f.1. 1970; Yadav & Gaikwad, Bull. Bot. Surv. 45: 57. 2003. *A. echinatus* Roxb., Fl. Ind. 2: 2110. 1832; FBI 6: 564. 1893; Matthew, Fl. Tamilnadu Carnatic 3: 1712. 1983.

A submerged or floating, aquatic, perennial herb; tuber ovoid, 3 x 2 cm. Leaves oblong-lanceolate, submerged leaves oblong-elliptic, 15 x 5 cm, floating leaves oblong, 4-16 x 1-5 cm, glabrous, base subcordate, margin entire, apex subacute, obtuse; submerged leaves petiolate, petiole *c.* 13 cm long. Spikes oblong, 15 cm long; peduncle nearly as long as petiole. Flowers patent, dense at base, apically lax; tepals 2, pinkish, obovate sub-orbicular as long as broad; stamens 6; ovaries 3, globose, ovules 2, style short. Fruit ovoid, 4-6 x 2 mm; seeds oblong, flat on back, turgid on the upper surface.

Common in ponds, tanks and ditches. Fl. & Fr.: January – April.

Molakalacheruvu (CTR), *KNR* & *DAM* 9831; Kotta cheruvu (CTR), *GVS* 45923 (MH); Horsley hills (CTR), *S.India Flora* 15464 (MH); Anantavaram (EG), *SS* 4990 (AU); Gandigam lake – Pennagaram (SLM), *EV* 22437 (MH); Vinukonda RF (GNT), *S.India Flora* 17401 (MH); Borapuram (MBNR), *BSS* & *BR* 32370 (BSID).

INDIA: Throughout India.

WORLD: Sri Lanka, Bangldesh.

Aponogeton natans (L.) Engl., Pflanzenr. IV. 13. 24: 11. 1906; Fischer 3: 1598. 1931; Yadav & Gaikwad, Bull. Bot. Surv. 45: 60. 2003; Matthew, Fl. Tamilnadu Carnatic 3: 1714. 1983. *Saururus natans* L., Mant. Pl. 227. 1771. *Aponogeton monostachyon* L. f., Suppl. Pl. 214. 1781; FBI 6: 564. 1893. *A. lucens* Hook.f., Fl. Brit. India *6: 564. 1893.*

A submerged or floating, glabrous, aquatic herb, tuber linear-oblong, 2.5-3.5 x 0.8-1.5 cm. Leaves mostly floating, oblong, 2-10 x 0.8-2 cm, glabrous, base subcordate-obtuse, margin entire, apex obtuse subacute. Spikes 2.5-15 cm long, usually dense-flowered; peduncle 10-45 cm, usually overtopping leaves, tepals 2, violetish, obovate-suborbicular, 2.5 x 1-5 mm, nearly twice as long as broad; stamens 6, anthers bluish-purple; ovaries 3, oblong. Fruit globose, 4 x 2 mm, smooth, sharply beaked.

Common in ponds, tanks, ditches throughout Eastern Ghats. Fl. & Fr.: December – February.

Kalasamudram RF (ATP), *TP* & *EC* 2870; Velugodu tank (KNL), *RVR* 14528; Tada (NLR), *MCK* 23509; Isukagundam (PKM), *BR* & *BSS* 30353; Sompalli reservoir (VSKP), *MHR* & *MCK* 14892; Diguvametta, Nallamalais (PKM), *JLE* 32387 (MH); Kambakam (CTR), *MCB* 45186 (MH); near Komaticheruvu (CTR), *GVS* 45851 (MH); Gudur (NLR), *KCJ* 18519 (MH); Repalle (GNT), *GVN* 16813 (MH); Amalapuram (EG), *KR* & *RG* 88815 (BSID); Srungavarapukota (VZN), *GVS* 21820 (MH); Yetapaka (KMM), *R.Rajan* 106064 (BSID); Egmore (MDS), *sine coll. s.n.* (MH); Kayappakam (CPT), *S.India Flora* 11233 (MH); Vandalur RF (CPT), *ANH* 45546 (MH); Tirupattur (NA), *KS* 6511 (MH); Sirupakkam RF (SA), *KRM* 50695 (MH);

INDIA: Throughout India.

WORLD: Sri Lanka, Bangladesh, China, Taiwan, Vietnam.

POTAMOGETONACEAE

1. Leaves all submerged, floating leaves absent, leaves acicular, not exceeding 0.2 cm across; stipular sheaths adnate to base of leaf blade for 2/3 or more length, blades channeled .... **Stuckenia**

1. Leaves both submerged and floating or all submerged, not acicular, exceeding 0.2 cm across, stipular sheaths of submerged leaves free from base of leaf blade or if adnate, then adnate portion less than ½ length of stipule, blades not channeled, flattened ....**Potamogeton**

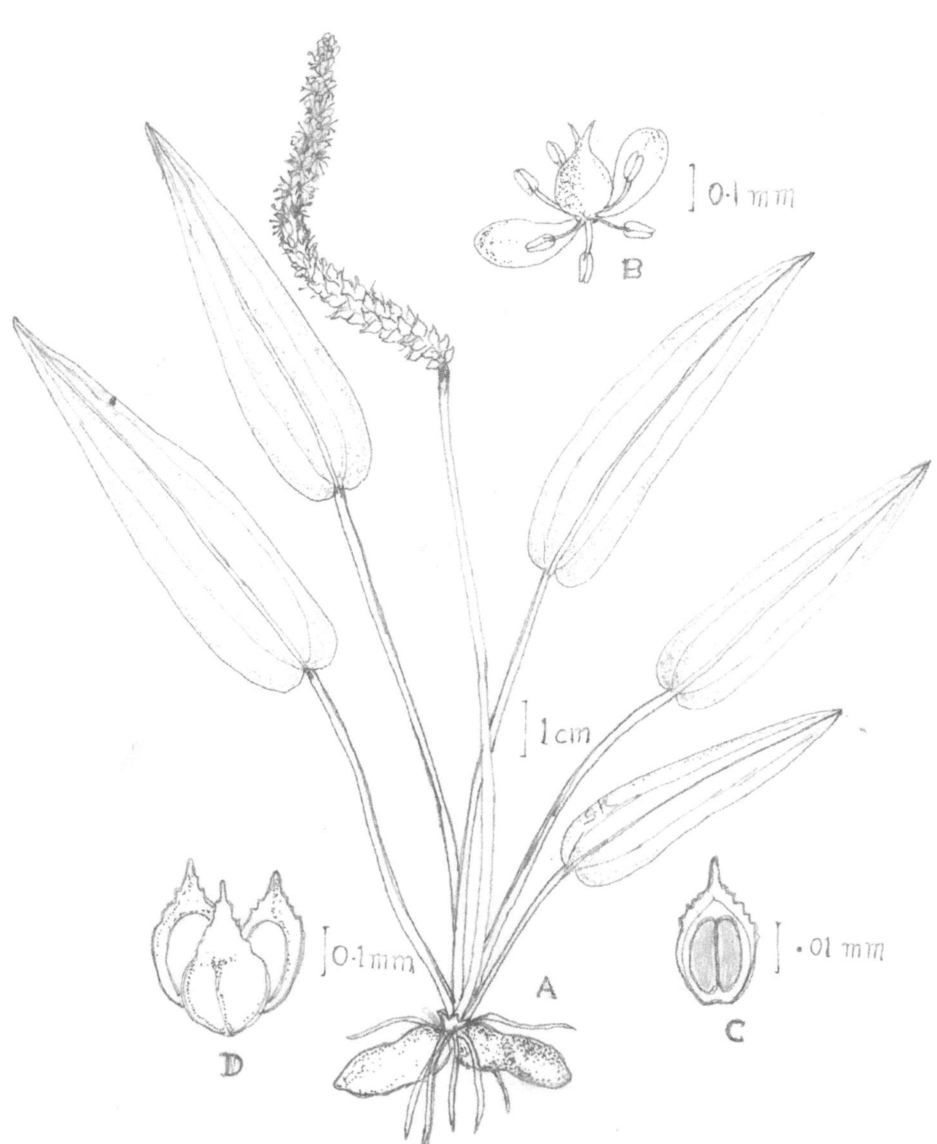

Figure 32. **Aponogeton natans** (L.) Engl.

A. Habit; B. Flower-entire; C. Ovary-l.s.; D. Capsule.

POTAMOGETON Linnaeus

1. Floating and submerged leaves present:
    2. Submerged leaves sessile; floating leaves not exceeding 3 cm in length... **P. octandrus**
    2. Submerged leaves petiolate; floating leaves exceeding 5 cm in length... **P. nodosus**
1. Floating leaves absent; leaves all submerged:
    3. Leaves 7-9-nerved, wavy, apex subulate-obtuse....................... **P. perfoliatus**
    3. Leaves 3-nerved, serrulate, apex rotund.. **P. crispus**

Potamogeton crispus L., Sp. Pl. 126. 1753; FBI 6: 566. 1893; Matthew, Fl. Taminadu Carnatic 3: 1715. 1983.

Submerged freshwater herb, stolons not more than 2 mm thick, much branched, reddish, lying flat below the surface of the substrate; stems 10-70 cm long, somewhat flattened and angular, usually branched, reddish. Leaves submerged, oblong-elliptic or oblanceolate, 1-5 x 0.3-0.8 cm, membranous, translucent, 3-nerved, glabrous, base slightly amplexicaul, subcordate, margin serrulate, apex rotund; petiole 0; stipule to 4 mm. Peduncle rigid, erect, emergent, up to 10 cm long; spikes with 2-10 flowers, dense, rarely more than 2 cm long. Drupelets globose, usually united below, 5-6 mm long.

In ponds (Matthew, 1983).

Satkosia Wild Life Sanctuary, *D.Hazra* & *D.Das s.n.* (BSID).

INDIA: Assam, Bihar, Uttar Pradesh, Maharashtra, Madhya Pradesh, Odisha, West Bengal, Tamil Nadu,

WORLD: Europe, Africa, Asia and Australia.

Potamogeton nodosus Poir. in Lam., Encycl. (Suppl. 4): 535. 1816. *P. indicus* Roxb., Fl. Ind. 1: 471. 1820; non Roth ex Roem. & Schult. 1818; FBI 6: 565. 1893; Fischer 3: 1600. 1831; Matthew, Fl. Tamilnadu Carnatic 3: 1716. 1983.

An aquatic submerged herb; stem terete, branched, smooth. Leaves heterophyllous, 4-12 x 2-3.5 cm, upper ones floating, broadly ovate-elliptic, sub-acute-obtuse at the ends, lower ones submerged, lanceolate, 7-9-nerved, base attenuate-cuneate, margin entire, apex acute-acuminate; stipules free, membranous, keeled, apex tapering. Spikes axillary, leaf-opposed, 2-5 cm long; peduncle 6 cm. Flowers sessile; tepals 4, suborbicular-ovate; stamens 4; ovaries 3, free, ovule 1. Drupelets obovoid, shortly beaked, 1-seeded.

Commong in ponds and ditches. Fl. & Fr.: July – December.

Kadiri – Kalasamudram (ATP), *NY* 654; Dadithota (ATP), *BR* & *ANS* 44805; Burakayalakota (CTR), *KNR* & *PSPB* 9833; Chintaparthi river (CTR), *MHR* 13377; Gujana River (KDP), *JLE* 14347 (MH); Horsley hills (CTR), *JSG* 15716 (MH); RR tank

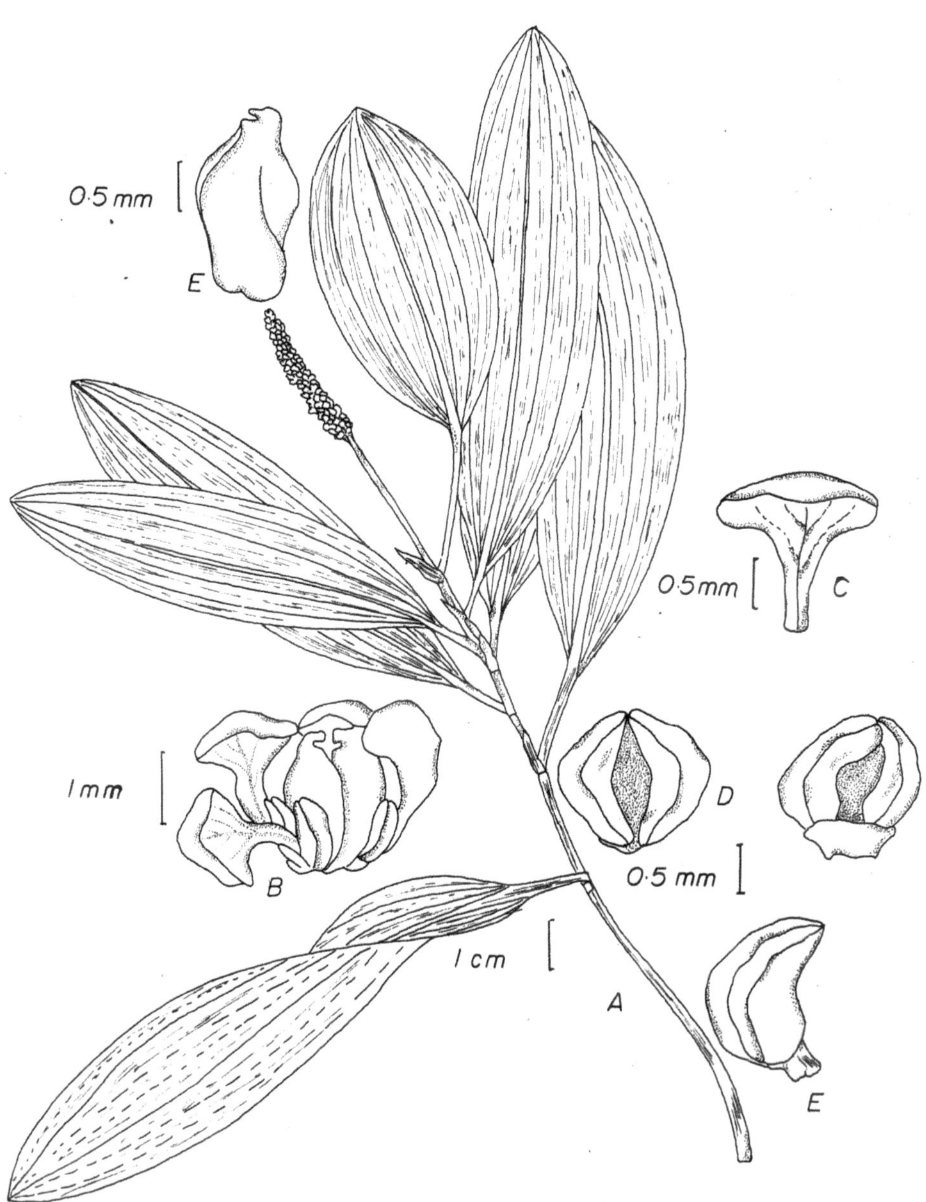

Figure 33. **Potamogeton nodosus** Poir.
A. Twig, B. Flower, C. Perianth lobe, D. Anthers, E. Pistil.

(PKM), *RKM* 0650 (CAL); Komarada (VZN), *MV* 2884 (AU); Palur (SA), *VNS* 4153 (MH); Sethiathope (SA), *KRM* 58107 (MH).

INDIA: Throughout India.

WORLD: Warm temperate and Tropical regions of Europe, Africa, Asia and America.

Potamogeton octandrus Poir. in Lam., Encycl. Suppl. 4: 517. 1816. *P. javanicus* Hassk., Acta Soc. Ind. –Neerl. 1, 8: 26. 1856; FBI 6: 566. 1892; Fischer 3: 1600. 1931.

A submerged or floating aquatic herb; stem very slender; submerged leaves very narrowly linear or filiform, without distinct petiole, floating leaves membranous, elliptic or ovate-oblong, acute, 1.2-4 cm long; petioles usually shorter; stipules free. Peduncle slender, axillary or leaf-opposed. Spikes 0.6-1.2 cm long, lax-flowered; tepals orbicular-obovate. Drupelets semiglobose with a hooked beak, ribs often toothed and tubercled.

Rare in ponds, tanks and ditches in Northern Eastrn Ghats. Fl. & Fr.: All seasons.

Araku valley (VSKP), *DDSR* 21389 (MH).

INDIA: Throughout India.

WORLD: Tropical Asia, Africa and Australia.

Potamogeton perfoliatus L., Sp. 126. 1753; FBI 6: 566, 1893; Fischer 3: 1600. 1931; Matthew, Fl. Tamilnadu Carnatic 3: 1717. 1983.

An aquatic herb; stem stout, terete, slightly branched. Leaves 2.5-6 x 0.9-1.5 cm, sessile, ovate-lanceolate, obtuse, membranous, transluscent, amplexicaul, base cordate, margin wavy, 5-9-nerved, stipules small, caducous. Peduncle axillary, rather stout, spike dense-flowered, 1.2-2.5 cm long, tepals elliptic-obovate, clawed. Drupelets compressed-globose with short curved beak, hardly keeled, smooth.

Common in ponds and ditches. Fl. & Fr.: December – February.

Gunjana River (KDP), *JLE* 14348 (MH); Ibrahimpatnam (KSN), *PV* 6055 (AU); Srungavarapukota (VSKP), *GVS* 21820 (MH).

INDIA: W. to E. India, W. Himalaya.

WORLD: Europe, Asia, Africa, N. America, Australia and Pakistan.

STUCKENIA Börner

Stuckenia pectinata (L.) Börner, Fl. Deut. Volk. 713. 1912. *Potamogeton pectinatus* L., Sp. Pl. 127. 1753; FBI 6: 567. 1893; Fischer 3: 1600. 1931; Saxena & Brahmam, Fl. Orissa 4: 2069. 1996; Matthew, Fl. Tamilnadu Carnatic 3: 1716. 1983.

An aquatic submerged herb; stem filiform, copiously distichously branched. Leaves all submerged, alternate, acicular, 2-8 x 0.2-0.5 cm, puberulous, apex acute, stipules adnate to the leaf-sheaths, the tips free. Spikes axillary, 2.5 cm long. Flowers a few in distant whorls, minute, green; tepals 4, obovate-oblong, base clawed, cuneate,

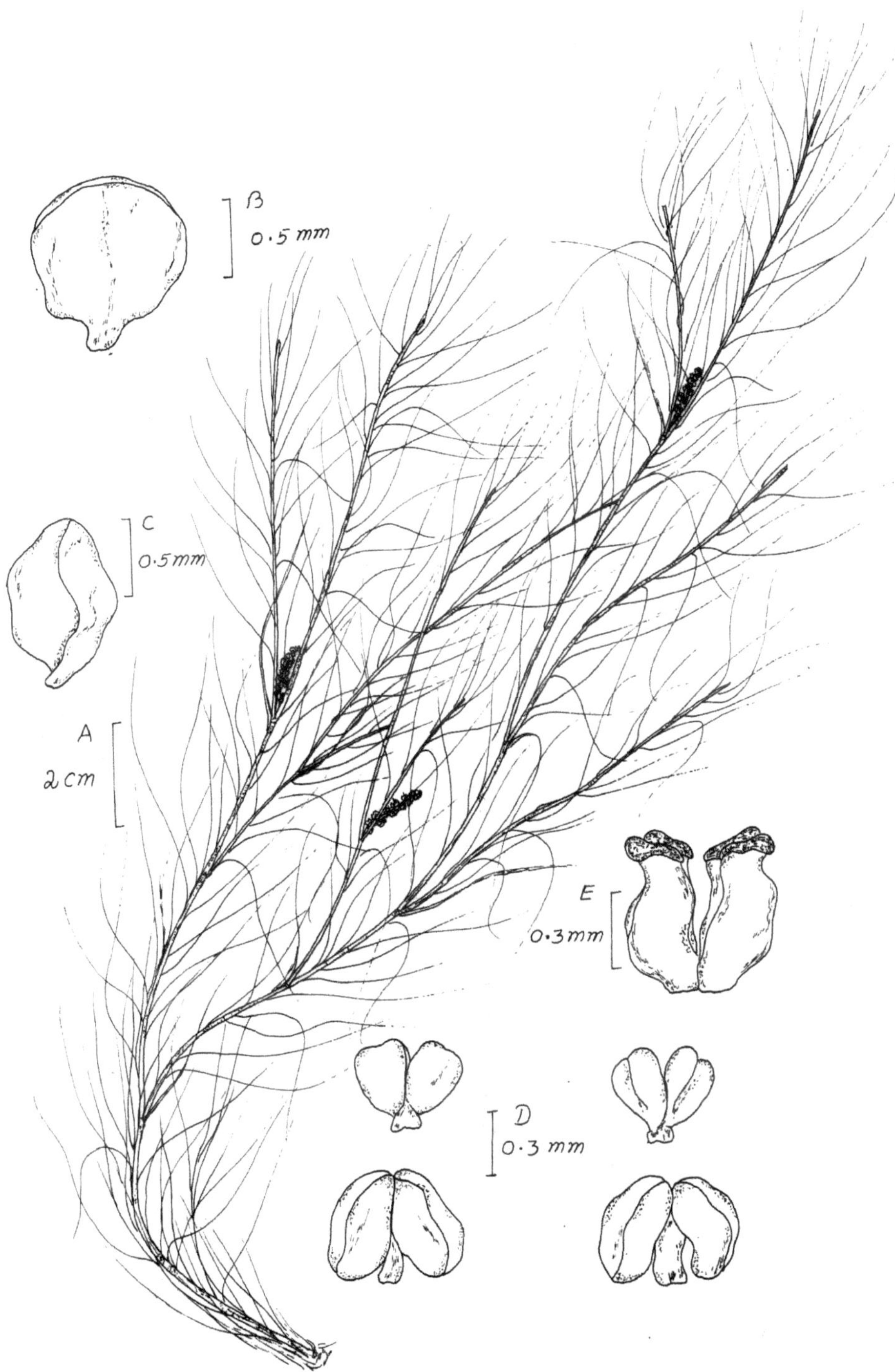

Figure 34. **Stuckenia pectinata** (L.) Börner, (Syn.: *Potamogeton pectinatus* L.)
A. Twig, B & C. Perianth lobes, D. Anthers, E. Carpels.

margin entire, apex rotund; stamens 4, filaments equal to anther lobes; ovaries 4, free, linear, stigma subsessile. Drupelets obovate, 3 x 2 mm, obscurely beaked.

Common in freshwater pools and low land water. Fl. & Fr.: December – March.

I. Sadum – Kokkanti (ATP), *MHR* 13926; Maddileti stream (KNL), *RVR* 3080; Sangalamadugu – Kondur (KDP), *MHR* 13301.

INDIA: Throughout India.

WORLD: North and South America, Europe, Africa, SW Asia, Russia, Pakistan, Sri Lanka, Myanmar, China, Philippines, Australia,

ERIOCAULACEAE

ERIOCAULON Linnaeus

1. Stem more than 10 cm long; leaves cauline, filiform .... **E. setaceum**
1. Stem absent or less than 5 cm long; leaves caespitose, not fliliform:
    2. Anthers black or dark brown:
        3. Spathe of male flowers 3-lobed:
            4. Female sepals unequal; odd sepals distinctly narrower than the lateral:
                5. Involucral bracts acuminate or cuspidate; floral bracts straw-coloured .... **E. xeranthemum**
                5. Involucral bracts obtuse or subacute; floral bracts blackish:
                    6. Involucral bracts exceeding the head, more than twice the length of floral bracts .... **E. heterolepis**
                    6. Involucral bracts not exceeding the head; less than twice the length of floral bracts:
                        7. Plant more than 6 cm high; floral bracts distinctly pubescent .... **E. richardianum**
                        7. Plant less than 3 cm high; floral bracts glabrous .... **E. echinulatum**
            4. Female sepals equal or subequal:
                8. Appendages arising from the middle of radial walls of each cell, they appear in vertical rows on the surface .... **E. luzulifolium**
                8. Appendages spread uniformly on the transverse radial walls of each cell, they appear to be in close transverse ring on the surface:
                    9. Heads straw-coloured; male and female petals eglandulose .... **E. oryzetorum**
                    9. Heads black or grey; male and female petals with black gland in each:

10. Male petals unequal; middle one larger than the others; receptacles villous .... **E. odoratum**

10. Male petals equal; receptacles glabrous or sparsely pilose .... **E. sollyanum**

3. Spathe of male flowers 2-lobed .... **E. thwaitesii**

11. Appendages of seeds ribbon-like or rectangular:

12. Male flowers with 3 sepals:

13. Male sepals free .... **E. longicuspis**

13. Male sepals connate in to a spathe:

14. Floral bracts densely hairy towards apex; leaves mostly turning into purplish when drying .... **E. quinquangulare**

14. Floral bracts glabrous or only sparsely hairy; leaves never turning into purplish on drying.. **E. conicum**

12. Male flowers with 2 sepals:

15. Cells of seed coat vertically elongated .... **E. truncatum**

15. Cells of seed coat laterally elongated .... **E. edwardii**

11. Appendages of seeds setiform:

16. Male sepals connate .... **E. edwardii**

16. Male sepals free .... **E. achiton**

2. Anthers white or yellow:

17. Leaves up to 30 cm long; male petals equal, middle one much longer than the others............................ **E. berviscapum**

17. Leaves not more than 15 cm long; male petals absent:

18. Involucral bracts acute; receptacles flat or convex.......... **E. leucomelas**

18. Involucral bracts obtuse; receptacles cylindrical:

19. Heads *c.* 6 mm across; seeds oblong ellipsoid .... **E. ritchieanum**

19. Heads *c.* 4 mm across; globose or ovoid .... **E. cinereum**

Eriocaulon achiton Körn, Linnaea 27: 630. 1854; FBI 6: 584. 1893; Mooney, Suppl. Bot. Bihar & Orissa 198. 1950; Saxena & Brahmam, Fl. Orissa 2079. 1996.

Erect herb, 2.5-10 cm high. Leaves subulate or enssiform, 5-18 m long, delicately 3-5-nerved with cross nervules (fenestrate), base dilated, scarcely diffeerentiated into a sheath, sparsely woolly. Peduncle mostly showingg a spiral twist, sheaths 7.5-15 mm, also often twisted, membranous above, Heads button-shaped to ovoid, 2.5-4.5 mm diam., pale, involucral bractss longer than the floral, ovate-lanceolate to narrowly lanceolate, acute or obtuse, floral bracts cuneate, obovate to oblanceolate, receptacle distinctly villous, columnar. Male sepals 2, distinct, linear, tapering to base; corolla stipitaate, very minute. Female flowers pedicelled, sepal very slender or 0; petals 3-0, linear, acute, with a few erect hairs; ovary usually slenderly stipitate. Seeds broadly ellpsoid with minute transverse aareoles.

Near Kutab, Bamra, in sandy soil by the side of stream; Keonjhar, Ranpur (Mooney *loc. cit.*). Fl. & Fr.: November – December.

INDIA: Goa, Karnataka, Maharashtra, Odisha, Meghalaya

WORLD: Bangladesh, Pakistan, Thailand, Vietnam.

Eriocaulon breviscapum Körn, Linnaea 27: 676. 1854; FBI 6: 575. 1893; Saxena & Brahmam, Fl. Orissa 4: 2080. 1996.

Aquatic herb with submerged leaves. Leaves linear, 15-37.5 cm long, flaccid, opaque, 7-11-nerved, nerves very slender. Scapes solitary, sheath 5-6.2 cm long, tip obtuse, membranous. Heads subglobose, rather broader than long, 1.2 cm diam., rather few-flowered, peduncles solitary, receptacle convex; involucral bracts few, green, dark floral bracts oblanceolate, shortly bearded. Male flowers: sepals densely bearded, petals villous, one much largest; stamens 6, anther white. Female flowers subsessile; sepals short, oblong, concave, tips trucate, villously bearded; petals oblanceolate, subequal, very minute, subglobose, pale yellowish, shining.

Occasional in flowing water in streams. Fl. & Fr.: November – February.

Bonai, *HFM* 2623 (DD).

INDIA: Karnataka, Maharashtra, Madhya Pradesh, Odisha. Endemic.

Eriocaulon cinereum R. Br., Prodr. 254. 1810; Ansari & N.P.Balakr., Eriocaul. India 172. 2009. *E. sieboldianum* Sieb. & Zucc. ex Steudel, Syn. Pl. Glum 2: 272. 1855; FBI 6: 577. 1893; Fischer 3: 1619. 1931.

A tufted, annual marsh herb; leaves capillary or very narrowly linear, finely acuminate, 1.5 x 0.15 cm, 5-nerved. Head spherical, whitish or purplish; peduncles numerous, slender, faintly ribbed, sheaths shorter than the leaves, acute; involucral bract obovate, floral bract obovate-cuneate, grey, receptacle columnar, sparsely pilose. Male floret: calyx lobes spathaceous, 3-partite, glabrous; petals short, glabrous, with apical black gland; stamens 3+3, white. Female floret: sepals 3, free, subequal, spathulate, glabrous; petals 3, free, linear – lanceolate, ciliate, ovary 3-lobed; style branches 3, filiform. Seed ellipsoid, transversly striate.

Common in open marshy grasslands and wet places. Fl. & Fr.: September – December.

Balapalle (KDP), *JLE* 15788 (MH & CAL); Gaddada (EG), *MM* 105099 (BSID); Vazeedu (KMM), *V.S.Raju* 670 (MH); Araku valley (VSKP), *Sine coll. s.n.* (MH); Girishchandpur, Rairakhl (SBP), *HFM* 2314 (DD); Baghmunda, Satkosia Tiger Reserve, *KCM* 7017 (BSID); Vandalur (CPT), *S.India Flora* 11576 (MH); Kattankulam lake, Guindy park (MDS), *ANH* 47173 (MH).

INDIA: Throughout India.

WORLD: Sri Lanka, Myanmar, China, Japan, Philippines and Australia.

Eriocaulon conicum (Fyson) C.E.C.Fischer, Fl. Madras 3: 1620. 1931. *E. dianae* var. *conica* Fyson, J. Indian Bot. 2: 260. 1921.

Glabrous herb. Leaves linear, up to 3.7 cm log, acuminate. Peduncles many, 6.2-15 cm long, usually 5-ribbed, sheaths close, shorter than the leaves, acute or obtuse. Heads conical with a flat base, 2-5-3.7 mm diam. and up to 5 mm high; involucral bracts obovate, rounded, pale brown, glabrous, not longer than the floral bracts and the flowers; floral bracts obovate, caudately cuspidate, not much longer than the flowers, outer glabrous, innermost white or yellow-puberulous near the apex, receptacle conical, pilose; male calyx spathaceous, split down one side, 3-lobed or –partite, lobes subacute, glabrous; corolla lobes equal or nearly so, none extruded beyond the floral bracts, glabrous or with a few white, papillose hairs at the tip, with or without a small apical black gland; female sepals 3, free, ligulate-spathulate, longer than the seapls, usually with a small apical black gland, glabrous or sparsely pilose; anthers black or dark green. Seeds oblong-subglobose, yellowish-brown, reticulated with transverse ridges.

Mahendragiri hills in Northern Eastern Ghats in Odisha (Fischer).

INDIA: Odisha, Karnataka, Tamil Nadu, Maharashtra. Endemic.

Eriocaulon echinulatum Mart., Pl. Asiat. Rar. 3: 29. 1832; Mooney, Suppl. Bot. Bihar & Orissa 197. 1950.

Stem 0. Leaves linear or linear-ensiform, 1.8-3.2 x 0.25-0.5 cm, finely acuminate, rather thick and fleshy when fresh, glabrous. Peducnles *c.* 4-10, 7.5-12.5 cm long, very slender, thinly hispidulous, faintly 5-ribbed; sheaths as long as or usually longer than the leaves, glabrous, distally somewhat inflated. Heads ovoid, longer than broad, 3.7-5 mm diam., pale and glistening or somewhat darker towards the apex, very echinate from the acuminate floral bracts; involucral bracts pale, scarious, shining, incurved, ovate-lnceolate, acuminate, glabrous, longer or paler than floral. Floral bracts dark, glistening, narrowly obovate or obovate-spathulatee, sharply suddenly acuminate, glabrous, hyaline, concave. Male flowers few, terminal; sepals 2, oblong, acute; petals 3, subequal, obscure, not produced beyond the floral bracts; stamens 6, anthers black. Female flowers numerous, outermost shortly pedicelled; sepals 2, very narrow, dark-winged down the back; petals 0; receptacle columnar, glabrous. Seeds smoth, yellow.

Motijharan, Sambalpur, commonly in muddy ground near spring (Mooney, *loc. cit.*).

Motijharan (SBP), *HFM* 2987 (DD).

INDIA: Odisha.

WORLD: Japan, Malaysia, Myanmar, Philippines, Thailand, Vietnam.

Eriocaulon edwardii Fyson, J. Indian Bot. 2: 313. 1921; Haines, Bot. Bihar & Orissa 3: 1070. 1924; Mooney, Suppl. Bot. Bihar & Orissa. 198. 1950.

Erect herb, 4.5-6 cm high. Leaves setaceous, 1-5 cm long, up to 6 mm broad at the base, apex acute, glabrous. Peduncle 6-20 cm long, slender; heads white, 3-5 mm diam., obconic, but finally globose, involucral bracts not reflexed 4 mm llong, scarious, glabrous, elliptic, acute, a little longer than the floral bracts; floral bracts obovte acute, grey, glabrous or sometimes minutely hairy; receptacle tall, villous. Male flowers: Sepals 2, connate into a spathe; petals 0; anthers 6, black. Female flower: Sepals 2, slender; petals 2, linear or 0.

Frequent in moist, sandy places near streams. Fl. & Fr.: Septeember – October.

Paniganda (GJM), *SX & MB* 1999 (RRL-B); Borapahad (SBP), *HFM* 3611 (DD); Sambalpur (SBP), *HFM* 2916 (DD); Kholgaon (SBP), *HFM* 2934 (DD).

INDIA: Sikkim, West Bengal, Bihar, Odisha, Madhya Pradesh. Endemic.

Eriocaulon heterolepis Steud., Syn. Pl. glumac. 2: 271. 1855. *E.dianae* Fyson, J. Indian Bot. 2: 259. 1921; Mooney, Supl. Bot. Bihar & Orissa, 266. 1950.

Submerged aquatic herb. Leaves flaccid, up to 17.5 cm x 6.2 mm. Peduncles several, slender, erect, up to 17.5 cm high, the head just emerging above water surface, sheaths lax, half as long as the leaves, obtuse. Heads hemispheric, 5-6.2 mm diam., involucral bracts acute or obtuse, straw-coloured; receptacle small, conical, pilose. Floral bracts obovate-cuneate, cuspidate, dark green, apex white papillose-hairy; corolla lobes white papillose-hairy at the tip with a dark gland; female sepals 3, free, dark, 2 narrowly cymbiform, white puberulous in the upper half, 1 flat, linear-spathulate, tip white hairy; petals 3, linear-oblanceolate, pilose, eglandular.

Near Borara, Sonabera, 700m in Odisha (Mooney *loc. cit.*). Fl.: October.

INDIA: Goa, Karnataka, Kerala, Maharashtra, Rajasthan, Odisha. Endemic.

Eriocaulon leucomelas Steud., Nomencl. Bot. ed. 2, 1: 585. 1840. *E. melaleucum* Mart., in Wall., Pl. Asiat. Rar. 3: 29. 1832; FBI 6: 574. 1893; Gamble 3: 1619. 1931; Saxena & Brahmam, Fl. Orissa 4: 2084. 1996.

Herb. Leaves linear, 1.2-7.5 cm long, apex acuminate, glabrous, 3-7-nerved, mouth of sheath truncate, or nearly so, rarely somewhat oblique, often lacerate, narrowly scarious. Peduncles 1-4, 2.5-20 cm high, glabrous, 5-8-ribbed, sheaths close, usually a little shorter than the leaves, glabrous, heads globose, black and densely snowy-white papillose-hairy, 4.5-8.8 mm diam.; involucral bracts broadly obovate, cuspidate, black, outermost nearly glabrous, inner with and inflexed and densely snowy papillose-hairy apex, receptacle conical or subglobose, glabrous. Male flowers: sepals 3, more or less united into a spathe, split down one side, obovate, concave, nearly black, apex densely white papillose-hairy; corolla lobes white papillose-hairy and with a black apical gland, one lobe larger and extruded beyond its floral bract. Female flowers: sepals free, elliptic to obovate, boat-shaped, acute, black, apex white-papillose-hairy; petals 3, linear, spathulate, pilose and with an apical black gland.

Occasional on sandy rocky bed of rivers in Eastern Ghats of Odisha (Panigrahi *et al.*, 1964). Fl.: February.

INDIA: Western Ghats.

Eriocaulon longicuspe Hook.f., Fl. Brit. India, 6: 573. 1893. *E. longicuspis* Hook.f., var. *polycephalum* (Hook.f.) Fyson, J. Indian. Bot. 2: 308. 1921; Gamble 3: 1618. 1931. *E. polycephalum* Hook.f., Fl. Brit. India 6: 573. 1893; Hajra in Sharma & Gupta, Glimp. Pl. Sci. 13: 1986; Matthew, Fl. Tamilnadu Carnatic 3: 1720.1983.

Scapigerous herb. Leaves oblong-lanceolate, to 10 x 0.4 cm, apex acute or acuminate; sheaths close, as long as the leaves, acute, glabrous, somewhat lacerate, mouth distinctly oblique. Heads globose, to 1 cm across; involucral bract obovate, glabrous, floral bract hyaline, obovate-cuneate, ciliate, cuspidate, receptacle globose, sparsely villous. Male florets: to 2 mm long, calyx spathaceous, 3-partite, apically ciliate; petals subequal, ciliate, glandular or not; stamens dark. Female florets: to 3 mm long, sepal 3, free, oblanceolate, ciliate, apically glandular. Seed ellipsoid, pale pink, striate.

Occasional on hills above 1400m.

INDIA: Central India, Peninsular India.

WORLD: Sri Lanka.

Eriocaulon luzulifolium Mart. in Wall., Pl. Asiat. Rar. 3: 28. 1832; FBI 6: 582. 1893.

A tufted, annual herb; leaves linear-lanceolate, to 8 x 0.3 cm, narrowed from the base to the tip, subacute, glabrous, flat, translucent, 10-12-nerved. Heads 5 mm across, hemispheric, hard, pale grey, pubescent; peduncles many, erect, glabrous, 6-25 cm long; sheaths 2.5-5 cm long, obliquely split, at length lacerate at the mouth, glabrous; involucral bracts oblong-obovate, obtuse, glabrous; floral bracts 2 mm long, spathulate-oblong, subacute, pubescent at the apex, receptacle conical or columnar, villous with long hairs. Male floret: sepals 3, oblong-obovate, concave, ciliate at the tips, free or two connate; petals 3, minute, equal without gland; stamens 3 + 3, anthers black. Female floret: sepals, linear-oblong, falcate, concave, acute, dorsally tipped with bristly hairs; petals 3, linear-lanceolate, ciliate without a gland; ovary subsessile, 3-lobed, style-branches 3, filiform. Seeds ellipsoid, orange-yellow, faintly ribbed.

Occasional in marshy localities in Anantagiri in Araku valley. Fl. & Fr.: September – December.

Anantagiri (VSKP), *GVS* 21758 (MH).

INDIA: Throughout India.

WORLD: China, Thailand.

Eriocaulon odoratum Dalzell, Hooker's J. Bot. Kew. Gard. Misc. 3: 280. 1851; FBI 6: 574. 1893; Fischer 3: 1618. 1931; Ansari & N.P.Balakr., Eriocaul. India 38. 2009. *E. collinum* Hook.f., Fl. Brit. India 6: 584. 1891; Fischer 3: 1620. 1931.

A tufted, annual herb. Leaves lanceolate, 6 x 0.3 cm, 7-nerved, apex acuminate. Heads subglobose, 6 mm across, grey; peduncles glabrous, 5-6-ribbed, slender, involucral bracts broadly obovate, rounded, glabrous, straw-coloured or pale brown; floral bracts obovate-cuneate, rounded or acuminate, yellowish-brown; sparsely hairy, receptacle villous. Male floret: sepals 3, free or connate, subequal; petals subequal, one much enlarged, exerted over floral bract, apically glandular;

stamens 3 + 3, dark. Female floret: sepals 2 or 3, dark, free, keeled, obovate, ciliate; petals 3, free, oblanceolate, ciliate, glandular; ovary 3-lobed, style branches 3, long filiform. Seed orange-yellow.

Occasional in marshy places. Fl. & Fr.: September – December.

Badanapalli (EG), *SS* 784 (AU); Gopalarayudupeta (VZN), *MV* 2136 (AU); Araku (VSKP), *NPBK* 10859 (MH); Kakkashola, Shevaroys (SLM), *EV* 77730 (MH); Brooklyn – Yercaud (SLM), *KS* 6577 (MH); Kaveri peak – Yercaud (SLM), *DBD* 31293 (MH);

INDIA: Andhra Pradesh, Karnataka, Kerala, Maharashtra, Tamil Nadu.

WORLD: Sri Lanka.

Eriocaulon oryzetorum Mart., Pl. Asiat. Rar. 3: 28. 1932; Haines, Bot. Bihar & Orissa 3: 1069. 1932; FBI 5: 579. 1893.

Acaulescent herbs; rootstock absent. Leaves rosulate, oblong-lanceolate, gradually narrowing towards tip, obtuse or truncate at apex, up to 12 x 0.8 cm, glabrous. Peduncles 1-3, rigid, up to 30 cm long; sheaths to 10 cm long, glabrous, obtuse, entire. Heads hemispherical, *c.* 7 mm across, straw-coloured. Receptacle ovoid, villous, involucral bracts obovate, obtuse, 3 x 2 mm, spreading, chartaceous, glabrous. Floral bracts cuneate, acuminate or cuspidate, 3.2 x 1.5 mm, chartaceous. Male flowers pedicellate; sepals obovate, connate into a spathe, 3-lobed, sparsely hoary towards apex, black; petals 3, unequal, lateral one minute, the odd one linear, up to 0.5 mm long, eglandulose; anthers 6, oblong, black. Female flowers pedicellate; sepals 3, free, oblanceolate, 2 mm long, lateral ones conduplicate, middle one flat, not keeled, acute or acuminate; petals 3, subequal, spathulate, 2 mm long, stipitate between sepals and petals; ovary stalked, ovoid; style 3-fid. Seeds oblong-ovoid or globose, obtuse or apiculate, 0.53 x 0.42 mm, purple.

Occasional in Keojhar. Fl. & Fr.: September – November.

Garjantoli (KJR), *HFM* 2803 (DD).

INDIA: Assam, Bihar, Himachal Pradesh, Madhya Pradesh, Meghalaya, Sikkim, Uttar Pradesh.

WORLD: Myanmar, Nepal, Singapore.

Eriocaulon quinquangulare L., Sp. Pl. 87. 1753; FBI 6: 582. 1893; Fischer 3: 1620. 1931; Ansari & N.P.Balakr., Eriocaul. India 97. 2009.

A tufted annual herb. Leaves lanceolate, purplish red, 3-5 x 0.2-0.3 cm, 7-nerved. Head globose, snow white, 6 mm across; peduncles numerous, twisted, 7-ribbed, puberulous; sheaths lax, shorter than the leaves, often purplish, obtuse; involucral bract obovate, obtuse, glabrous; floral bract hyaline, obovate, cuneate, ciliate, cuspidate; receptacles globose, sparsely villous. Male floret: calyx lobes oblong, apex white papillose hoary, petals subequal, ciliate, glandular or not; stamens 3 + 3, dark. Female floret: sepals 3, free, oblanceolate, densely ciliate; petals 3, subequal, pilose, apex with a black gland; ovary stipitate or subsessile, 3-lobed, style branches 3, filiform. Seed oblong, ellipsoid, pale brown, smooth or striate.

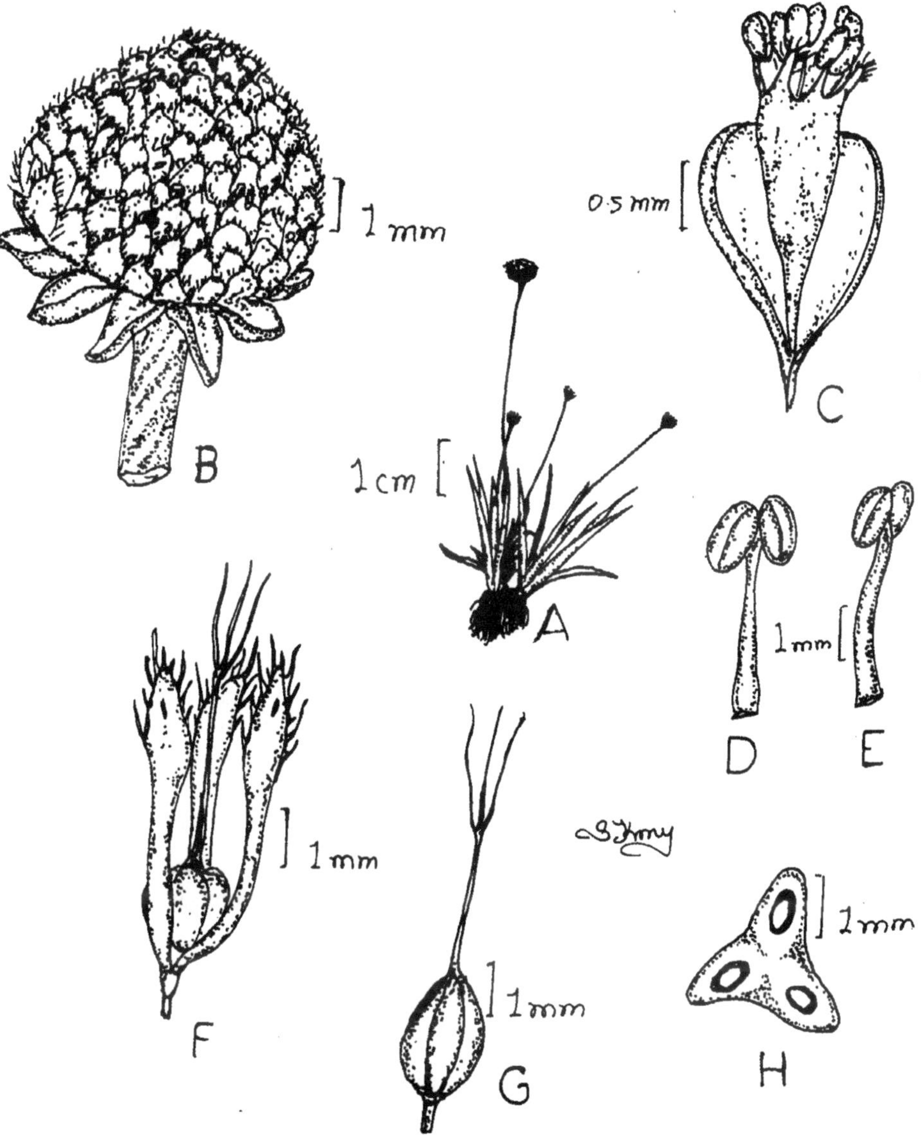

Figure 35. **Eriocaulon quinquangulare** L.
A. Habit; B. Head; C. Male flower; D. Female flower; G. Pistil; H. Ovary c.s.

Occasional in marshy localities. Fl. & Fr.: August – December.

Garugudukona (ATP), *TP* 955; Kalasamudram RF (ATP), *NY* 1187; Chelama (KNL), *KNR* & *DAM* 10630; Rollamadugu (KDP), *BR* & *BSS* 30303; Guvvalacheruvu RF (KDP), *R.V. Reddy* 8146; Talakona (CTR), *KNR* 9847, *GVS* 46958 (MH); Macherla RF (GNT), *VRK* 3291; Kambakkam (CTR), *MCB* 45177 (MH); Horsley hills (CTR), *DRC* 1116A (MH); Srisailam (KNL), *JLE* 16895 (MH); Balapalle (KDP), *JLE* 15743 (MH); Devaramadugula, near Addatigala (EG), *GVS* 68577 (MH & CAL); Tanjavanam (VSKP), *GVS* 42624 (MH); Araku (VSKP), *NPBK* 10859 (MH); Nellipaka RF (KMM), *RCS* 99084 (BSID); Baghmundo, Satkosia Tiger Reserve, *KCM* 7016 (BSID); Katrang, Satkosia Tiger Reserve, *KCM* 6794 (BSID); Noana, Similipal (MBJ), *sine. coll.* 12299 (BSID); Madanancheri (NA), *MBV* 761 (MH); Veerambakkam (NA), *MBV* 706 (MH); Karai RF – Gingee (SA), *KRM* 52856 (MH); Arambakkam (CPT), *DN* 220 (MH); Shevaroy hill slopes, Yercaud (SLM), *AVNR* 26759 (MH); Tickepalli to Linepada (GJM), *CAB* 1215 (MH); Krishnagiri (KGR), *S.India Flora* 13888 (MH); on the way to Kalchipadi (DMP), *TRS* 95556 (MH).

INDIA: Throughout India.

WORLD: Sri Lanka, Nepal, Myanmar.

Eriocaulon richardianum (Fyson) R.Ansari & N.P.Balakr., Eriocaul. India 51. 1994 & Eriocaula. India (Rev.). 46. 2009. *E. dianae* Fyson var. *richardiana* Fyson, J. Indian Bot. Soc. 2: 260. t. 14. 1921; Fischer 3: 1620. 1931.

A tufted annual herb; leaves oblong-lanceolate, to 12 x 0.7 cm, 7-nerved, apex acute. Head subglobose, 7 mm across, grey; peduncles many, striate, 6-20 cm long; sheaths lax, half as long as the leaves, apex obtuse; involucral bract lanceolate or ovate, glabrous; floral bract oblanceolate – cuneate, apically ciliate; receptacle conical, villous. Male floret; 1.8 mm long, calyx spathaceous, split down one side, lobes 3, small, rounded, apex white papillose-hairy, petals 3, ciliate, apically glandular; stamens 3 + 3. Female floret: 2 mm long; sepals 3, free, dark, subequal, 1.5-2 mm, petals 3, free, oblanceolate, keeled, 1.8 mm; eglandular; ovary 3-lobed, style branches 3, filiform. Seeds transversely ellipsoid.

Occasional in marshy localities. Fl. & fr.: August – November.

Palakonda hills (KDP), *CS* & *DAM* 7680.

INDIA: Kerala, Karnataka, Andhra Pradesh. Endemic.

Eriocaulon ritchieanum Ruhl. in Engl., Pflanzenr. 13: 73. 1903; Fischer 3: 1618. 1921; Ansari & N.P.Balakr., Eriocaul. India 161. 2009. *E. horsley-kondae* Fyson, J. Indian Bot. Soc. 3: 13. 1924.

A tufted annual herb. Leaves linear, acuminate, 2.5-7.5 x 0.2-0.3 cm, 4-7-nerved. Heads subglobose, 5 mm across, black, the apex with white indumentum; sheaths close, usually a little longer than the leaves, acute, often divided into 2-5 segments; involucral bracts suborbicular or obovate-oblong, rounded; floral bracts obovate-lanceolate, boat-shaped, acute or cuspidate, white papillose-hairy at the apex on the back; receptacle conical or sub-hemispheric, glabrous. Male floret: calyx-lobes spathaceous, split down one side, obscurely 3-lobed, blackish upwards, apex more or less densely white papillose; petals white papillose-hairy and with an apical black gland; stamens 3 + 3. Female floret: 2 mm long; sepals 3, free, oblong or obovate, boat-shaped, greenish black, truncate or rounded, denticulate, apex white papillose hairy; petals 3, linear-spathulate, longer than the sepals, pilose with a large apical black gland; ovary 3-lobed. Seeds sub-globose, brown, angular.

Rare in marshy localities in Horsley hills and Simlipal forests. Fl. & Fr.: August – December.

Horsley hills (CTR), *JSG* 20985 (CAL), *CECF* 4439 (CAL), *KP* 6484 (BSID; Simlipahar (MBJ), *SX* & *MB* 4445, 4991, 5053, 5546 (RRL-B).

INDIA: Odisha, Andhra Pradesh, Madhya Pradessh, Maharashtra, Gujarat, Karnataka, Tamil Nadu. Endemic.

Eriocaulon setaceum L., Sp. Pl. 129. 1753; FBI 6: 572. 1893; Fischer 3: 1118. 1931; Hajra in Sharma & Gupta, Glimp. Pl. Sci. 74. 1986. *E. capillus-naiadis* Hook.f., Fl. Brit. India 6: 572. 1893.

Submerged aquatic herb; stems clothed with short capillary leaves, densely disposed like the hairs of a tail. Leaves very slender, 3.7-5 cm long, 1-nerved. Tip of the stem floating near the water surface and bearing very many, slender, umbellately fascicled peduncles. Peduncles 3.7-7.5 cm long, 6-ribbed, exserted from the water, sheaths 1.8-2.5 cm long, with membranous tip. Heads small, *c.* 2.5 mm diam., pale grey or nearly black; receptacles convex or conical, villous; involucral bracts very small, oblong or cuneate, glabrous; floral bracts cymbiform, all or outer glabrous, coriaceous and black or membranous and whitish and the inner with long white dorsal hairs. Male florests: sepals usually connate into a split spathe; corolla-tube with 3 minute lobes with white hairs; stamens 6, anthers black. Female florets: sepals obovate-cymbiform, hairy on the back at apex; petals 3, narrow, ciliate, usually with a black gland. Seeds oblong smooth, dark brown.

Eastern Ghats of Odisha (Hajra *loc. cit.*). Fl.: August – September.

INDIA: Bihar, Gujarat, Karnataka, Maharashtra, Meghalaya, Odisha, Uttar Pradesh, West Bengal, Kerala.

WORLD: Sri Lanka.

Eriocaulon sollyanum Royle, Ill. Bot. Himal. Mts. 409. T. 97. F.1. 1839. *E. trilobum* Buch.-Ham. ex Körn, Linnaea, 27: 645. 1856; FBI 6: 583. 1893; Haines, Bot. Bihar & Orissa. 3: 1068. 1924; Saxena & Brahmam, Fl. Orissa 2087. 1996.

Small herbs, 6-20 cm high. Leaves linear or ensiform, mostly erect, 5-6.2 cm long, many-nerved, fenestrate. Heads small, globose, 2.5-5 mm diam., grey or dark coloured, involucral bracts oblong or obovate, obtuse, nearly glabrous, usually excceding the radius of the head, spreading or reflexed, floral bracts obcuneate, with triangular, obtuse tip or suddenly acuminate, triangular tip with short white hairs at the back; receptacle globose, villous. Male floret: sepals more or less connate with rounded hairy tips; petals not longer than the stamens and almost concealed by the hairs, usually with black gland; anthers black. Female floret: sepals oblanceolate, concave and more or less keeled at the tip, with a few white hairs; petals linear, slightly broader upwards, with a few white hairs. Seeds amber-coloured with close minute lines.

Occasional in Eastern Ghats of Odisha. Fl. & Fr.: October – May.

Joranda, Similipahar (MBJ), *SX* & *MB* 4373 (RRL-B); Chandragiri (KHD), *HFM* 1722 (DD).

Note: Ansari and Balakrishnan (1994) consider *E. sollyanum* as a *nomen nudum*, and give the name for this species as *E. trilobum* Körn. Contrary to this statement, *E. sollyanum* is validly published on the coloured Plate 97 of the work of Royle (Royle, Ill. Bot. Himal. Mts.: tab. 97, fig. 1. 1840) (Stearn, 1943). This illustration with an analysis conforms to the ICN.

Eriocaulon thwaitesii Körn., Linnaea, 27: 627. 1856; FBI 5: 583. 1893. *E. mariae* Fyson, Fl. Pulney Hill Tops. 331. 1914. *E. gamblei* C.E.C.Fischer, Fl. Pres. Madras 3: 160. 1930.

Tufted marshy, acaulescent herbs. Root stock absent. Leaves oblong-lanceolate, 5-9 x 0.2-0.4 cm, acute. Head to 6 mm, globose; involucral bract obovate-oblong, glabrous; floral bract narrowly oblanceolate, margin ciliate, apex obtusely caudate; receptacle densely villous. Male floret to 2 mm long; sepals 2, occasionally 3, free, oblanceolate, to 1.5 mm, glabrous; petals 3, ciliate, glandular; stamens dark. Female floret to 2 mm long; sepals 2, sometimes 3, dark, ciliate; petals 3, minute, with apical tuft of long pilose haris. Seed ellipsoid, transeversely striate.

Common in slopes and plateaus of Eastern Ghats of Odisha and Sirumalai and Kolli hills in Tamil Nadu above 1200 m altitude. Fl. & Fr.: October – January.

Ooradi, Sirumalai hills (DGL), *SKS* 924 (SGH).

INDIA: Kerala, Tamil Nadu.

WORLD: Nepal, South East Asia, Sri Lanka.

Eriocaulon truncatum Buch.-Ham. in Wall., Pl. Asiat. Rar. 3: 29. 1832; FBI 6: 578. 1893; Fischer 3: 1619. 1931; Ansari & N.P.Balakr., Eriocaul. India 110. 2009. *E. truncatum* var. *malaccense* Hook.f., Fl. Brit. India 6: 578. 1893.

A tufted annual herb; leaves oblong-lanceolate, to 6 x 0.25 cm, 7-nerved, acute. Head 4 mm across; peduncles many, shallowly 5-8-ribbed, 5-20 cm; involucral bract oblong-obovate, 2.5 mm, glabrous, apex obtuse, pale-straw-coloured; floral bract obovate, concave, glabrous, receptacle narrowly conic, glabrous. Male floret: sepals 2, dark, obovate; petals 3, glabrous; stamens 3 + 3, dark. Female floret: sepals 2, free, linear-oblanceolate; petals 3, narrowly oblanceoate, glabrous, with or without an apical gland; ovary 3-lobed; style branches 3, filiform. Seed ellipsoid, brown, longitudinally ribbed.

Occasional in marshy localities. Fl. & Fr.: August – December.

Kalasamudram (ATP), *KRKS* 40106; Guvvalacheruvu (KDP), *R.V. Reddy* 7997; Srisailam (KNL), *BSN* 4355; Kailasakona (CTR), *GVS* 32022 (MH); Near Akkagarigudi (CTR), *GVS* 45949 (MH); Talakona Water fall (CTR), *LR* & *M.S.Rao* 2558 (BSID); Kamayyapalem – Polavaram Agency (WG), *DCSR* 272 (CAL); Gaddada (EG), *MM* 102636 (BSID); Araku valley (VSKP), *NPBK* 520 (CAL); Cherukonda (VSKP), *GVS* 42741 (MH); Thakurmunda (MBJ), *SX* & *MB* 5389 (RRL-B); Motijharan (SBP), *HFM* 2911 (DD); Barapahad (SBP), *HFM* 2623 (DD).

INDIA: Peninsular to E. & NE. India.

WORLD: Sri Lanka, Maynmar, China, Malay Isles, Philippines.

Eriocaulon xeranthemum Mart. in Wall., Pl. Asiat. Rar. 3: 29. 1823; FBI 6: 584. 1893; Fischer 3: 1610. 1931; Ansari & N.P.Balakr., Eriocaulaceae in India 54. 2009.

A tufted annual herb. Leaves caespitose, 2-4 x 0.25-0.4 cm, linear-lanceolate, acute, flat, glabrous, translucent, 7-11-nerved. Heads globose, sheaths reaching 2.5 cm long, obliquely split, glabrous; involucral bracts longer than the flowering head, lanceolate, glabrous, pale yellow or nearly white; floral bracts oblong-obovate, hairy at the apex, receptacle glabrous. Male florets: obovate, hairy at the apex, petals minute, glabrous; stamens 3 + 3, yellow. Female floret: 2 mm long, subsessile; sepals 2 (rarely 3, the third capillary), unequal, lanceolate, hairy at the tip; petals 1.6 mm long, linear or oblanceolate, hairy at the tip and with a minute gland, yellow; ovary sessile, 3-lobed, style-branches 3, filiform; seeds narrowly oblong, yellow, papillose.

Occasional in marshy localities. Fl. & Fr.: August – November.

Similipahar (MBJ), *SX* & *MB* 4791 (RRL-B); Garjantolia (KJR), *HFM* 2802 (DD); Barapahad (SBP), *HFM* 3626 (DD).

INDIA: Andaman Islands, Andhra Pradesh, Assam, Bihar, Goa, Kerala, Karnataka, Madhya Pradesh, Maharashtra, Meghalaya, Manipur, Tamil Nadu, West Bengal.

WORLD: Tropical Africa, central Himalayas, Bangladesh, Indonesia, Myanmar, Nepal and Thailand.

CYPERACEAE

1. Florets unisexual:
    2. Nut enclosed in a utricle.. **Carex**
    2. Nut not enclosed in a utricle:
        3. Nut enveloped by two glumes and falling together with them.. **Diplacrum**
        3. Nut not enveloped by two glumes .... **Scleria**
1. Florets bisexual:
    4. Glumes on spikelets distichous:
        5. Stigmas 2; nutlets 2-sided with 1 margin facing spikelet axis:

6. Spikelets with more than 2 glumes; spikelets axis and glumes persistent.. **Pycreus**
6. Spikelets with 1 or 2 glumes; spikelet axis deciduous, spikelets falling whole:
    7. Spike sessile; lower glumes bisexual, upper male/sterile..**Kyllinga**
    7. Spike stalked, all flowers bisexual.......................... **Queenslandiella**

5. Stigmas 3, rarely 2; nutlets trigonous, rarely bixconvex with 1 side facing spikelet axis:
    8. Annual or perennial; glumes not winged.................................. **Cyperus**
    8. Annual; glumes prominently winged ....**Courtoisina**

4. Glumes on spikelets spiral:
    9. Hypogynous scales present ....**Lipocarpha**
    9. Hypogynous scales absent:
        10. Style base dilated and constricted or articulated above the nut:
            11. Hypogynous bristles present, nut beaked:
                12. Leaves absent, style persistent ....**Eleocharis**
                12. Leaves present, style not persistent...................... **Rhynchospora**
            11. Hypogynous bristles absent, but not beaked:
                13. Style base persistent; if falling not leaving a tumour on the nut.. **Fimbristylis**
                13. Style deciduous leaving a tumour on the nut...........**Bulbostylis**
        10. Style base not dilated, continuous with the nut:
            14. Nut tightly enclosed in a corky thickened rhachilla .... **Remirea**
            14. Nut not enclosed in a corky thickened rhachilla:
                15. Leaves reduced to sheaths (except *S. maritimus*):
                    16. One involucral bract over 1.5 cm, longest involucral bract erect, culmlike and apparently continuous with culm; inflorescence appearing lateral:
                        17. Plants colonial from elongate rhizomes, perennial, usually more than 0.8 m high; hypogynous bristles present ....**Schoenoplectus**
                        17. Plants caespitose, annual, less than 0.8 m high; hypogynous bristles absent.. **Schoenoplectiella**
                    16. Atleast 2 involucral bracts over 1.5 cm, longest involucral bract lealflike, erect or spreading...**Bolboschoenus**

15. Leaves well developed:

    18. Bract leaves equal or slightly longer than the inflorescence..**Fuirena**

    18. Bract leaves three times longer than the inflorescence... **Actinoscirpus**

ACTINOSCIRPUS (Ohwi) Haines & Lye

Actinoscirpus grossus (L.f.) Goetgh. & D.A. Simpson, Kew Bull. 46: 171. 1991. *Scirpus grossus* L.f., Suppl. Pl. 104. 1782; FBI 6: 659. 1893; Fischer 3: 1666. 1931. *Schoenoplectus grossus* (L.f.) Palla, Allg. Bot. Z. Syst. 17: Beibl. 3: 1911; Saxena & Brahmam, Fl. Orissa 4: 2204. 1996; Pullaiah & Hanumanthappa in Pullaiah, Fl. Andhra Pradesh 3: 1107. 1997.

Perennial, stoloniferous herb; stolons producing tubers. Culms to 200 cm tall, sharply angled, faces concave, septate-nodose, with a corm like hard enlargement at base. Leaves equaling the culm, 1-ribbed, flattish-plicate, septate-nodose, sheaths spongy, prominently septate-nodose, tightly surrounding the culm base. Inflorescence a large terminal corymb with primary, secondary and tertiary rays; rather dense, to 12 cm long; involucral bracts 4, foliaceous, exceeding the inflorescence, the longest 40 cm long; spikelets solitary, ellipsoid, to 10 mm long, densely manny-flowered, light brownish, rust-coloured. Achenes broadly obovate, subcompressed-triantular, stramineous brown.

Rare in marshy fields in Southern Eastern Ghats. Fl. & Fr. Agusust – December.

Kalasamudram (ATP), *KRKS* 39139; Near Kanigiri reservoir (NLR), *MCK* 22885; Palur (SA), *VNS* 4136 (MH);

INDIA: Throughout India.

WORLD: S.C.Asia, S.China, Malesia, Tropical Australia.

BOLBOSCHOENUS (Ascherson) Palla

Bolboschoenus maritimus (L.) Palla in Koch., Syn. Deutsch. Fl. ed. 3. 2532. 1904. subsp. **maritimus**. *Scirpus maritimus* L., Sp. Pl. 75. 1753; FBI 6: 658. 1893; Fischer 3: 1666. 1931; Saxena & Brahmam, Fl. Orissa 4: 2208. 1996. *Schoenoplectus maritimus* (L.) Lye, Blyttia 29: 145. 1971; Pullaiah & Hanumanthappa in Pullaiah, Fl. Andhra Pradesh 3: 1109. 1997. *Scirpus tuberosus* Desf., Fl. Atlant. 1: 50. 1798; Verma & Veenachandra, Rec. Bot. Surv. India 21: 263. 1981; Karthikeyan & *al.*, Florae Indicae Enumeratio Monocotyledonae 71. 1989.

Perennial, erect, rhizomatous creeping herb with tubers, to 75 cm high; culms triquetrous. Leaves flat, linear, to 30 cm long, acuminate, ligulate, sheaths long. Inflorescence terminal umbel, rayed, unequal. Involucral bracts 3-5, leaflike. Spikelets ovoid to cylindric, acute; glumes pale straw-coloured, 5-7 mm long, nearly flat, with a strong excurrent midrib produced into an awn, puberulous, margins membranous; stigmas 2. Achenes obovate, trigonous, smooth, to 4 mm long.

Occasaional in moist waste lands. Fl. & Fr.: October – March.

Kalasamudram (ATP), *NY* 1181 (SKU & MH), *MHR* 14921; Chintalapally Road (ATP), *KH* 7474; Saidapuram (NLR), *PMR* 23067; Rayachoty (KDP), *KS* 7812 (MH); Nagarjunakonda valley (GNT), *KT* 9817 (CAL); Mylavaram (KSN), *CAB* 8164 (MH); Rathamhutta hills (KMM), *RCS* 99054 (BSID); Narsingpur (CTK), *HFM* 1682 (DD); Sitalpani, Satkosia Tiger Reserve, *KCM* 8376 (BSID).

INDIA: Throughout India except NE. India.

WORLD: Cosmopolitan.

BULBOSTYLIS Kunth *nom. cons.*

1. Inflorescence capitate; achenes smooth.
   2. Stems thick, rigid, curved, shallowly grooved, up to 10 cm long .... **B. subspinescens**
   2. Stems slender, striate, up to 30 cm long .... **B. barbata**
1. Inflorescence of loose or crowded anthelas; achenes transversely wrinkled:
   3. Glumes glabrous, muticuous .... **B. densa**
   3. Glumes pubescent, mucronate .... **B. puberula**

Bulbostylis barbata (Rottb.) Kunth ex C.B. Clarke in Hook. f., Fl. Brit. India 6: 651. 1893. *Scirpus barbata* Rottb., Progr. 27. 1772. *Stenophyllus barbata* (Rottb.) Cooke, Fl. Pres. Bombay 2: 887. 1908; Fischer 3: 1662. 1931.

1. Bracts as long as or shorter than the heads; glumes 2–2.2 mm with a short, straight mucro at apex .... subsp. **barbata**
1. Bracts much longer than the heads; glumes 3 mm with a recurved mucro at apex .... subsp. **pulchella**

Bulbostylis barbata (Rottb.) Kunth ex C.B. Clarke subsp. **barbata**

Annual, densely tufted erect herb, to 30 cm high, culms slender, striate, leaved at base. Leaf blades filiform, 4 to 8 cm long, sheaths membranous, brownish. Inflorescence a terminal head of 2 -20 sessile spikelets, 5-15 mm across, brownish. Involucral bracts 3, spreading, to 2 cm long, setaceous. Spikelets sessile, lanceolate, 3-8 mm long, angular, subdensely 7 to 15-flowered, pale brownish; glumes ovate, to 2 mm long with recurved short awn, membranous, yellowish-green and becoming rusty brown, prominently keeled, ciliolate; stamen 1; style base depressed-globular, style filiform, stigmas 3. Achenes obovate-orbicular, trigonous, pale rounded at apex.

This is a very common species found everywhere. Usually seen in open dry especially sandy areas, rocky surfaces, waste lands, road sides, and as a weed in cultivated lands. Fl. & Fr.: August - December.

Gani RF (KNL), *RVR* 1508; Guvvalacheruvu (KDP), *RVR* 7846, *KS* 6402 (MH); Tirumala (CTR), *MHR* 13360; Vijayapuri south (GNT), *VRK* 5820; Mahanandi (KNL), *JLE* 25482 (MH), *VBH* 86676 (BSID); Seshachlam RF (KDP), *A.M.Rao* 324

(BSID); Nagapatla RF (CTR), *KS* 6856 (MH); Emmalur hills (NLR), *P.Venu* 110111 (BSID); Udayagiri (NLR), *V.V.Seshagiri Rao* 4496 (BSID); Diguvametta (PKM), *JLE* 32544 (MH); Nagarjunakonda (GNT), *J.Swamy & S. Nagaraju* 7254 (BSID); Kondapalle (KSN), *CAB* 7979 (MH); Rampachodavaram (EG), *MM* 102652 (BSID); Near Patel river, Araku (VSKP), *NPBK* 10860 (MH); Nattavaram (VSKP), *CAB* 1834 (MH); Sithanur (CPT), *S.India Flora* 11218 (MH); Parangipettai (SA), *KRM* 64147 (MH); Komatiyur (NA), *KS* 7476 (MH); Tippukadu RF (NA), *KRM* 16601 (MH); Hogainakkal (KGR), *EV* 20513, 20514 (MH); Kallar river (SLM), *R. Ansari* 50013 (MH); Pampasar RF, Satkosia Tiger Reserve, *KCM* 5116 (BSID); Satkosia Wild Life sanctuary, *D.Hazra & D.Das* 18111 (BSID).

INDIA: Throughut India.

WORLD: Widespread in the Old World Tropics and subtropics from N.Africa to China and Japan, south to Indonesia and Australia.

Bulbostylis barbata (Rottb.) Kunth ex C.B. Clarke subsp. **pulchella** (Thw.) T. Koyama, Bot. Mag. (Tokyo) 93: 341. 1980. *Isolepis pulchella* Thw., Enum Pl. Zeyl. 350. 1864. *Bulbostylis barbata* Kunth ex C.B. Clarke var. *pulchella* (Thw.) C.B.Clarke in Hook. f., Fl. Brit. India 6: 652. 1893.

Plants more rigid than subsp. *barbata*. Head 10-15 mm across, more densely bearing numerous spikelets. Glumes 3 mm long, wholly subdensely pilose with often tubercle-based brownish hairs, the midrib thick, not clearly nerved, projecting beyond the glume apex forming a recurved awn-like cusp.

Not common. Fl. & Fr.: August – November.

Palakonda hills (KDP), *CS* 9381; K.K.Kottala (KDP), *KRKS* 37663; Macherla (GNT), *VRK* 3524.

INDIA: South India.

WORLD: Sri Lanka, Vietnam.

Bulbostylis densa (Wall.) Hand.-Mazz. in Karsten & Schenk., Veg. 20. 7: 16. 1930. *Scirpus densa* Wall. in Roxb., Fl. Ind. 1: 231. 1820. *Bulbostylis capillaris* (L.) Kunth ex C.B.Clarke var. *trifida* (Nees) C.B.Clarke in Hook. f., Fl. Brit. India 6: 652. 1893. *Isolepis trifida* Nees in Wight, Contrib. Bot. India 108. 1834; Fischer 3: 1662. 1931.

Annual, tufted, glabrous, slender herb, to 20 cm high; culms capillary, leaved at base. Leaf blades filiform, to 10 cm long, light green; sheaths pale, membranous, loosely pilose with white hairs at orifice. Inflorescence simple to sub compound, occasionally contracted nearly head like, bearing 2 to 8 spikelets, to 4 cm long; rays 1-5, capillary, 1 to 5 cm long. Involucral bracts setaceous, to 1.5 cm long. Spikelets solitary, oblong-ovate, to 6 mm long, somewhat angular, *c*. 10-flowered, chestnut brown. Achenes obovate-orbicular, trigonous, pale and eventually becoming greyish, transversely wrinked, distinctly buttoned.

Common in open wet grassy places. Fl. & Fr.: September – December.

SEDS – Penukonda road (ATP), *BR* & *ANS* 35356; Batrepalli (ATP), *KRKS* 40332; Kongampallam, Yercaud (SLM), *SKK* 26849 (MH); Banigoccha, Satkosia Tiger Reserve, *KCM* 7099 (BSID); Katrang, Satkosia Tiger Reserve, *KCM* 6793 (BSID).

INDIA: Throughout India.

WORLD: Africa, Sri Lanka, China, Japan.

Note: This taxon can easily be recognized by its less tufted habit, capillary leaves and stems and the inflorescence with 3 spikelets.

Bulbostylis puberula (Poir.) Kunth ex C.B.Clarke in Hook. f., Fl. Brit. India 6: 652. 1893. *Scirpus puberulus* Poir. in Lam., Encycl. 6: 767. 1804. *Isolepis gracilis* Nees in Wight, Contr. Bot. India 109. 1834; Fischer 3: 1662. 1931.

Annual, tufted, slender erect herb, to 30 cm high. Leaves basal, much shorter than the culms; blades filiform, puberulent; sheaths pubescent, straw brownish. Inflorescence simple, to 1.5 cm long and as broad, open but crowded to almost headlike with up to 5 short rays and, to 15 spikelets. Involucral bracts 2-4, filiform, dilated at membranous base, the lowest to 4 cm long. Spikelets solitary, ovate-oblong, to 7 mm long, angular, to 14-flowered; glumes broadly ovate, to 2.5 cm long, apex with recurved mucro, strongly keeled on back, pale-greenish to light brown, densely pubescent; stamen 1; style base conical-globular, stigmas 3. Achenes broadly obovate, trigonous, to 1 mm long.

Usually thrives in sandy areas near river banks. Fl. & Fr.: October – December.

Gani RF (KNL), *RVR* 1523; Anakapalle (VSKP), *sine coll. s.n.* (MH).

INDIA: Peninsular India.

WORLD: Tropical Africa, Madagascar, Sri Lanka, Myanmar, China, Malaysia, Laos, Cambodia, Thailand, Vietnam.

Bulbostylis subspinescens C.B. Clarke in Hook. f., Fl. Brit. India 6: 652. 1893; Fischer 3: 1662. 1931.

Annual, tufted herb, to 10 cm high. Culms thick, hairy, rigid, curved, shallowly grooved. Leaves basal, about one-third to half the length of stem, numerous, channeled, hispid-puberulous; sheath very broad, striate. Spikelets 3.7-6.2 mm long, in dense, terminal, almost prickly, globose heads. Involucral bracts 3, as long as the head. Spikelets to 1.8 cm long, stellately spreading; glumes 3-3.5 mm long; keel brown (when dry), somewhat rounded, scarcely excurrent, sides scarious, finely pubescent; styles slender, *c.* 1.2 mm. Achenes broadly obovoid, obtusely trigonous, pale brown.

Occasional in sandy areas. Fl. & Fr.: August – December.

Visakhapatnam (VSKP), *KH* 10988; Guvvalachervu (KDP), *KS* 6402 (CAL); Araku (VSKP), *NPBK* 10860 (CAL); Srisailam (KNL), *JLE* 22086 (CAL).

INDIA: Odisha, Andhra Pradesh. Endemic.

CAREX Linnaeus

1. Inflorescence panicled in the axils of all or atleast the lower involucral bracts:
   2. Spikelets almost all 2.5 to 10 cm long:
      3. Utricles strongly inflated, broadly obovoid; to subglobose, red on maturity .... **C. baccans**
      3. Utricles not inflated, ellipsoid-oblong, pale to castaneous:
         4. Male spikelet equal to female; urticle equal to glume ......**C. myosurus**
         4. Male spikelete shorter than female; urticle not equal to glume .... **C. filicina**
   2. Spikelets almost all 0.4 to 1.5 cm long:
      5. Utricles glabrous; glumes straw or dirty white coloured .... **C. stramentitia**
      5. Utricles short hairy; glumes reddish brown .... **C. cruciata**
1. Inflorescence spicate with a solitary clusters in the axils of each involucral bract or spikelets solitary, terminal:
   6. Stigma 2; glumes mucronate:
      7. Spikelets 10-14 in a cluster, androgynous:
         8. Urticle planoconvex, glabrous .... **C. nubigena**
         8. Urticle trigonous, pubescent .... **C. lindleyana**
      7. Spikelets solitary, unisexual .... **C. phacota**
   6. Stigma 3; glumes obtuse:
      9. Leaves scattered, leaf-sheath pinkish .... **C. hebecarpa**
      9. Leaves subbasal, leaf-sheath not pinkish:
         10. Urticles not winged, glabrous, beak deeply bifid:
             11. Male and female flowers equal in part of spike .... **C. jackiana**
             11. Male part shorter than the female part of spike .... **C. rara**
         10. Urticles winged, ciliate, beak shortly bifid .... **C. speciosa**

Carex baccans Nees in Wight, Contr. Bot. India 122. 1834; FBI 6: 722. 1894; Fischer 3: 1687. 1931.

Perennial, loosely tufted, rhizomatous herb; culms to 110 cm high, acutely 3-angled, clothed at base with purple or greyish brown sheaths, slightly disintegrating into fibres. Leaves cauline, dense on the lower portion, leaf blades widely linear, upper ones overtopping the culms, to 9 mm wide, deeply green, coriaceous, stiff, 3-costate, flattish, scabrid on upper surface; sheaths to 15 cm long, tightly surrounding the culm internode, upper ones greenish, lower ones reddish brown, eventually spitting in to oblique fires on ventral side. Inflorescence a compound panicle, interrupted and contiguous above; partial panicles 6, elliptical, to 20 cm long, single at each node, further branched into *c.* 15 spikes; upper peduncles enclosed in the involucral bract sheath, the lower ones exserted. Lower involucral bracts leaf like, overtopping the inflorescence, the sheathing base to 5 cm long. Spikes suberect, cylindrical, to 5 cm long, androgynous; pistillate part subdensely many-flowered; male portion linear, dark red when young. Achenes loosely enveloped, short beaked at subdensely contracted apex, elliptic, to 3 mm long, acutely 3-angled, dark brown.

This is a rare species usually seen in wet grassy slopes in forests. Fl. & Fr.: July – August.

Anantagiri – Galikonda (VSKP), *TP & EC* 7399; Sunkarimetta, Galikonda (VSKP), *GVS* 21633 (MH); Minumuluru (VSKP), *GVS* 28276 (CAL); way to Gudem, Chintapalli (VSKP), *JLE* 37125 (MH).

INDIA: Peninsular & C. India, E. to C. Himalaya.

WORLD: Sri Lanka, S.China, Taiwan, Thailand, Indo-China, Malesia.

Carex cruciata Wahlenb., Vet. Akad. Handl. Stockh. 24: 149. 1803; FBI 6: 715. 1894.
C. candensata Nees in Wight, Contr. Bot. India 123. 1834; FBI 6: 716. 1894.

Perennial herb with a short, stout, woody rhizome which is clothed with fibrous remains of leaf-sheaths. Culms trigonous, smooth, glabrous. Leaves sub-basal, often 1-2 cauline, as long as or longer than stem, to 0.8 cm broad; sheaths trigonous. Panicle slender, compound, to 25 mm long with 8-10, solitary, 7 cm long secondary panicles. Involucral bracts foliaceous, becoming smaller upwards. Spikelets cylindrical, linear-oblong, to 1.5 cm long. Bracteoles glumaceous, aristate. Male florets: to 5; glumes ovate-oblong, aristate, reddish brown. Utricle inflated, ellipsoid, faintly trigonous, many-ribbed, with a hairy margin, thinly pilose, narrowed into a 2-fid beak. Female florets: glumes longer than the male glumes; shortly aristate.

Shady localities of ravines and forest edges in Northern Eastern Ghats. Fl. & Fr.: August – October.

Sunkarimetta (VSKP), *GVS* 19603 (MH & CAL); Purunakote, Satkosia Tiger Reserve, *KCM* 5257 (BSID).

INDIA: Peninsular, C. & NE. India, Himalaya, Andaman & Nicobar Islands.

WORLD: India through SE. Asia to S. Asia, Taiwan, the Ryukyu and Queensland.

Carex filicina Nees in Wight, Contrib. Bot. India 123. 1834; FBI 6: 717. 1894; Fischer 3: 1686. 1931; Mattew, Fl. Tamilnadu Carnatic 3: 1727. 1983.

Densely tufted herb, up to 90 cm high, culms trigonous. Leaves as long as or shorter than the culm, linear, 2-8 mm wide, persistent sheaths purplish red or blackish. Involucral bracts leafy, longer than inflorescence branches, single, rarely binate. Inflorescence a slender, interrupted or continous, decompound, fuscous panicle, up to 60 cm long. Spikelets numerous, bisexual, androgynous, oblong, male part shorter than female part, 3-7-flowered; female part 2-16-flowered; glumes brown or reeddish brown, ovate-lanceolate, apex acuminate; style slightly thickened, stigmas 3. Utricle triquetrous, not inflated. Nutlets yellowish brown at maturity, ellitpic, trigonous.

It is found growing on swamps and marshes, in shady areas of moist forests. Fl.: August.

Valamuru, Maredumilli (EG), *MM* 105052 (BSID); Shevaroy Bauxite hill – Yercaud SLM), *AVNR* 26776 (MH); Similipahar (MBJ), *SX* 3793 (RRL-B).

INDIA: Odisha, Andhra Pradesh, Karnataka, Goa, Kerala, Tamil Nadu.

WORLD: Sri Lanka, Bhutan, Nepal, China, Indonesia, Malaysia.

Carex hebecarpa C.A. Mey., Petersb. Mem, Sav. Entrang. 1: 223, t. 12. 1831; FBI 6: 747. 1894; Fischer 3: 1686. 1931. var. **ligulata** (Nees) Kuekenth. in Engl. Pflanzenr. 38: 745. 1909; Mattew, Fl. Tamilnadu Carnatic 3: 1728. 1983. *C. ligulata* Nees is Wight, Contrib. Bot. Ind. 127. 1834.

Perennial, very loosely tufted, rhizomatous herb, to 75 cm high; rhizome short, prostrate, knotty, covered tightly with reddish-brown scales; culms erect from ascending base, lower half clothed with bladeless sheaths and leaved only above the middle, the internodes triquetrous, becoming yellowish; basal sheaths bladeless, reddish-brown below. Leaves subdensely alternate, equaling to or surpassing the inflorescence; leaf blades linear-lanceolate, to 35 cm long, to 12 mm wide, deeply green above, whitish green beneath, roughened above, the sheathing portion to 5 cm long, sharply trigonous, greenish, ligule brown, membranous, to 1.5 cm long. Involucral bracts foliaceous, all surpassing the inflorescence, the lower 2-4 long sheathing, the upper 2-3 short sheathing. Spikes 6, terminal spike staminate, linear, to 2.5 cm long, lightly brown, almost equaling the uppermost pistillate spike; lateral spikes pistillate, cylindrical, to 4 cm long, subdensely many-flowered, glaucous green; peduncles to 3 cm long. Achenes tightly enclosed, elliptic, triquetrous, 3 mm long.

Occasional in wet forests as undergrowth. Fl. & Fr.: September – January.

Sunkarimetta to Galikonda (VSKP), *GVS* 19604 (MH); Orchidarium area – Yercaud (SLM), *AVNR* 18246 (MH);

INDIA: Central to E. Himalaya.

WORLD: Bhutan, Nepal, China.

Carex jackiana Boott, Proc. Linn. Soc. London 1: 260. 1845; FBI 6: 735. 1894; Fischer 3: 1686. 1931; Matthew, Fl. Tamilnadu Carnatic 3: 1729. 1983.

Herb, up to 50 cm high. Leaves several, to 30 cm high, to 5 mm wide. Spikes 3-5, distant, to 1.5 cm, terminal one linear. Male flowers: lateral ones ovoid or oblong, scape radical, to 10 cm; involucral bracts nearly equal to spikes. Famale flowers: glumes ovate, acute, aristate; style 3-fid. Utricle ellipsoid, lanceloate, glabrous, nerves prominent, beak elongate, deeply 2-fid. Nut obovoid, triquestrous.

In Tamil Nadu Carnatic (Matthew, 1983).

INDIA: Sri Lanka, Yunnan, Australia, Malesia.

Carex lindleyana Nees in Wight, Contrib. Bot. India 121. 1834; Fischer 3: 1687. 1931. *C. glaucina* Boeckeler, Linnaea 40: 353. 1876; Matthew, Fl. Tamilnadu Carnatic 3: 1728. 1983.

Eerect, rigid herb, to 1 m high. Leaves as along as stem, 0.5-1 cm wide, flat or canaliculate. Panicles oblong, distant, to 10 cm long, 2 or 3 at node, unequal, rachis appressed-pubescent; involucral bracts overtopping; spiklets 1-2.5 mm. Male flowers: at apex, rhachilla pubescent; glumes imbricate, ovate, to 2.5 mm, persistent, mucronate, sides purple or brown, obscurely nerved, puberulous, keel yellow, 3-nerved, acute. Utricle trigonous, to 3.5 mm, exceeding glume, ribbed, pubescent, beak 2-fid. Nut trigonous, to 2 mm.

Hills above 1200 m in Eastern Ghats of Tamil Nadu (Matthew, 1983).

INDIA: Peninsula.

Carex myosurus Nees in Wight, Contrib. Bot. Ind. 122. 1834; FBI 6: 723. 1894; Fischer 3: 1687. 1931; Mattew, Fl. Tamilnadu Carnatic 3: 1729. 1983.

Perennial, herb, to 90 cm high with a short, stout, woody rhizome which is clothed with persistent, fibrous remains of leafsheaths. Culms erect, obtusely trignous, minutely hairy on faces just a little below the sheath. Leaves subbasal, as long as or longer than the stem, to 1 cm broad. Panicle simple, to 50 cm long, with secondary panicles reduced to solitary spikes. Involucral bracts foliaceous. Spikelets oblong-cylindric, terete, 7 x 0.8 cm, male at apex, many-flowered. Utricle equal to glume, ellipsoid-oblong, ribbed, scabrid, beak short, 2-fid, pale to castaneous, stipitate; achene ellipsoid, trignous, to 2 mm long, castaneous.

Usually found along wet forest slopes, shady places of ravines and in mountain sides. Fl. & Fr.: September – December.

Ebul RF, north of Gudem (VSKP), *VNS* 586 (CAL).

INDIA: Thtoughut India.

WORLD: Myanmar, Nepal, China, Vietnam.

Carex nubigena D. Don., Trans. Linn. Soc. London 14: 326. 1824; FBI 6: 702. 1893; Fischer 3:1165. 1931; Mattew, Fl. Tamilnadu Carnatic 3: 1730. 1983. *C. nubigena* D.Don var. *fallax* Clarke, J. Linn. Soc. Bot. 36: 301. 1904.

Perennial herbs, to 70 cm high. Leaves to 20 cm, 2 mm wide. Inflorescence spicate; involural bracts 1-3, overtopping. Spikelets 10 in a cluster, ovioid-oblong, condensed, 3 x 0.5 cm, hardly interrupted, subsessile, 7-10-fid, apically with male flowers, basally with female flowers. Male glume ovate-lanceolate; keel nerved; stamens 2. Female glume ovate, 3-4 mm; keel nerved; style bifid. Urticle plano-convex, ellipsoid, 4 mm, exceeding glume; beak 2-fid, nearly equal in inflated base. Nut biconvex, ovoid, to 1.5 mm, yellow, gibbous on one side.

Occasional on stream banks in high altitudes of Sirumalai hills of Tamil Nadu. Fl. & Fr.: November – February.

Ooradi, Sirumalai hills (DGL), *SKS* 518 (SGH).

INDIA: Himalaya and Peninsualr India.

WORLD: Malaysia, Taiwan, Java, Philippines, Indonesia and Sri Lanka.

Carex phacota Spreng., Syst. Veg. 3: 826. 1826; FBI 6: 708. 1894; Fischer, 3: 1686. 1931; Matthew, Fl. Tamil Nadu Carnatic 3: 1730. 1983; Brahmam & Saxena, Fl. Orissa 2101. 1996.

Herb, stem erect, to 2 m high, to 4 mm wide, rigid, stout, striate-sulcate. Leaves subbasal or occasionally a few higher up, as long as the stem or shorter, to 7 mm wide, scaberulous. Inflorescence spicate involucral bracts overtopping; spikelet solitary, terete, elongate, drooping, terminal one female, lateral one male. Male flowers: glume obovate, 4 mm, retuse, awn antrorsely scabrid, sides purplish, nerveless; keel strong, 3-nerved; stamens 2 or 3, anthers oblong, to 2 mm. Female flowers: glume narrow, linear, to 3 mm, awn erect, stiff; keel strong, yellow; style 3-fid, 3 mm. Utricles biconvex, ellipsoid, plano-convex, to 2.5 mm, not exceeding glume, castaneous, densely scaberulous, shortly beaked, obscurely 2-fid or not. Nut plano-convex, to 2 mm, yellow, glossy, obtuse, cuneate.

Hills above 1400m, in marshy places, rising above the herbs in Eastern Ghats of Odisha and Tamil Nadu. Fl. & Fr.: March – April.

Opposite to Marapalam, Nagalur road – Yercaud (SLM), *AVNR* 27465 (MH); Similipahar (MBJ), *SX* 422 (RRL-B).

INDIA: Himalayas, Assam Nilgiri and Pulney hills.

WORLD: Sri Lanka, Thailand, China, Korea, Japan, Malesia.

Carex rara Boott, Proc. Linn. Soc. London 1: 284. 1845. subsp. **patanicola** T.Koyama, Bull. Natl. Sci. Mus. 17(1): 65. 1974; Veeranjaneyulu *et al.*, J. Econ. Taxon. Bot. 35: 652. 2011.

Perennials, densely tufted in large clumps; rhizome very short; culms slender, 15-45 cm high, acutely triquetrous, smooth, leaves only at base. Leaves longer than the culm, crescentiform in transverse section, smooth-margined, abaxially 1-costate, tip subabruptly sub-acute; sheaths 1-8 cm long, membranous. Inflorescence a single terminal spike, without bract; spike androgynous, erect, oblong-cylindrical, 0.7-2 x 3-3.5 cm; staminate part linear, shorter than the pistillate part, 3-7 x 1 mm; pistillate part 0.4-1.5 cm, many-flowered; pistillate glumes ovate-elliptic, shallowly boat-shaped, 1.5-2 x 1-1.2 mm, membranous, the costa 3-nerved forming a mucronate apex. Utricles spreading, broadly ovate, 2-2.7 x 1 mm, obtusely trigonous, slightly inflated, glabrous, 7 to 9-veined; achenes ovate, triquetrous, 1.2-1.5 x 0.8 mm, apex subacute; style-base not thickened, stigmas 3.

Rare in edges of moist streams in Nallamalais (Veeranjaneyulu *et al.*, 2011). Fl. & Fr.: October – December.

Gundlabrahmeswaram (KNL), *BR* & *DV* 39340.

INDIA: E. to C. Himaaya, Andhra Pradesh.

WORLD: Bhutan, Nepal, China, Japan, Korea.

Carex speciosa Kunth, Enum. Pl. 2: 504. 1837; FBI 6: 729. 1894; Fischer 3: 1686. 1931; Mattew, Fl. Tamilnadu Carnatic 3: 1731. 1983.

Annual, tufted, rhizomatous, erect herb, to 30 cm high. Culms slender, winged. Leaves basal, exceeding stem, to 3 mm wide, multistriate. Involucral bracts overtopping. Spike 1, oblong, terete. Spikelets male at apex, female at base. Male flowers: glume ovate, to 1.5 mm long, margins hyaline; stamens 3, filaments unequal in length, connate at base. Female flowers: glumes broadly ovate, ovary and style enclosed in utricle, ovary triquestrous, obovoid, style dilated at base. Utricle ovoid-ellipsoid, base trigonous, to 4 mm long, pale, ribbed margins winged, exceeding glume, beak conical, truncate, 2-fid with ciliate margins; achenes ellipsoid, trigonous with prominent angles, stipitate, pale brown with darker markings, to 1 mm long.

Occasional along moist shady slopes of forests. Fl. & Fr.: September – December.

Maredumilli (EG), *KH* & *BR* 10904; Rampa hill (EG), *MSR* 1543 (CAL) 1543 (CAL); Forest near Sunkarimetta (VSKP), *NPBK* 10935 (MH); Chintapalli (VSKP), *GVS* 28205 (MH & CAL); Kankanapalli (VZN), *MV* 2594 (AU); Kongampallam – Yercaud (SLM), *SKK* 26850 (MH);

INDIA: Peninsular, C. & E. India, E. to C. Himalaya.

WORLD: Bhutan, Nepal, China, Laos, Cambodia, Indonesia, Thailand, Vietnam.

Carex stramentitia Boott ex Boeck., Linnaea 40: 351. 1876; FBI 6: 717. 1894.

Perennial, rhizomatous herb; rhizome woody, clothed with persistent, fibrous remains of leaf sheaths. Culms to 100 cm high, trigonous; nodes hairy. Leaves sub-basal, nearly as long as the culms, plicate. Panicle dense, compound, to 25 cm long with 3 secondary panicles. Involucral bracts foliaceous. Spikelets cylindric, straw-coloured, 1-3 cm long. Male florets: glumes ovate-oblong, aristate, straw-coloured, 3 mm long. Utricle ellipsoid-rhomboid, trigonous, greenish, glabrous, with subentire, glabrous beak, many-nerved.

Rare in wet open places in forests in coastal districts. Fl. & Fr.: December – January.

Ebul RF, north of Gudem (VSKP), *VNS* 589 (CAL); Dummakonda (EG) *VNS* 222 (CAL); Similipahar (MBJ), *SX* 409 (RRL-B), *SX, MB* & *Prabhakar Rao* 4661 (RRL-B), *SX* & *MB* 5046 (RRL-B).

INDIA: C. to E. India, E. to C. Himalaya.

WORLD: Myanmar, S.China, Thailand, Laos, Malaysia.

COURTOISINA Soják

Courtoisina cyperoides (Roxb.) Soják, Cass. Nar. Mus. Odd. Prir. 148: 193. 1979 publ. 1980. *Kyllinga cyperoides* Roxb., Fl. Ind. 1: 187. 1820. *Mariscus cyperoides* (Roxb.) A. Dietr., Sp. Pl. 2: 348. 1832. *Cyperus pseudokyllingioides* Kuk., in Engler, Pflanzenreich IV. 20, (Ht.101): 501. 1936. *Indocourtoisia cyperoides* (Roxb.) Bennet & Raizada, Indian Forester 107: 432. 1981; Saxena & Brahmam, Fl. Orissa, 4: 2193. 1996. *Courtoisia cyperoides* (Roxb.) Nees, Linnaea, 9: 286. 1834; FBI 6: 625. 1893; Fischer 3: 1645. 1931.

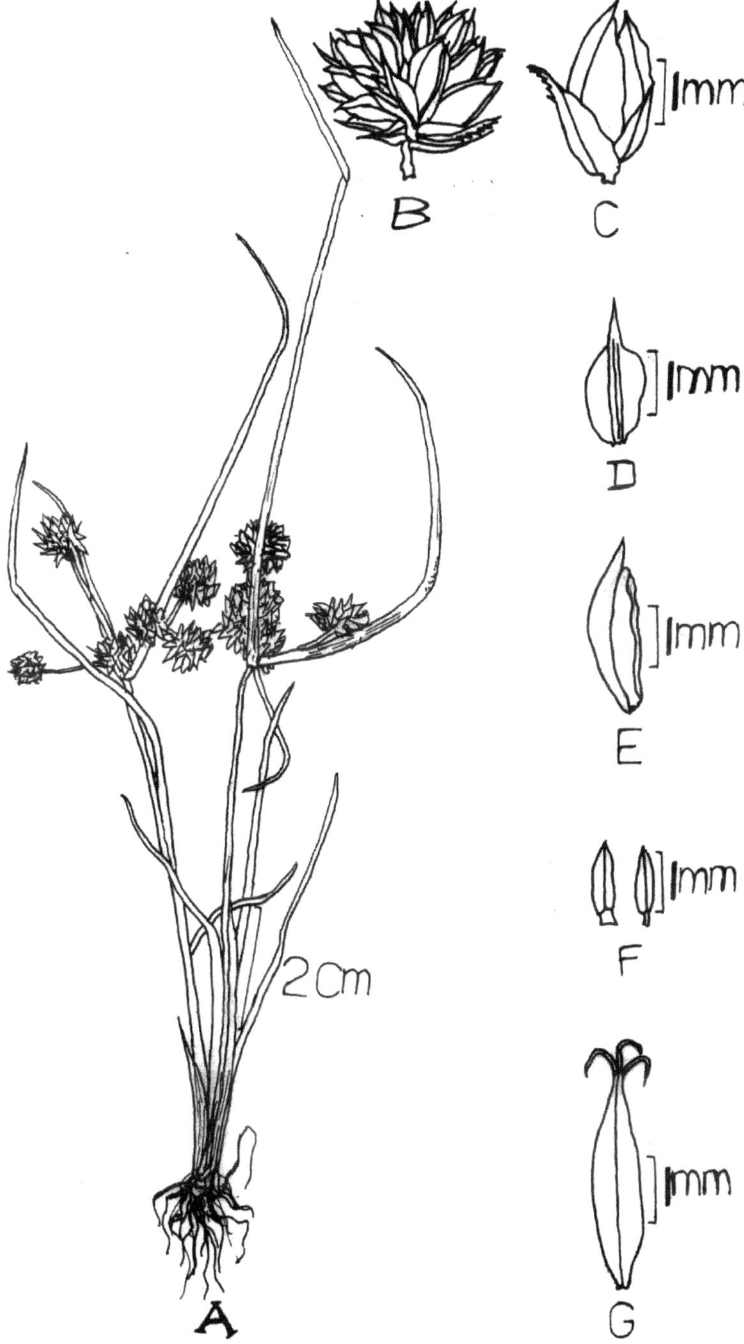

Figure 36. **Courtoisina cyperoides** (Roxb.) Soják
[*Indocourtoisia cyperoides* (Roxb.) Bennet & Raizada]
A. Habit, B. Spikelet, C. Rhachilla, D. Opened glume, E. Closed glume, F. Stamens, G. Achene with stigmas.

Annual, erect, tufted, non rhizomatous herb to 30 cm high, culms trigonous, glabrous. Leaves few, basal, as long as the stem, gradually acuminate; sheaths 2.5 cm; yellow, membranous. Inflorescence compound, branches ending in spikes. Involucral bracts 6, lower one much exceeding the inflorescence, spreading. Spike globose, 6 mm long; spikelets many, 4 mm long, clustered, compressed, 1-flowered, broadly elliptic, golden-brown-coloured, shining, rhachilla slender, narrowly winged. Achenes trigonous, linear oblong, 2.8 mm long, apiculate, golden brown, strongly puncticulate, about ¾ of the length of the glume.

Occasional in rice fields as a weed and also found in waste lands. Fl. & Fr.: November – January.

Kalasamudram RF Tank (ATP), *NY* 1186; SEDS – Penukonda Road (ATP), *BR & ANS* 35341; Batrepalli (ATP), *KRKS* 37852; Rudrakod (KNL), *BR & BSS* 30175; Nallamalais (KNL), *JLE* 32512 (MH); Anantagiri (VSKP), *GVS* 19568 (MH); Kavalur, Javadi hills (NA), *MBV* 856 (MH); Hogainakkal (KGR), *KCJ* 18011 (MH);

INDIA: Throughout India except NW. Region.

WORLD: Myanmar, Indo-China, Malesia.

CYPERUS Linnaeus

1. Florets unisexual .... **C. mollipes**
1. Florets bisexual:
    2. Rhachilla not articulated, persistent:
        3. Stigmas 3, nut trigonous:
            4. Rhachilla not winged; stamen 1 .... **C. leucocephalus**
            4. Rhachilla winged; stamens 1 to 3:
                5. Spikelets spicately disposed on conspicuously elongated rhachis thus forming spikes:
                    6. Rhachilla of spikelets winged with base of glumes, which is decurrent along the rhachilla internode:
                        7. Spikes cylindrical, rhachis 3-7 cm long; spikelets numerous:
                            8. Rhizome short; spikes umbelled or corymbose; spikelets distinctly compressed: involucral bracts 3-6 .... **C. exaltatus**
                            8. Rhizome woody, creeping; spikes digitate; spikelets terete; involucral bracts 3-12 .... **C. digitatus**
                        7. Spikes broadly elliptic, rhachis 1-4 cm long; spikelets occasionally up to 30:
                            9. Plants with creeping rhizomes or stolons (except in *C. bulbosus* which bears tunicate bulb at base of the culm):

10. Culms thick, more than 5 cm wide and 60 cm high, clothed at base with bladeless sheaths:

    11. Culms obtusely trigonous to subterete; involucral bracts much shorter than the corymb:

        12. Culms not septate........................**C. corymbosus**

        12. Culms transversely septate.......... **C. articulatus**

    11. Culms acutely trignous; involucral bracts much surpassing the corymb.... **C. pangorei**

10. Culms slender, 1-2 mm wide and less than 60 cm high, leaves with elongated blade:

    13. Culms arising from bulb like tuber; stolons filiform, soon disappearing.....**C. bulbosus**

    13. Stolons conspicuous and persistent, culm base forming corm like enlargement:

        14. Glumes 2-2.5 mm long; leaves as long as the culm..........**C. stoloniferus**

        14. Glumes 3-4 mm long; leaves much shorter than the culm:

            15. Culms 10 to 30 cm high; longest involucral bracts to 15 cm long...................................**C. rotundus**

            15. Culms 30-60 cm high, longest involucral bract to 10 cm long.. **C. bifax**

9. Plants without creeping rhizome or stolons:

    16. Spikelets linear to lanceolate, 1.5-2 mm wide, weakly flattened with two acute edges; glumes close together.................**C. tenuiculmis**

    16. Spikelets filiform, subterete, less than 1 mm wide; glumes spaced:

        17. Spikelets spreading; glumes rounded, obtuse at apex bearing narrow hyaline margins towards apex ....**C. distans**

        17. Spikelets erect patent; glumes shallowly emarginated at apex, bearing very broad hyaline margin towards apex... **C. nutans**

6. Rhachilla of spikelets without conspicuous wings, base of glumes not decurrent along the rhachilla internode:

18. Plants perennial with stolons:

19. Rhachis of spikes glabrous; glumes obtuse at white hyaline apex .... **C. procerus**

19. Rhachis of spikes densely hispid; glumes acute at apex with broad hyaline apex .... **C. pilosus**

18. Plants annuals with fibrous roots only:

20. Spikelets 1-2.5 cm long, 2-3 mm wide; glumes 2-3.5 mm long, acute at apex .... **C. compressus**

20. Spikelets 1 cm long, 1-1.5 mm wide; glumes 1-1.5 mm long.

21. Rhachilla of the spike glabrous; keel 3-5-nerved, smooth, not winged.................. **C. iria**

21. Rhachilla of the spike scabrid on the angles, keel 7-nerved, winged with serrate margins at apex .... **C. alulatus**

5. Spikelets digitately disposed or capitately congested at apices of inflorescence rays, thus forming heads without elongated rhachis:

22. Inflorescence with open elongated rays:

23. Large sized perennials with conspicuous rhizomes; atleast some leaves and involucral bracts more than 10 mm wide; culms 30-90 cm high .... **C. platystylis**

23. Small to medium sized annuals, leaves and bracts almost 5 mm wide; culms up to 30 cm high:

24. Leaves flattish; glumes muticous; achenes obovate:

25. Spikelets digitate in clusters of 3-15; glumes acute at apex:

26. Glumes sublaxly imbricate and slightly spaced; thus achenes exposed between glumes; plants tufted with fibrous roots only .... **C. tenuispica**

26. Glumes densely imbricate and close together so that the achenes not shown between the glumes; rhizomes frequently elongated .... **C. halpan**

25. Spikelets many, congested on globose heads .... **C. difformis**

24. Leaves canaliculated; glumes awned or mucronate at apex; achenes oblong:

27. Glumes darker in colour, long with short erect mucro at apex............................ **C. amabilis**

27. Glumes awned at apex:

28. Spikelets chestnut black; glumes with straight awn .... **C. castaneus**

28. Spikelets ferrugineus; glumes with recurved awn:

29. Involucral bracts 5; style shorter than the achenes .... **C. cuspidatus**

29. Involucral bracts 3; style longer than the achens .... **C. rubicundus**

22. Inflorescence congested in a singled head:

30. Annual, tufted herb .... **C. meeboldii**

30. Perennial rhizomatous herbs:

31. Spikelets 3- or 4-flowered; nutlets with spongy base of style extending downward along angles of nutlet .... **C. cephalotes**

31. Spikelets 8-flowered or more; nutlets without enlarged spongy apex:

32. Culms usually tall; spikelets several to more than 20, laxly fascicled; glumes conspicuously distichous, *c.* 4 mm, apex obtuse and muticous .... **C. niveus**

32. Culms dwarf; spikelets numerous, very densely arranged; glumes spirally imbricate or obscurely distichous, *c.* 2 mm, apex curved mucronate......... **C. michelianus**

3. Stigmas 2; nutlets biconvex:

33. Inflorescence umbelled, stems robust up to 120 cm tall.. **C. alopecuroides**

33. Inflorescence capitate, stems densely tufted, up to 25 cm high .... **C. laevigatus**

2. Rhachilla articulate, deciduous, leaving a knob above the two lowest glumes:

34. Spikelets flattened, 3 to many-flowered, glumes folded on a conspicuous keel:

35. Glumes awned .... **C. squarrosus**

35. Glumes not awned:

36. Plants 40-60 cm tall .... **C. compactus**

36. Plants 15-30 cm tall .... **C. dubius**

34. Spikelets terete, 1 or 2-flowered, glumes not or scarcely keeled:

37. Base of the stem hardly enlarged:

38. Spikelets shortly cylindric, oblong or ovoid, 1.2-1.8 x 0.8-1.2 cm, with sides narrowed toward base; anthers 1.5 mm; nutlets 0.6-0.9 mm wide; leaf blade margin scabrous:

39. Spikelets 1 or 2-fruited, greenish straw-coloured .... **C. cyperinus**

39. Spikelets 1-fruited, whitish-green .... **C. paniceus**

38. Spikes cylindric, 2-3.5 x 0.6-1 cm, with parallel sides; anthers 0.8-1 mm; nutlets 0.5 mm wide; leaf blade margin smooth .... **C. cyperoides**

37. Base of the stem bulbous .... **C. clarkei**

Cyperus alopecuroides Rottb., Descr. Pl. Rar. Progr. 20. 1772; Mattew, Fl. Tamilnadu Carnatic 3: 1733. 1983. *Juncellus alopecuroides* (Rottb.) C.B.Clarke in Hook. f., Fl. Brit. India 6: 595. 1893; Fischer 3: 1629. 1931.

Perennial, erect, stout, rhizomatous herb, to 120 cm high, culms rounded below, trigonous above. Leaves many, flat, as long as the stem, coriaceous, acute; sheaths dark. Inflorescence compound umbel with 4-8 primary rays, to 15 cm long; secondary rays 3-7, to 2 cm long, bearing clusters of oblong spikes, densely covered with small spikelets. Spikelets spicate, to 1 cm long, 15-30-flowered, persistent, elliptic-lanceolate. Involucral bracts 5, far overlapping to 40 cm long, leaf like. Achenes broadly elliptic, light brown, stipitate, apiculate.

Usually found in marshy low lands, along the banks of canals, ponds, lake margins and in swamps. Fl. & Fr.: September – February.

Paikaticheruvu (ATP), *NY* 1219; Tungabhadra river (KNL), *KH* 10942; Kolleru lake (WG), *KH* & *BR* 10908; Atapaka (KSN), *PV* 5348 (AU); Sorada (GJM), *SX* 2891 (RRL-B); Lanjigarh (KHD), *MB* & *Rout* 7070 (RRL-B).

INDIA: Widely distributed in India.

WORLD: Africa from the Mediterranean to the tropics.

Cyperus alulatus Kern, Reinwardtia 1: 464. f. 1. 1952; Saxena & Brahmam, Fl. Orissa 4: 2113. 1996. *C. iria* var. *rectangularis* Kuckenth. in Engl., Pflanzenr. 101: 152. 1935. *C. rectangularis* (Kuekenth.) Bennet, Indian Forester 95: 692. 1969.

Annual, tufted herb, to 40 cm high, culms triquetrous. Leaves as long as the culms, linear, serrulate margin near apex. Inflorescence terminal, lax, compound umbel. Involucral bracts 8, unequal, the longest one up to 30 cm long. Spikes to 3 cm long; rhachis flexuous, scabrid on the angles; rays 8, unequal, to 15 cm long. Spikelets greenish yellow, patent; glumes *c.* 2 mm long, scarcely imbricate, concave, orbicular, emarginated at apex, keel prominent, curved, winged with serrate margin at the top, 7-nerved; stamens 22, styles 3-fid Achenes obovoid, triquetrous, reddish-brown, apiculate.

Occasional in dry rice fields, ditches and puddles (V.S.Raju *et al.*, 2008). Fl. & Fr.: August – November.

Palaspal (KJR), *GP* 8561 (CAL).

INDIA: Throughout India.

WORLD: Nepal.

Cyperus amabilis Vahl, Enum. Pl. 2: 318. 1806; FBI 6: 598. 1893.

Annual slender, tufted herb; culms up to 30 cm tall, trigonous. Leaves 2, sub-basal, half as long as culms. Inflorescence a compound umbel; rays 9-10 cm long, involucral bracts 2, leaf like, longest up to 15 cm; spikelets linear, golden-brown. Achenes obovate, brown, faintly rugose.

Common in marshy places. Fl. & Fr.: August – October.

Japalitheertham (CTR), *KI* 21562.

INDIA: Assam, Bihar, Madhya Pradesh, Odisha, Andhra Pradesh, Kerala.

WORLD: Tropical America, Africa.

Cyperus arenarius Retz., Observ. 4: 9. 1786; FBI 6: 602. 1893; Fischer 3: 1640. 1931; Matthew, Fl. Tamilnadu Carnatic 3: 1733. 1983.

Perennial, trufted, rhizomatous, stoloniferous herb, whole plant glaucous green, rhizome creeping, clothed with chestnut brown sheaths, roots thick; culms to 10 cm long, erect, terete, stout, clothed below with long withered sheaths. Leaves to 6 cm long, terete, coriaceous, sulcate, scaberulous, acute. Heads solitary on the summit of the stems, globose; involucral bracts 2, overtopping. Spikelets radiating, 10-15 in a cluster, oblong, brown-coloured, 8-10-flowered; rhachilla stout, not winged. Achenes triquetrous, smooth, brown, apiculate.

Coastal species, but reported from inland also. Fl. & Fr.: July – December.

Kanekal (ATP), *BR* & *ANS* 448818; Velgode (KNL), *BR* & *BSS* 29344; Chelama (KNL), *BR* & *BSS* 32248; Tummalapenta (NLR), *JSG* 12395 (MH); Somasila (NLR), ASR 4949 (BSID); Gollabandh (GJM), *SX* 1301, 1305 (RRL-B).

INDIA: Peninsular, C, N & NW. India.

WORLD: Iran, Pakistan, Sri Lanka, Cochinchina.

Note: This species can be easily identified by its single dense spikelets and long rhizomes. Usually this species forms large communities along sandy sea-shores.

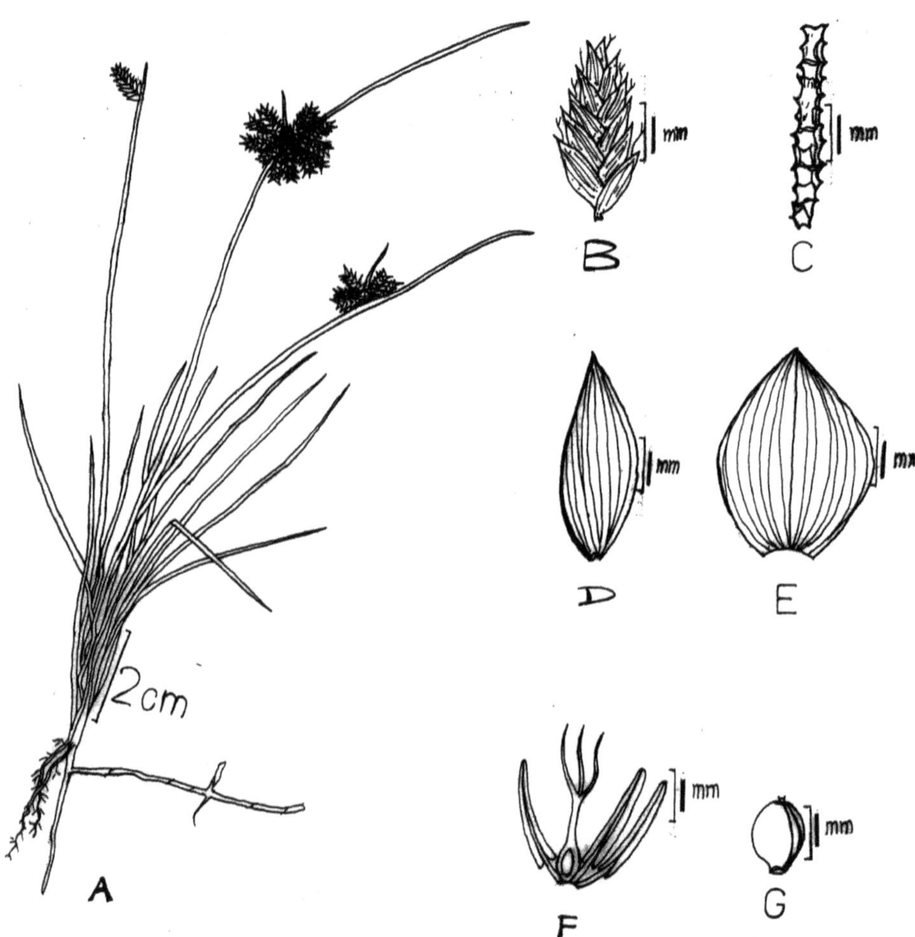

Figure 37. **Cyperus arenarius** Retz.
A. Habit, B. Spikelet, C. Rhachilla, D. Closed glume,
E. Opened glume, F. Pistil with stamens, G. Achene.

Cyperus articulatus L., Sp. Pl. 44. 1753; FBI 6: 611. 1893; Fischer 3: 1641. 1931; Matthew, Fl. Tamilnadu Carnatic 3: 1734. 1983.

Perennial, erect, glabrous, stout herb, culms to 1.2 cm wide, terete, transversely septate. Leaves reduced to 7-20 cm long, brown to grey purple-coloured sheaths and covers the base of the stem. Inflorescence compound umbel; involucral bracts 5, much shorter than the corymb, to 1 cm long, primary rays 5-8, to 9.5 cm. Spikes shortly racemose, to 4 cm long, 2-15-spikeletted, spikelets oblong, narrow, linear, compressed, acute, *c*. 35-flowered; rhachilla 2-winged. Achenes oblong, trigonous, maturing brown.

A water loving sedge, usually found in standing waters of ponds, canals and along river banks in large colonies. Fl. & Fr.: March – September.

Gangalakunta (ATP), *KH* 7477; Rangapuram RF (KNL), *RVR* 1501; Pulipadu tank – Darsi (PKM), *MCK* 22852; Venkatagiri (NLR), *BSN* 3417 (BSID); Somavarapupadu (PKM), *RKM* 0672 (CAL); K. Kota (KSN), *PV* 5810 (AU); Chettaraparu (WG), *KS* 5118 (MH & CAL); Hogainakkal (KGR), *KCJ* 17993 (MH); Ganjam (GJM), *JSG s.n.* (CAL).

INDIA: S. to E. India.

WORLD: Tropical Africa, Mediterranean, Sri Lanka, Indo-China, SE. USA to C. & S. America, Neotropics.

Cyperus bulbosus Vahl, Enum. Pl. 2: 342. 1806; FBI 6: 611. 1893; Fischer 3: 1641. 1931.

Perennial, stoloniferous herb, to 30 cm high; tuber ovoid-globose, clothed with coriaceous, black-brown, striate coat splitting into several segments; stolons filiform, to 10 cm long, terminated by tuber. Culms triquetrous arising from bulb like tuber. Leaves to 20 to a culm, slightly over-topping the culm; blades to 20 cm long, incurved plicate, scabrid on upper margins; sheaths to 6 cm long, broader than the blade, membranous, cinnamon-coloured, the inner ones pale white. Inflorescence simple, to 3 cm long, spicately bearing 2 subdigitate clusters of spikelets on very short axis. Involucral bracts 3, patent, the lowest surpassing the inflorescence, the longest to 10 cm long. Spikelets subdigitate-spicate on very short rhachis, patent linear, to 2 cm long, to 20-flowered, sanguineous brown; rhachilla flexuous, the internodes winged. Achenes ellipsoid, trigonous, maturing black, apiculate at obtuse apex.

Occasional in sandy beaches, river sides, and marshy areas. Fl. & Fr.: September – December.

Nemakal (KNL), *KH* 10975.

INDIA: Throughout plains of India.

WORLD: West Africa, Sahara, Africa north of Sahara, East Africa, Iran, Pakistan, Sri Lanka, Malayais, Vietnam, Australia.

Note: *C. bulbosus* can easily be distinguished by the black-coated bulbils which give rise to new plants. In the absence of the bulbils *C. bulbosus* may easily be confused with *C. rotundus*. But the former can be distinguished by spaced lower bracts, and more distinctly nerved glumes.

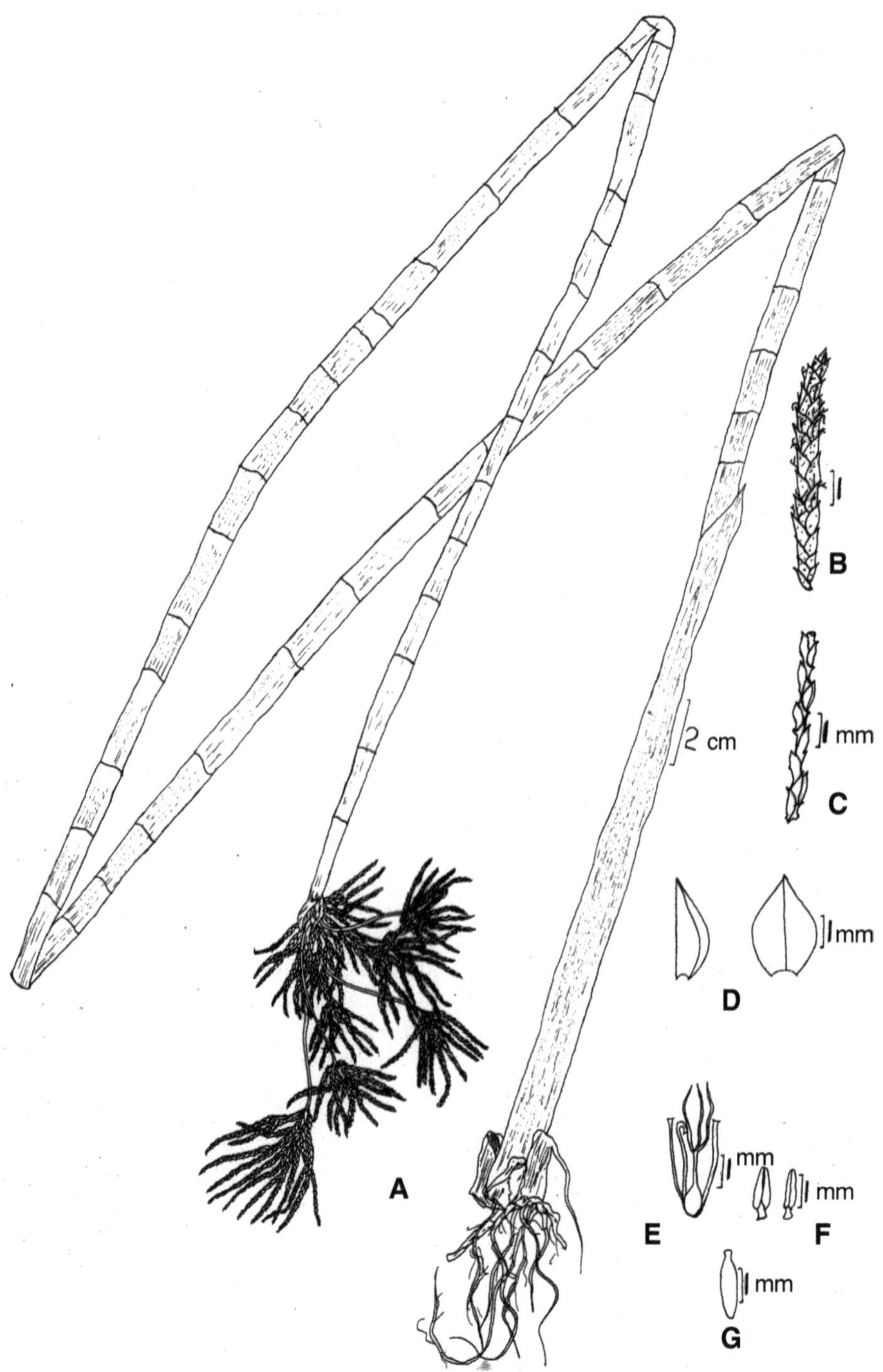

Figure 38. **Cyperus articulatus** L.
A. Habit, B. Spikelet, C. Rhachilla, D. Glumes closed and opened, E. Pistil, F. Stamens, G. Achene.

Cyperus castaneus Willd., Sp. Pl. 1: 278. 1797; FBI 6: 598. 1893; Fischer 3: 1639. 1931.

Annual, tufted, glabrous herb to 20 cm high, culms slender, triquetrous, 10-leaved at base. Leaves 9 cm long; blades setaceous, weakly scabrid, reddish brown. Inflorescence simple, to 8 cm long, frequently congested in a single head. Involucral bracts 4, surpassing the inflorescence; rays 5, to 5 cm long. Spikelets stellately clustered in groups of 5 to 10, slenderly linear, to 10 mm long, flattened, 20-40-flowered, chestnut black; rhachilla striaghtish, wingless. Achenes oblong with parallel sides to oblong-obovoid, brown, puncticulate.

Common in open wet lands, grassy place, rice fields, and in swamps *etc.* Fl. & Fr.: August – December.

Kalasamudram (ATP), *MHR* 14920; Batrepalli (ATP), *KRKS* 37861; Rollamadugu (KDP), *BR* & *BSS* 33101; Salur (VZN), *MCK* 25210; Srisailam (KNL), *JLE* 22115 (MH); Near Komativaricheruvu (CTR), *GVS* 46831 (MH & CAL); SNG Palem (KSN), *PV* 5799 (AU); Y. Ramavaram (EG), *SS* 693 (AU); Tippukadu RF (NA), *KRM* 17666 (MH); on the way to Kalachipadi (DMP), *T. Ravishankar* 95554 (MH); Karchuli (GJM), *MB* 2842 (RRL-B).

WORLD: Cochinchina to Tropical Australia, Malesia.

Cyperus cephalotes Vahl, Enum. Pl. 2: 311. 1805; FBI 6: 597. 1893; Fischer 3: 1639. 1931. *Anosporum cephalotes* (Vahl) Kurz, J. Asiat. Soc. Beng. Pt. 2, Nat. Hist. 39(2): 84. 1870.

Perennial, rhizomatous, stoloniferous herb; stolons to 15 cm long, brownish, rooting at nodes; culms to 40 cm high, triquetrous at least distally. Leaves 2 to several, half as long as the culm, blades narrowly linear, to 50 cm long, 1-costate, stiffish, scabrid, sheaths to 70 cm long, the dorsal hyaline part eventually brownish, the lower ones cinnamon-coloured. Inflorescence on ovoidal, conical, lobed head, bearing 4 glomerules of 8 spikelets. Involucral bracts foliaceous. Spikelets ovoid, to 8 mm long, densely to 20-flowered, brown to fuscous with yellowish margins. Achenes ovoidal, trigonous and dorsally flattened, brown, minutely puncticulate, the wings developing into yellowish corky ridges.

Occasional in wet or flooded rice fields, in tanks *etc.* Fl. & Fr.: July – December.

Makkuva (VZN), *MV* 6932 (AU).

INDIA: West Bengal to Tamil Nadu.

WORLD: Myanmar, S. China, Indo-China, Malesia, N. Australia.

Cyperus clarkei Cooke, Fl. Bombay 2: 873. 1908. *Mariscus clarkei* (Cooke) T. Koyama, J. Jap. Bot. 51: 313. 1976; Matthew, Fl. Tamilnadu Carnatic 2: 1770. 1983. *M. bulbosus* C.B.Clarke in Hook. f., Fl. Brit. India 6: 620. 1839. non Steud.; Fischer 3: 1644. 1931.

Perennial, rhizomatous, stoloniferous herb; stolons to 6 cm long, the internodes to 5 mm long, clothed with pale brownish membranous scales, culms solitary, to

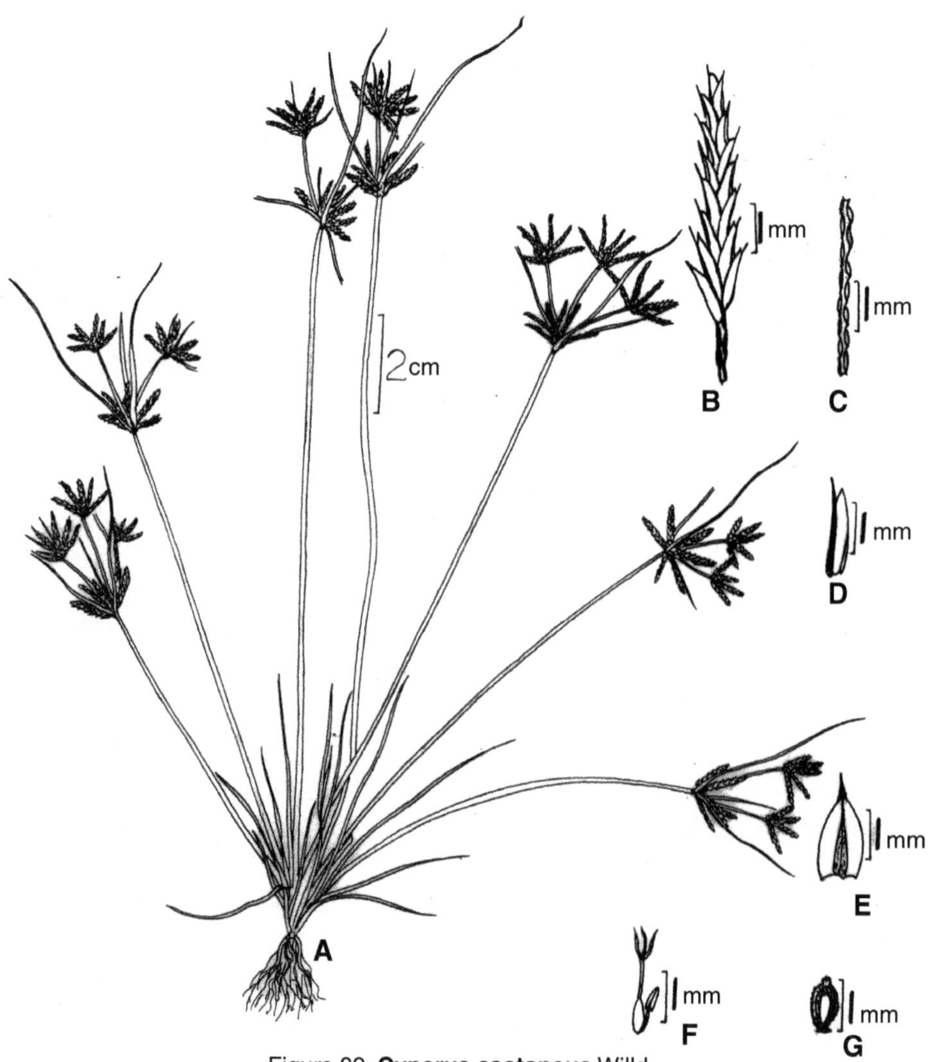

Figure 39. **Cyperus castaneus** Willd.
A. Habit, B. Spikelet, C. Rhachilla, D. Closed glume, E. Opened glume, F. Pistil with stamens, G. Achene.

22 cm high, triquetrous, the base thickened, to form a lanceolate enlargement, to 4 cm long, clothed with dusky brown, membranous sheaths. Leaves 6 to a culm, shorter than the culm; blades 18 cm long, herbaceous, 1-costate, plicate with incurved margins, glaucous green; sheaths 4 cm long, membranous. Dusky brown inflorescence simple, spikes cylindrical, subsessile, 2 cm long, obtuse. Achenes oblong-elliptic, trigonous, 1.5 mm long, maturing dark brown.

Rare on wet shallow soil over rock outcrops. Fl. & Fr.: September – December.

Kalasamudram RF (ATP), *TP* & *EC* 2816; North Dhone RF (KNL), *RVR* 3198; Guvvalacheruvu (KDP), *RVR* & *R.V.Reddy* 7847; near Duggamakonda temple (CDP), *KS* 6354 (MH); Seshachalam RF (KDP), *A.M.Rao* 319 (BSID); Nagaptla RF (CTR), *KS* 6855 (MH); Devarapalli – Maredumilli (EG), *MM* 105067 (BSID); Kottaiyur RF, Yelagiri hills (NA), *MBV* 839 (MH).

INDIA: Rajsthan, Maharshtra, Andhra Pradesh, Kerala, Tamil Nadu.

WORLD: Sri Lanka, Myanmar, Vietnam.

Cyperus compactus Retz., Observ. Bot. 5: 10. 1788. *C. spinulosus* Roxb., Fl. Ind. 1: 203. 1832. *Mariscus microcephalius* C. Presl, Rel. Haenk. 1: 182. 1830; FBI 6: 624. 1893. *M. compactus* (Retz.) Bold., Zakfl. Java 77. 1916; Fischer 3: 1645. 1931; Matthew, Fl. Tamilnadu Carnatic 2: 1771. 1996.

Perennial, rhizomatous herb, rhizome corm like; culms solitary, erect, to 50 cm high, obtusely trigonous, septate nodulose. Leaves 3, longer than the culm; blades elongated, linear, thickish, canaliculated, glaucous green, markedly septate nodose; sheaths elongated, cylindrical, septate nodose, purple brown. Inflorescence compound, ample, 7 cm long and as wide. Involucral bracts 8, patent, the lower ones much surpassing the inflorescence, the longest 60 cm long; primary rays 6, patent, the longer one up to 7 cm long; secondary corymbs with 5 raylets; spikes nearly globose, bearing many stellately arranged spikelets; rhachis very short. Spikelets linear, red brown, jointed at base, congested in a globsoe head, 12 mm long, pale green and tinged with sanguineous brown; rhachilla flexuose with pale white hyaline wings. Achenes linear-oblong, trigonous, to 2 mm long, yellow brown.

Occasional along ravines and flowing streams. Fl. & Fr.: March – September.

Mahanandi (KNL), *RVR* 14536; Kolambarathi (KNL), *SS* & *BR* 20146; Peccheruvu (KNL), *BR* & *BSS* 30167 (SKU & BSID); Maredumilli (EG), *KH* & *BR* 9739; Near Cheediplem (EG), *GVS* 68618 (MH); Tatipudi (VZN), *MCK* 18865; Balapalle (KDP), *JLE* 15767 (MH & CAL); Krishnanandi (KNL) *JLE* 25529, 42317 (MH & CAL); Gokavaram (EG), *CAB* 8268 (MH); Bhogapuram (VZN), *MV* 3971 (AU); Purunakote RF, Satkosia Tiger Reserve, *KCM* 5290 (BSID).

INDIA: Throughout India.

WORLD: Southern Continental China, Taiwan and Malesia.

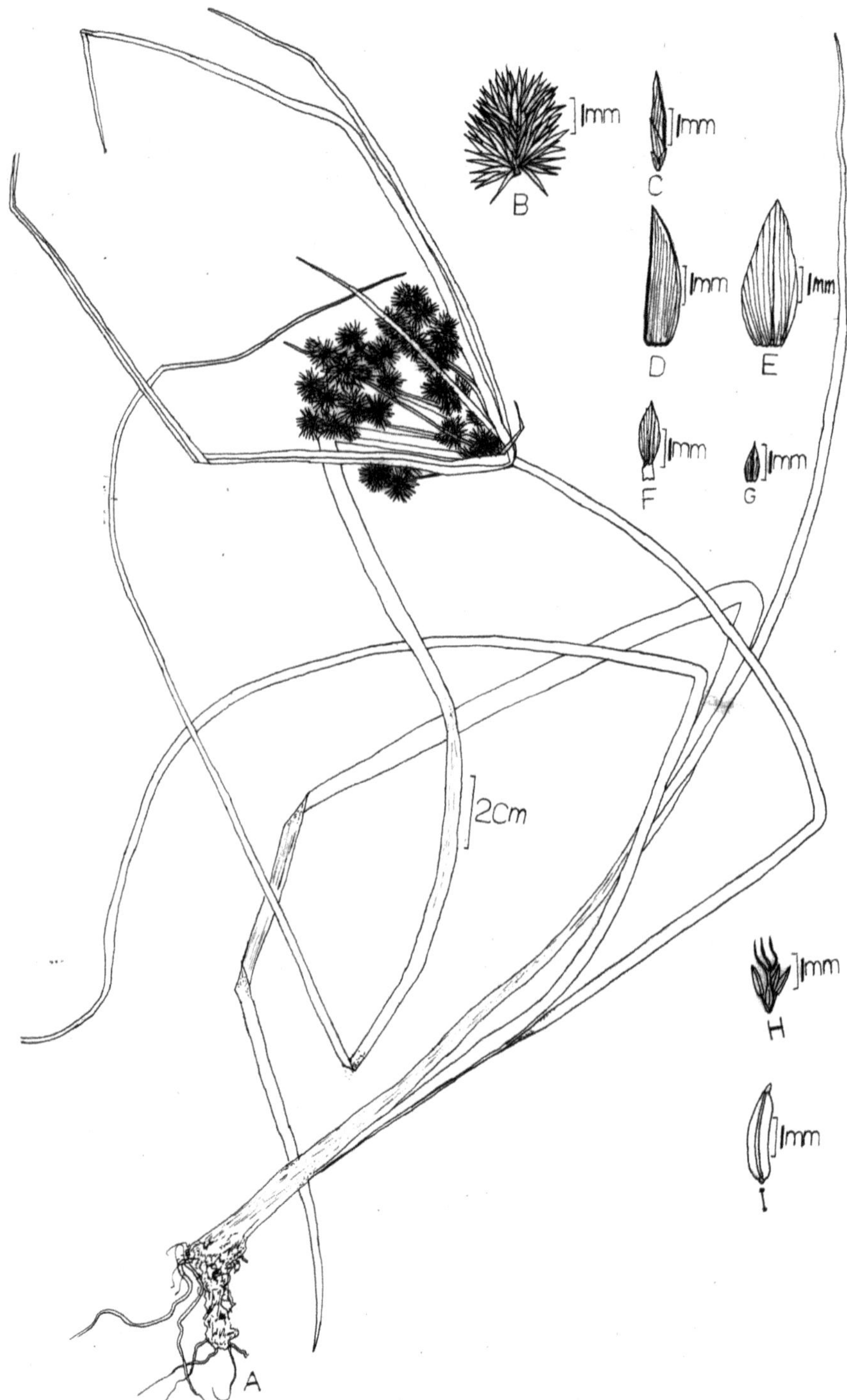

Figure 40. **Cyperus compactus** Retz. [Syn.: *Mariscus compactus* (Retz.) Bold]
A. Habit, B. Spike, C. Spikelet, D – G. Glumes, H. Pistil with stamens, I. Achene.

Cyperus compressus L., Sp. Pl. 46. 1753; C.B. Clarke in Hook. f., Fl. Brit. India 6: 605. 1893; Ficher 3: 1640; Matthew, Fl. Tamilnadu Carnatic 3: 1734. 1983. *C. pectiniformis* Roem. & Schult., Mant. 2: 128. 1824. *C. compressus* L., var. *micranthus* T.Koyama, Gard. Bull. Singaporee 30: 139. 1977

Annual, tufted herb. Culms patent, to 30 cm high, 3-sided. Leaves 4 to a culm; basal; blades linear, shorter than the culms, light green; sheaths membranus, pale brownish, striate. Inflorescence to 10 cm long; involucral bracts 7, foliaceous, unequal, the longest one to 20 cm long; rays 8, patent, to 10 cm long. Spikes bearing to 10 spikelets on abbreviated axis. Spikelets lance-oblong, 1-2.5 x 0.2-0.3 long, *c.* 20-flowered, compressd, greenish and stramineous at maturity; rhachilla flexuose. Achenes 1.8 mm long, broadly obovate, 3-sided, dark brown, shiny.

Common weed of cultivated fields, waste lands, in open grass lands, along the sides of water courses, along paths on moist sandy soils. Fl. & Fr.: September – February.

Kuppagal RF (KNL), *KH* 10967; Rachakuntapalli (KDP), *KRKS* 37643; Kailasakona (CTR), *MHR* 14911; Mellavagu (GNT), *VRK* 5722; Mahanandi (KNL), *JLE* 25483 (CAL); Kolletikota (KSN), *PV* 5828 (MH); Rajavomangy (EG), *GVS* 68651 (MH & CAL); Muzumamidivalasa, Maredumilli (EG), *MM* 102521 (BSID); Salur (SKLM), *NPBK* 954 (CAL); Koriguttalu (KMM), *R.Rajan* 106050 (BSID); Lakshmigudem (KMM), *RCS* 104264 (MH); Nellippakka RF (KMM), *RCS* 99079 (BSID); Mamillagudem (KMM), *R.Rajan* 105916 (BSID); Kayapakam (CPT), *S.India Flora* 11228 (MH); Tippukadu RF (NA), *KRM* 17615 (MH); Pennagaram RF (SLM), *EV* 22427 (MH); Hogainakkal RF (KGR), *EV* 20661 (MH); Way to Kuttathur (NA), *KS* 7451 (MH); Berhampore (GJM), *VNS* 4716 (MH).

INDIA: Throughout India.

WORLD: Cosmopolitan.

Cyperus corymbosus Rottb., Descr. Icon. Rar. 42. t. 4. 1773; FBI 6: 612. 1893; Fischer 3: 1641. 1931; Matthew, Fl. Tamilnadu Carnatic 3: 1735. 1983.

Perennial, glabrous, glaucous-green, rush-like herb; rhizome creeping, clothed with dark brown scales; culms to 90 cm long, 6 mm wide, terete below, trigonous above. Leaves reduced to sheaths. Umbel compound, to 14 cm long; involucral bracts 3, longest to 13 cm long, green; lowest involucral bract much shorter than the corymb; primary rays 6-9, very unequal, the longest reaching 10 cm long; secondary rays terminated by spikes or corymbs of 4-8 very slender spikelets. Spikes *c.* 10, 6-9-spikeletted. Spikelets spicate, very variable in length, 2.5 cm long, rusty brown, rhachilla winged, 15-20-flowered. Achenes obovoid, tapering towards the base, trigonous.

Common in marshy areas near river banks, canals, along the sides of the ponds, in ditches *etc.* Fl. & Fr.: August – December.

Kalasamudram RF (ATP), *NY* 1185; Batrepalli (ATP), *KRKS* 39005; Palakonda hills (KDP), *CS* 8271; Leguntapadu (NLR), *PMR* 16429; Macherla RF – Ethipothala

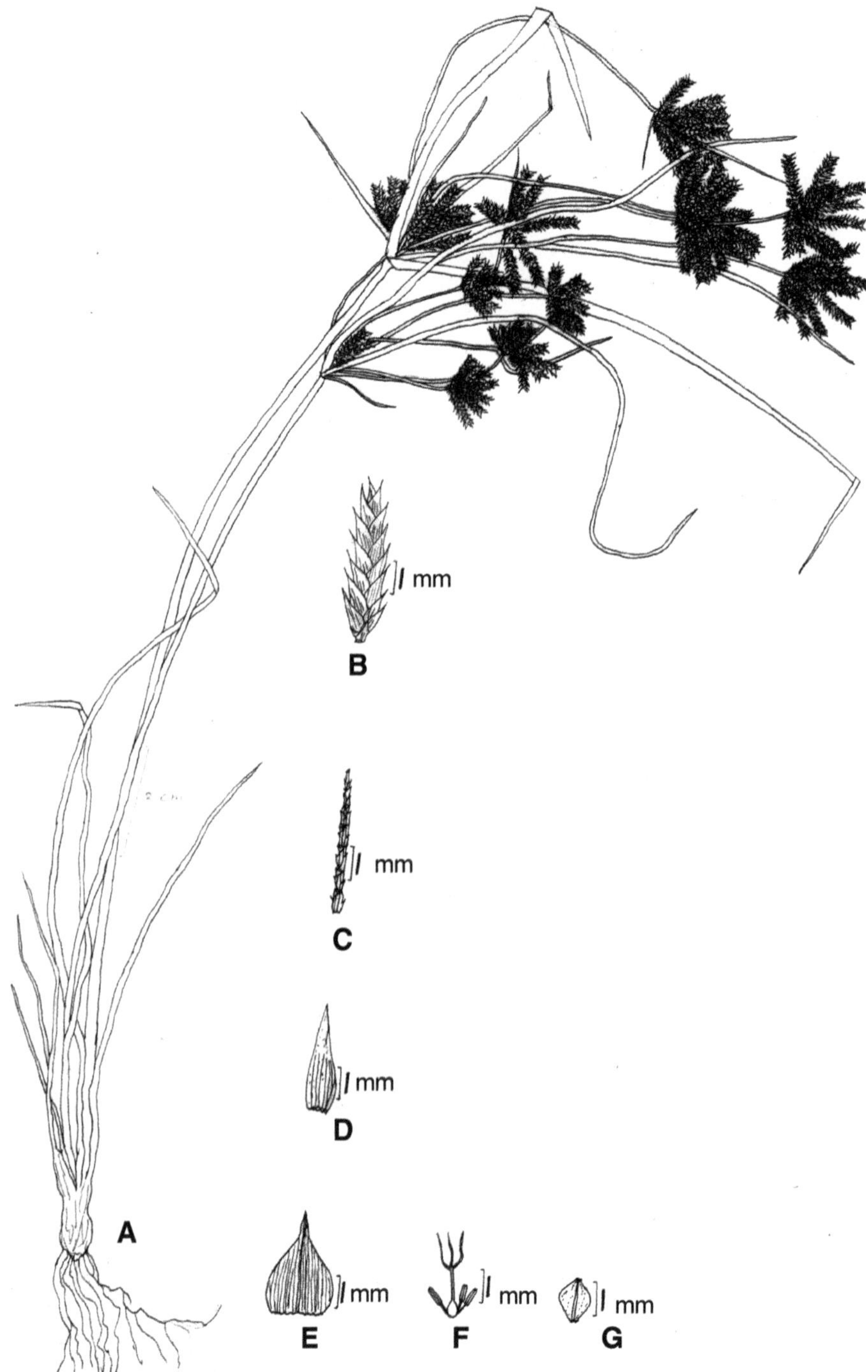

Figure 41. **Cyperus compressus** L.
A. Habit, B. Spikelet, C. Rhachilla, D. Closed glume, E. Opened glume, F. Pistil with stamens, G. Achene.

falls (GNT), *VRK* 3283; Rampa hill (EG), *KH* & *BR* 9730; Jalipinaidupeta (SKLM), *MCK* 18890.

INDIA: Throughout India.

WORLD: Pantropical.

Cyperus cuspidatus Kunth, Nov. Gen. Pl. 1: 204. 1815; FBI 6: 598. 1893; Matthew, Fl. Tamilnadu Carnatic 3: 1735. 1983. *C. uncinatus* auct. non Poiret: Fischer 3: 1639. 1931.

Annual, erect, tufted herb, 9 cm high. Leaves setaceous, as long or half as long as culms, filiform, sheaths purple. Umbel simple, consisting of a central sessile head and few or many filiform spreading unequal rays, each terminated by a head of stellately spreading spikelets; involucral bracts 5, longest to 5 cm long, filiform. Spikes obovoid, digitately and stellately spikeletted. Spikelets 5-15 in a cluster, compressed, 12-24-flowered, 1 cm long, reddish-brown, rhachilla persistent. Achenes obovoid, triquetrous, winged, dark-brown.

Common in moist or damp sandy soil or cultivated and fallow fields. Fl. & Fr.: July – December.

Guvvlacheruvu (KDP), *RVR* 7968; Araku (VSKP), *GVS* 21549 (MH); Similipahar (MBJ), *SX* & *MB* 5175 (RRL-B).

INDIA: Throughout India.

WORLD: Pantropical.

Cyperus cyperinus (Retz.) Suringar, Cyperus 154. 1898. *Kyllinga cyperina* Retz., Observ. Bot. 6: 21. 1791. *Mariscus cyperinus* (Retz.) Vahl, Enum. Pl. 2: 377. 1806; FBI 6: 621. 1893; Fischer 3: 1644. 1931; Matthew, Fl. Tamilnadu Carnatic 2: 1771. 1983. *M. tenuifolius* Schrad. ex Nees, Fl. Bras. 2(1): 46. 1842.

Perennial, rhizomatous herb, rhizome woody, clothed with brown fibres, culms few, erect, to 40 cm high, triquetrous, leaves at thickened base. Leaves 7, rarely equaling the culm; blades narrowly linear, long attenuate, 2.5 mm wide, herbaceous, plicate; sheaths pale green and stained with purplish pink, the basal sheaths dark red brown. Umbel simple, terminal, open with short rays, rays 9, to 2 cm long each terminated by a spike; spikes cylindrical, narrowed to base, 2 cm long, greenish, densely bearing up to 60 spikelets. Involucral bracts 8, patent, the lower few surpassing the anthela, the longest up to 16 cm long. Spikelets patent, greenish straw-coloured, linear-lanceolate, 5 mm long, bearing 4 glumes and 2 achenes; rhachilla with lanceolate wings. Achenes oblong, trigonous, 2.5 mm long, brownish.

Occasional in marshy areas, banks of streams, and in open grassy places in hilly areas. Fl. & Fr.: July – October.

Talkona (CTR), *BR* & *BSS* 29334; Maredumilli RF (EG), *KH* & *BR* 9740; D.M. Bhavi (KSN), *PV* 5650 (AU); Chelama (KNL), *JLE* 16733 (MH); Chelama (KNL), *JLE* 16771 (MH); Kondavidu fort (GNT), *S.India Flora* 17500 (MH); Anantagiri (VSKP), *NPBK* 11002 (MH & CAL), GVS 19478 (MH); Dharakonda (VSKP), *sine coll. s.n.*

(MH); Bhadrachalam RF (KMM), *RCS* 104313 (BSID); Mahendragiri (GJM), *VNS* 5738 (MH); Pampasar RF, Satkosia Tiger Reserve, *KCM* 5102, 5129 (BSID).

INDIA: Peninsular, NW. & E. India.

WORLD: S.W.Asia, Sri Lanka, Maynamar, Bangladesh, Bhutan, China, Indonesia, Philippines, Thailand, Vietnam, NE Australia, Pacific Islands.

Cyperus cyperoides (L.) Kuntze, Revis Gen. Pl. 3(3): 333. 1898. *Scirpus cyperoides* L., Mant. Pl. 2: 181. 1771. *Mariscus sumatrensis* (Retz.) Raynal, Adansonia 15: 110. 1975; T.Koyama, Gard. Bull. Singapore 30: 154. 1977. *Kyllinga sumatrensis* Retz., Observ. Bot. 4: 13. 1786. *Mariscus sieberianus* Nees (Linnaea 9: 286. 1835, nom. nud.) ex C.B.Clarke in Hook.f., Fl. Brit. India 6: 122. 1893. *Cyperus biglumis* C.B. Clarke, J. Linn. Soc., Bot. 21: 199. 1884. *C. konkanensis* T. Cooke, Fl. Bombay 2: 874 1908.

Perennial herbs, culms up to 60 cm high, acutely triquetrous. Leaves shorter than culms, leaf sheath 20-75 cm, brown to reddish brown, leaf blade 3-6 mm wide, gradually long acuminate, scabrid in the upper part. Involucral bracts 5-8, obliquely spreading, leaflike, longer than inflorescence. Inflorescence open, simple, rays 3-15; spikes cylindrical, spikelets numerous, at first obliquely erect, finally at right angles to the raachis, linear to linear-lanceolate; glumes lanceolate-oblong to ovate-oblong; stamens 3; stigmas 3. Nuts linear-oblong, straw-coloured at first but dark brown when mature.

Rare in marshy areas along streams.Fl. & Fr.: August – November.

Near Lankamala water falls (KDP), *MHR* & *KI* 14904; Upper Ahobilam (KNL), *JLE* 25566 (MH); Anantagiri (VSKP), *GVS* 19568 (MH); Shevaroy hills, Yercaud (SLM), *AVNR* 26773 (MH).

INDIA: Throughout India.

WORLD: Tropics and subtropics of Africa, Asia and Australia, West Indies.

Cyperus difformis L., Cent. Pl. II. 6. 1756; FBI 6: 598. 1893; Fischer 3: 1640. 1931; Matthew, Fl. Tamilnadu Carnatic 3: 1736. 1983.

Annual, glabrous, tufted herb with many reddish fibrous roots; culms to 40 cm high, weak, triquetrous towards the top. Leaves few, shorter than the stem, acuminate, flaccid, to 13 cm long; sheaths stramineous or brown. Spikes in a simple or compound or contracted into a head, to 1.5 cm across; involucral bracts 3, overtopping, longest to 20 cm long, leaflike; rays of the umbel 3-6, to 3 cm long. Spikelets many in dense heads, dusty or brown, oblong, 10-20-flowered, compressed, rhachilla not winged. Achenes ellipsoid, trigonous, yellow or brown.

Very common species found as a weed in rice fields, and cultivated lands, often associated with *C. haspan*, *C. iria* and *Pycreus sanguinolentus*. Fl. & Fr.: Throughout the year.

Bukkapatnam (ATP), *JSG* 20926 (MH); Pennahobilam (ATP), *NY* & *TP* 376; Kalasamudram (ATP), *NY* 1170 (SKU & MH); Gani RF (KNL), *RVR* 1656; Leguntapadu (NLR), *PMR* 16433; N.S. right canal (GNT), *VRK* 3548; Near

Padmapuram gardens (VSKP), *MHR* & *MCK* 14839; Lakkavarapu reservoir (VZN), *MHR* & *MCK* 14868; Bukkapatnam (ATP), *JSG* 15157 (DD); Balapalli (CDP), *JLE* 14311 (MH); Guvvalacheruvu (KDP), *KS* 7790, 7792 (CAL); Penchalakona (NLR), *BSN* 5127 (BSID); Polavaram agency (WG), *DCSR* 117 (CAL); Tiger camp to Valamuru, Maredumilli (EG), *MM* 105041 (BSID); Rangamatigedda (SKLM), *GVS* 62346 (MH & CAL); Koikonda road (MBNR), *S.R.Srinivasan* 104511 (BSID); Lakshmigudem (KMM), *RCS* 104266 (MH & BSID); Dhanaiyaigudem (KMM), *R.Rajan* 108508 (BSID); Nayakkanur Javadi hills (NA), *MBV* 992 (MH); Vazhiyur – Polur (NA), *EV* 55605 (MH); Yelagiri hills (NA), *MBV* 807 (MH); Gingee RF (SA), *KMS* 12328 (MH); Tulka, Satkosia Tiger Reserve, *KCM* 6505 (BSID).

INDIA: Throughout India.

WORLD:Tropical, Subtropical and temporate regions of both hemispheres.

Cyperus diffusus Vahl, Enum. Pl. 2: 321. 1806; FBI 6: 603. 1893; Fischer 3: 1639. 1931.

Perennial, tufted herb with short corm like rhizome clothed with dark brown scales and their fibrous remnants; culms to 84 cm high, rigid and coarse, trigonous. Leaves flattish, 3-costate, scabrous on margins, sheaths pale green eventually turning rusty brown. Inflorescence decompound, diffuse; primary rays many, spreading, 15 cm long; secondary rays to 5 cm long; involucral bracts 12, leafy, longer than the corymb, the longest up to 30 cm long. Spikelets sessile, compressed, digitate in groups of 3-5 at apices of secondary and tertiary rays, linear oblong, 8 mm long, 12-flowered, green tinged with straw-colour, internodes winged. Achenes elliptic, 3-sided, slightly surpassing the glume, 2 mm long, brownish.

Common in wet low lands, along canals and as a weed in paddy fields. Fl.& Fr.: December – August.

Papanasanam (CTR), *MHR* 13955; Punyagiri hill – S. Kota (VZN), *KH* & *BR* 9714; Near Japali (CTR), *GVS* 45905 (MH); Tiger camp to Valamuru (EG), *MM* 105041 (BSID); Sesharayi (EG), *SS* 621 (AU); Jamparakota-Palakonda range (VSKP), *KCJ* 17240 (MH); Near Papanasanam falls (CTR), *KS* 6957 (CAL); Saptasajya (DKL), *SX* 2111 (RRL-B), *HFM* 2045 (DD); Similipahar (MBJ), *SX* 3778 (RRL-B), *SX* & *MB* 4598 (RRL-B); Goriska forest (KJR), *MB* & *NKD* 8450 (RRL-B); Singare forest (KHD), MB & *NKD* 7684 (RRL-B); Sorisiapada (DKL), *NKD* & *Rout* 8167 (RRL-B); Panigada (GJM), *SX* & *MB* 1933 (RRL-B); Purunakote RF, Satkosia Tiger Reserve, *KCM* 5167 (BSID).

INDIA: Throughout India (except in dry West).

WORLD: S. China, Taiwan, Solomon Islands, Malesia.

Cyperus digitatus Roxb., Fl. Ind. 1: 209. 1820; FBI 6: 618. 1893; Fischer 3: 1642. 1931. *C. digitatus* Roxb. var. *hookeri* (Boeck.) C.B.Clarke in Hook. f., Fl. Brit. India 6: 618. 1893.

Perennial, erect, stout, robust herb with woody creeping, rhizome, culms triquetrous, to 100 cm high. Leaves as long as the stem, coriaceous, multistriate, flat, with scaberulous margins; sheaths purplish. Umbel very large, broad, to 15 cm long; involucral bracts 3-12, to 40 cm long, leaflike; primary rays 3-6, rigid, trigonous, to 9 cm long, terminated by stellately spreading sessile or cylindric spikes of unequal length, loosely set all round with innumerable cylindric spikes of unequal length, loosely set all round with innumerable horizonatally spreading spikelets; secondary rays 1-3. Spikes digitate, terete, 7-15 in a cluster, to 2.5 cm long, sessile, 10-20-flowered. Spikelets terete, 3 x 0.5 cm, spicate, at right angles to rhachis, acute, yellowish-brown, rhachilla winged. Achene ellipsoid, trigonous, yellowish.

Occasional near water logged places, at the sides of ponds, channels ditches, and in marsh areas. Fl. & Fr.: July – December.

Penakacherla (ATP), *NY* 592 (SKU & MH); Tungabhadra river (KNL), *KH* 10941; Duggeru (VZN), *MV* 6894 (AU); Kavri peak, Yercaud (SLM), *DBD* 31306 (MH); Yercaud (SLM), *DBD* 31417 (MH);

INDIA: NW. to E. India and Southwards.

WORLD: Pantropics and subtropics.

Cyperus distans L. f., Suppl. 103. 1781; FBI 6: 607. 1893; Fischer 3: 1640. 1931; Matthew, Fl. Tamilnadu Carnatic 3: 1737. 1983.

Key to varieties

1. Spikelets distant, spreading at maturity .... var. **distans**
1. Spikelets crowded, erect at maturity .... var. **pseudonutans**

Cyperus distans L. f., var. **distans**

Perennial, erect, robust, leafy herb; culms not tufted, to 100 cm high, trigonous. Leaves flat, as long as stem, scaberulous; sheaths purplish. Inflorescence brown, decompound, to 25 cm long. Involucral bracts 5-7, leaflike, very much elongated, primary rays 7-9, secondary rays 1-5. Spikes ovate, with 15-20 patent spikelets, at right angles to rhachis. Spikelets distant, spreading, narrowly linear, 2 cm long, purplish, 13-15-flowered; rhachilla persistent. Achenes ellipsoid, triqetrous, grey apicualte.

Occasional on the margins of ponds, ditches, rivers, rice fields and in wet places. Fl. & Fr.: September – December.

Bukkapatnam RF (ATP), *NY* 642; Omkaram (KNL), *BSS* & *MSR* 29348; Talakona (CTR), *BR* & *BSS* 29335; Gundlamotu reservoir (PKM), *BR* & *BSS* 33150; Chinnamantanala (KNL), *JLE* 42290 (MH); Maredumilli RF (EG), *KH* & *BR* 10902, *MM* 105051 (BSID); Tiger camp, Maredumilli (EG), *MM* 105029 (BSID); Tyada forest (VSKP), *KH* & *BR* 9719; Javakulavagu river side (KNL), *JLE* 25487 (MH & CAL); Agiripalle (KSN), *PV* 5230 (MH); Anchetti (SLM), *KCJ* 18044 (MH); Krishnagiri (SLM), *S.India Flora* 14903 (MH); Pachidya (GJM), *VNS* 5829 (MH); Purunakotee RF, Satkosia Tiger Reserve, *KCM* 5175 (BSID).

INDIA: Throughout India.

WORLD: Pantropic.

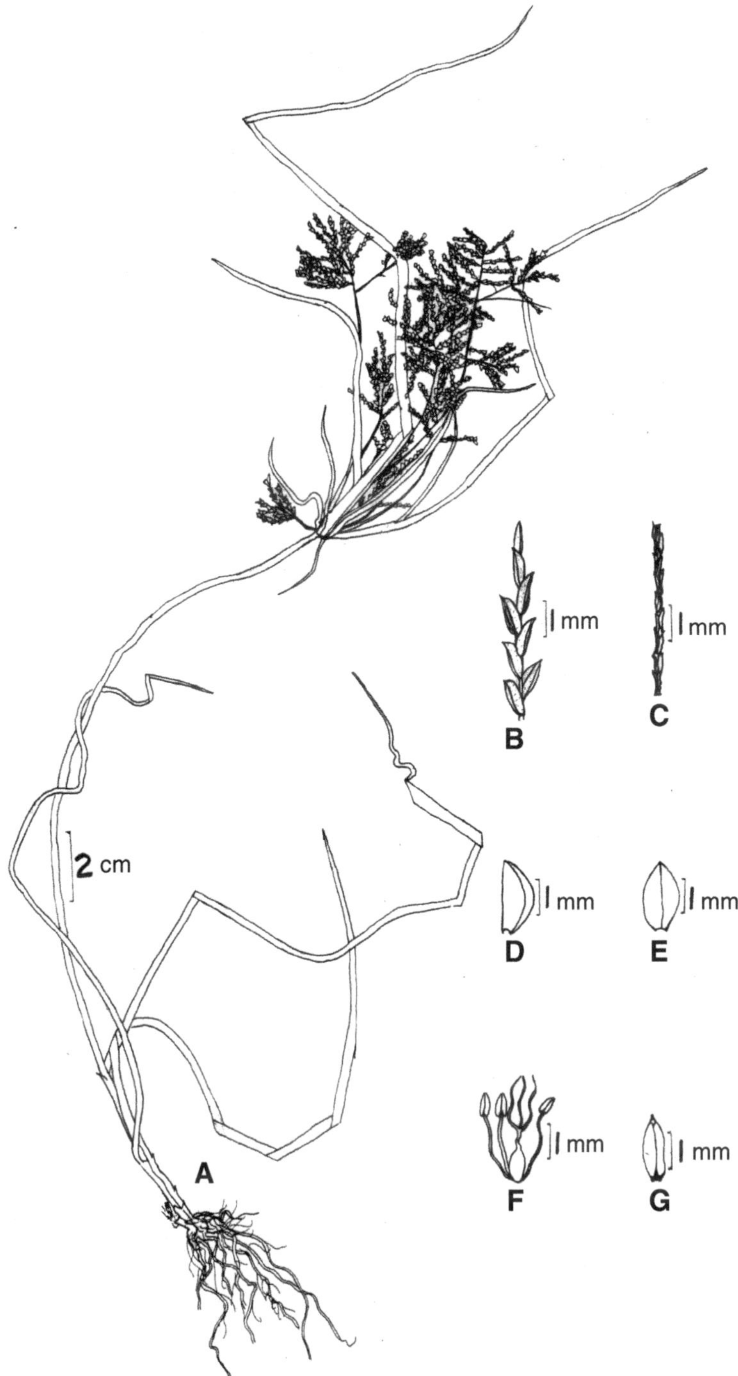

Figure 42. **Cyperus distans** L. var. **distans**
A. Habit, B. Spikelet, C. Rhachilla, D. Closed glume, E. Opened glume, F. Pistil with stamens, G. Achene.

Cyperus distans L. f. var. **pseudonutans** Kuekenth. in Engl., Pflanzenr. 101: 140. 1935. *Cyperus natans* auct. non Vahl: FBI 6: 607. 1893.

Perennial, rhizomatous herb, culms to 90 cm high. Spikes rather densely spikeletted. Spikelets crowded, erect or suberect at maturity.

Occasional in wet places and along streams. Fl. & Fr.: September – December.

Kalasamudram (ATP), *MHR* 14927; Batrepalli (ATP), *KRKS* 37850; Omkaram (KNL), *DV* & *BR* 29351; Velugodu – Yoganandakonda (KNL), *BR* & *BSS* 29336; Nulakamaddi (EG), *GVS* 24533 (MH); On the way to Punyagiri hills – S. Kota (VZN), *KH* & *BR* 5597; Gundlabrahmeswaram, Nallamalais (KNL), *JLE* 16913 (MH); Nulakamaddi (EG), *GVS* 24533 (MH); Mundanchola, near Sunkarimitta (VSKP), *NPBK* 10951 (MH).

INDIA: NW. to NE. India and southwards.

WORLD: Sri Lanka.

Cyperus dubius Rottb., Descr. Icon. Rar. 20. t. 4. f. 5. 1773. *Mariscus dubius* (Rottb.) Kukenth. ex Fischer, Fl. Madras 3: 1644. 1931; Matthew, Fl. Tamilnadu Carnatic 2: 1772. 1983. *M. dregeanus* Kunth, Enum. Pl. 2: 120. 1837; FBI 6: 620. 1893.

Perennial, tufted herb with short rhizome; culms thickish, stiffly erect, to 20 cm high, triquetrous, with a narrowly ovoidal bulbose enlargement with thickened basal sheaths, 30 mm long, clothed with brown outer sheaths and their dusky brown fibrous remnants. Leaves 7 to a culm; blades linear surpassing the culm, soft, herbaceous, scabrid on upper margins; sheaths to 5 cm long, scarious, whitish. Inflorescence a dense, hemispherical, ovoid, deltoid head, to 12 mm long and as wide. Involucral bracts 4, spreading, the longest 13 cm long; spikes broadly ovoidal, densely bearing 20 spikelets, light greenish; rhachis subulate, 2 mm long. Spikelets patent, ovate, 5 mm long, bearing 6 glumes, 6-flowered, rhachilla straightish; wings lanceolate, white-hyaline. Achenes elliptical, 1.5 mm long, trigonous, apiculate, light brown.

Occasional in open rocky ground. Fl. & F.: August – December.

Ahobilam (KNL), *BR* & *BSS* 29327 (SKU & BSID); Tirumala – Papanasanam (CTR), *MHR* 13970; Kambakkam (CTR), *BR* & *MVS* 32116; Simhachalam hill (VSKP), *KH* & *BR* 5590; Ummithela - Kodur (KDP), *JLE* 14334 (MH); Gundlabrahmeswaram, Nallamalai (KNL), *JLE* 32670 (MH); Nagapatla RF (CTR), *KS* 6857 (MH & CAL); Nallathippa, Kambakkam range (CTR), *MCB* 45172 (MH); Near Rollapenta (PKM), *VBH* 83968 (BSID); Kondavidu (GNT), *S.India Flora* 17501 (MH); Rampachodavaram (EG), *VNS* 231 (CAL), *MM* 105110 (BSID); Dharakonda (VSKP), *S.India Flora* 17027 (MH); Near Anantagiri (VSKP), *NPBK* 11010 (MH); Salur (VZN), *NPBK* 940 (CAL); Billy guttalu (KMM), *R.Rajan* 105979 (BSID); Pennagaram RF (SLM), *EV* 20743 (MH); Muthanur – Harur (DMP), *EV* 51863 (MH); Tirupporur RF (CPT), *ANH* 45577 (MH); Gingee RF (SA), *KRM* 13032 (MH); Sirupakkam RF (SA), *KRM* 50694 (MH); Puthari, Inner Javadi RF (NA), *KS* 6049 (MH); Tirupattur (NA), *KS* 6512 (MH); Tippukadu

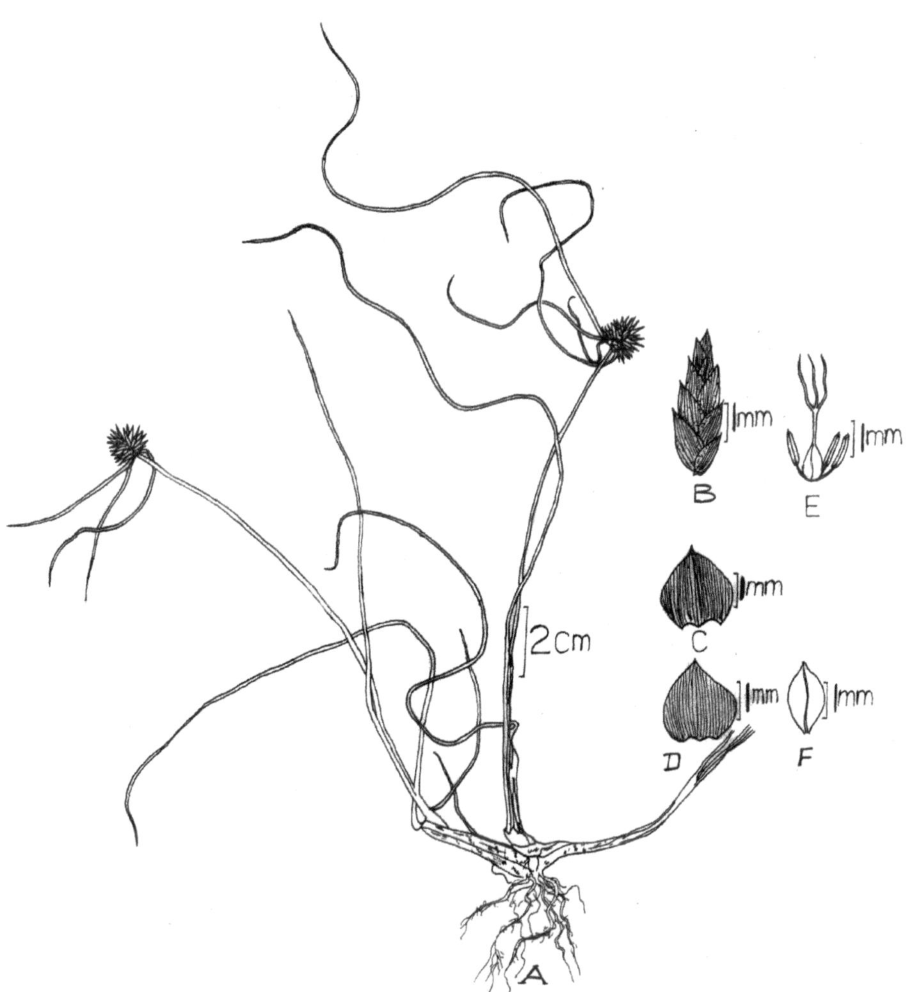

Figure 43. **Cyperus dubius** Rottb. [Syn.: *Mariscus dubius* (Rottb.) Kuekenth.]
A. Habit, B. Spikelet, C & D. Stamens, E. Pistil with stamens, F. Achene.

RF (NA), *KRM* 17667 (MH); Krishnagiri (SLM), *S. India Flora* 14948 (MH); Illalur RF, Tirupporur (CPT), *ANH* 45582 (MH); Nirmaljhar- Salapa (GJM), *MB* 2592 (RRL-B).

INDIA: Through out India.

WORLD: Widespread in Old World Tropics from Africa to N. Australia.

Cyperus elatus L., Cent. Pl. 2: 301. 1756 & Sp. Pl. ed. 2. 67. 1762; FBI 6: 618. 1893; Fischer 3: 1642. 1931.

Perennial, stout, robust herb, to 120 cm high; culms above bluntly trigonous, the angles smooth. Leaves nearly as long as culm. Inflorescence compound umbel. Involucral bracts 10, to 80 cm long; primary rays to 10, to 20 cm long; secondary rays to 5 cm long; spikes narrowly cylindric, clustered at the ends of the rays and few sessile at the base of the partial umbels, to 5 cm long; spikelets densely spirally inserted on the angular rhachis, erect, linear, wings of rhachilla lanceolate, yellow, caducous. Achenes trigonous, ellipsoid, pale-brown.

Occasional on river banks and swamps. Fl. & Fr.: August – December.

Kuneru – Komarada (VZN), *MCK* 25256; Porumamilla (KDP), *JSG* 10886 (DD).

INDIA: Uttar Pradesh, Madhya Pradesh, Andhra Pradesh, Tamil Nadu, Kerala.

WORLD: Bangladesh, Cambodia, China, Indonesia, Laos, Thiland, Vietnam.

Cyperus esculentus L., Sp. Pl. 45. 1753; FBI 6: 616. 1893; Fischer 3: 1641. 1931; Saxena & Brahmam, Fl. Orissa 4: 2127. 1996.

Perennial; stolons several, slender, clothed with pale scales, ending in tubers; tubers ovoid to globose, *c.* 1 cm diam., transversely zoned when young and covered with grey tomentum when ripe; stolons disappearing after tuber formation; stems slender, rigid, trigonous or triquetrous, smooth, 15-30 cm long. Leaves several, somewhat shorter or longer than the stems, rather rigid, gradually acuminate, 3-6 mm wide, lower sheaths stramineus to reddish brown. Inflorescence usually compound, *c.* 5-10 cm long, involucral bracts 3-5, often shorter than the inflorescence, primary rays 3-8, unequal, up to 10 cm long, secondary rays 3, up to 3.7 cm long. Spikes ovoid, with few to numerous spikelets; rachis glabrous. Spikelets distinctly alternate at the ends of the rays, oblong-linear, 5-20 x *c.* 2mm, subcompressed, obtuse, obtuse, 8-16-flowered, rachilla slightly flexuous, broadly winged; glumes laxly imbricate, ovate to elliptic, 2.5-3 mm long, membranous, obtuse, sometimes minutely mucronuate, yellow to pale brown with whitish hyaline margins, distinctly 7-9-nerved over their whole breadth; stamens 3; stigmas 3. Nut trigonous, obovoid to oblong-obovoid, 1.5 mm long, obtuse, hardly apiculate.

Horsley hills at 1300 m (Gamble, cf. Fischer *loc cit.*).

INDIA: From the Gangetic plain to Nilgiri.

WORLD: Widely distributed from the Mediterranean region to S. Africa and through India to N. Queensland, abundant in America, also cultivated.

Cyperus exaltatus Retz., Observ. Bot. 5: 11. 1789; FBI 6: 617. 1893; Fischer 3: 1642. 1931; Matthew, Fl. Tamilnadu Carnatic 3: 1737. 1983.

Perennial, stout, glabrous, tufted erect herb with a short rhizome; culms to 110 cm high, rigid, trigonous. Leaves folded, to 75 cm long, thick, to 8 mm wide, nerves prominent above, multistriate; sheaths brown to purple, spongy. Umbel compound, to 20 cm long; involucral bracts 3-6, overtopping, longest to 55 cm long, leaf like; primary rays 7-10, unequal, to 20 cm long, spreading; secondary rays 3-5, to 5 cm long, bearing spikes. Spikes umbelled or corymbose, terete, loose, to 4 x 1 cm long, 30-40-spikeletted, rhachis visible, flexuous. Spikelets spicate, at right angles to rhachis, oblong, ascending chestnut brown, much compressed, 1 cm long, acute, 10-20-flowered; rhachilla winged. Achenes ellipsoid, tapering to both ends, trigonous, brown.

Common in wet fields, swamps, marshy areas, along streams and canals. Fl. & Fr.: August – December.

Hottabetta RF (ATP), *BR* & *ANS* 35571; Kanigiri reservoir (NLR), *MCK* 22886; Tadepalligudem (WG), *MCK* 22914; Jalipinaidupeta (SKLM), *MCK* 18891; Nunna (KSN), *PV* 5925 (AU); Madugukota (EG), *SS* 9910 (AU); Gokavaram (EG), *CAB* 8267 (MH); Chintapalli (VSKP), *GVS* 28212 (CAL); Barmakonda RF (VSKP), *KCJ* 17148 (MH); Salur (VZN), *NPBK* 929; Duggeru (VZN), *MV* 6862 (AU); Malkangiri-Chitrakonda (KPT), *SX* & *MB* 6827 (RRL-B); Brudhakol (GJM), *MB* 2784 (RRL-B); Kalasanapur (GJM), *VNS* 4654 (MH); Singarapet (NA), *KS* 6578 (MH);

INDIA: NW. to NE. India and southwards.

WORLD: Tropical Africa, Pakistan, Sri Lanka, Bangladesh, Myanmar, Nepal, China, Thailand, Vietnam, Papua New Guinea, Australia.

Cyperus haspan L., Sp. Pl. 45. 1753 (*halpan*); FBI 6: 600. 1893; Fischer 3: 1640. 1931; Matthew, Fl. Tamilnadu Carnatic 3: 1738. 1983.

Key to Subspecies

1. Rhizome long, creeping, stem spaced; glumes 1.5 – 1.8 mm; stamens 3 .... subsp. **juncoides**
1. Rhizome short, stem tufted; glumes 1 mm long; stamens 1 .... subsp. **haspan**

Cyperus haspan L. subsp. **haspan**

Perennial, tufted, rhizomatous herb, culms to 50 cm high, acutely triquetrous, leaved at base. Leaves 2 to a culm, blades linear, to 15 cm long, herbaceous, light green; sheaths to 4 cm long, lower ones membranous, red purple. Umbel compound, dense to subloose, to 10 cm long. Involucral bracts 2, the lowest one surpassing the inflorescence, the longest one to 8 cm long; primary rays to 10, unequal, patent, up to 7 cm long; secondary rays to 2 cm long bearing 3-5 stellately spreading spikelets. Spikelets linear-oblong, to 10 mm long, subdensely to 28-flowered, chestnut brown, the wingless rhachilla hidden by densely disposed glumes. Achenes broadly obovate, trigonous, cream-yellow.

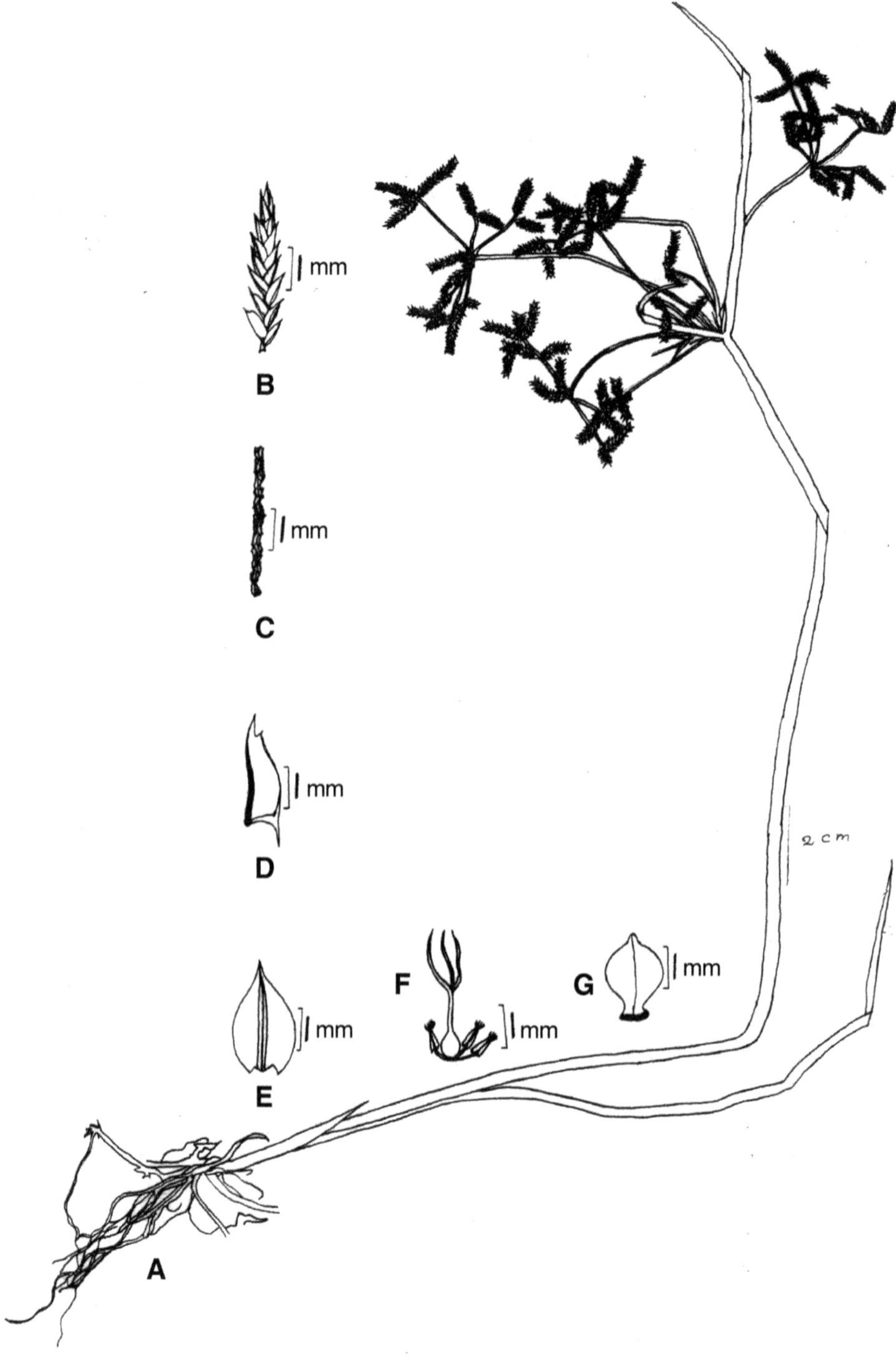

Figure 44. **Cyperus haspan** L. subsp. **haspan**
A. Habit, B. Spikelet, C. Rhachilla, D. Closed glume, E. Opened glume, F. Pistil with stamens, G. Achene.

Common weed in rice fields, abundant in wet and marshy areas. Fl. & Fr.: August – December.

Batrepalli (ATP), *KRKS* 40351; Ahobilam (KNL), *BR* & *BSS* 29332; Korrapadu (KDP), *AL* & *BSS* 28634 B; Kota (NLR), *PMR* 19513; Maredumilli (EG), *KH* & *BR* 10906 A; Owk RF (KNL), *RVR* 3035; Seethalam river bed (CTR), *MCB* 45093 (MH); Isukagundalu, GBMLS (PKM), *J.Swamy* & *L.Rasingam* 14430 (BSID); Minumuluru (VSKP), *GVS* 29747 (MH & CAL); Srisailam (KNL), *JLE* 16894 (CAL); Ahobilam (KNL), *JLE* 32610 (MH); Nunna (KSN), *PV* 5923 (AU); Maredumilli towards Kutravada (EG), *GVS* 68542 (MH); Khotnighada area (VSKP), *GVS* 19506 (MH); Majhipada, Satkosia Tiger Reserve, *KCM* 7049 (BSID).

INDIA: Throughout India.

WORLD: Tropical and warm temperate regions of both hemispheres.

Cyperus haspan L. subsp. **juncoides** (Lam.) Kuekenth., Feddes Rep. 23: 184. 1926. *C. juncoides* Lam. Tabl. Encycl. 1: 147. 1791. *C. halpan* var. *evoluta* Valck., Sur. Gesl. Cyp. Mal. Arch. 92. 1898. *C. halpan* var. *flaccidissimus* Kuekenth. in Engl., Pflanzenr. 101: 248. 1936.

Perennial herb with long creeping rhizome, the culms are disposed in a row in a spaced manner.

Occasional in marshy areas and in wet grass lands. Fl. & Fr.: September – November.

Owk RF (KNL), *RVR* 3035; Guvvalacheruvu (KDP), *R.V.Reddy* 7966; Pedatheerthala kona (CTR), *BR* & *AMR* 26090; Maredumilli (EG), *KH* & *BR* 10906 B; S.Kota reservoir (VZN), *MCK* 18810; Seethampeta tank (SKLM), *MCK* 25289; Kumaradhara (CTR), *GVM* & *JHFB* 118903 (MH); Nellippkka RF (KMM), *RCS* 99090 (BSID).

INDIA: N. India, Andhra Pradesh.

WORLD: Pantropical.

Cyperus iria L., Sp. Pl. 45. 1753; FBI 6: 606; Fischer 3: 1640. 1931; Matthew, Fl. Tamilnadu Carnatic 3: 1738. 1983. *C. iria* var *paniciformis* (Franch. & Sav.) C.B.Clarke in Hook. f., Fl. Brit. India 6: 607. 1893.

Annual, glabrous, erect, tufted, marshy herb; culms to 50 cm long, triquetrous, striate, rigid. Leaves flat or channeled, to 30 cm long, 6 mm wide, scaberulous, multistriate, acuminate, flaccid; sheaths purplish. Umbel decompound, 5-15 cm long; primary rays 5-7, to 11 cm long; secondary rays 3-5, to 8 cm long, bearing irregularly fascicled umbellules, formed of narrow interrupted spikes of 15-20, small, few-flowered spikelete. Rhachilla of the spike glabrous. Spikelets spicate, oblong, 8 x 1 mm, to 15-flowered. Achenes obovoid, triquetrous, black or brown.

Common weed in wet and flooded paddy fields. Fl. & Fr.: July – December.

Chelama (KNL), *BR* & *BSS* 32245 (BSID); Palakonda hill (KDP), *CS* 7699; Darsi (PKM), *MCK* 22827; S.Kota reservoir (VZN), *MCK* 18812; Diguvametta (PKM), *JLE*

32515 (MH); Kambakkam hill (CTR), *MCB* 45189 (MH); Venkatagiri hill (NLR), *BSN* 4137 (VV); Malakondapenta RF (PKM), *RKM* 0506 (CAL); Marripalem (EG), *GVS* 27415 (MH); Rampachodavaram (EG), *MM* 105103; Araku (VSKP), *NPBK* 10886 (CAL); Rangamati gedda near Singapuram (SKLM), *GVS* 62338 (MH); Rathamhutta hills (KMM), *RCS* 104227 (MH & BSID); Parnasala RF (KMM), *RCS* 99053 (MH & BSID); Krishnagiri (SLM), *S.India Flora* 14905 (MH); Kottur RF, Yelagiri hills (NA), *MBV* 802 (MH); Tippukadu RF (NA), *KRM* 17650 (MH); Muthanur – Harur (DMP), *EV* 51859 (MH); Mahendragiri (GJM), *VNS* 5747 (MH); Pachidya (GJM), *VNS* 5849 (MH); Tulka RF, Satkosia Tiger Reserve, *KCM* 8326 (BSID); Purunakote RF, Satkosia Tiger Reserve, *KCM* 5296 (BSID).

INDIA: Throughout India.

WORLD: Tropical and temperate regions of both hemispheres.

Cyperus laevigatus L., Mant. Alt. 179. 1771; Matthew, Fl. Tamilnadu Carnatic 3: 1738. 1983. *Juncellus laevigatus* (L.) C.B.Clarke in Hook. f., Fl. Brit. India 6: 596. 1893; Fisher 3: 1629. 1931.

Perennial, tufted, rhizomatous herb, rhizome creeping horizontally with solitary distant stems, their bases enclosed by imbricate, shining, chestnut red scales. Culms to 30 cm high, round, fleshy. Leaf blades undeveloped. Spikelets 30 in one apparently lateral head, straw-coloured. Involucral bracts 2, the lower as though a continuation of stem, to 7 cm long, the other much shorter. Spikelets ovate, compressed but thick, straw-coloured dotted with brown spots. Achenes obovoid, brown, stipitate.

Occasional in moist localities, salt marshes, sandy beds of rives and muddy soil. Fl. & Fr.: September – March.

Guttur kona (ATP), *NY* 983 (SKU & MH); Kalasamudram (ATP), *KRKS* 39535; Kurichedu (PKM), *MCK* 22818; Gooty (ATP), *JSG* 11091 (DD); way to Guvvalacheruvu (KDP), *KS* 7789 (MH).

INDIA: Western India, Andhra Pradesh.

WORLD: Most warm regions.

Cyperus leucocephalus Retz., Observ. Bot. 5: 11. 1789; FBI 6: 602. 1893; Fischer 3; 1640. 1931. *Sorostachys leucocephalus* (Retz.) Lye, Nord. J. Bot. 1: 190. 1981.

Perennial, tufted, rhizomatous herb; rhizome very short, woody; culms to 30 cm high, very slender, nodose at the base, each carrying one head. Leaves shorter than the stem. Heads dense, globose, white, containing 40 spikelets; involucral bracts 3, the longest reaching to 6 cm long, long acuminate. Spikelets compressed, elliptic, to 18-flowered. Achenes oblong-ellipsoid, trigonous, shortly apiculate, black.

Occasional in marshy and swampy areas. Fl. & Fr.: July – October.

Guvvalacheruvu (KDP), *KS* 6382 (MH & CAL); Isakagundam (PKM), *VBH* 83998 (BSID); Pottangi (KPT), *HFM* 3946 (DD).

INDIA: S., C. & E. India.

WORLD: W Africa, Sudan, S Asia and Australia.

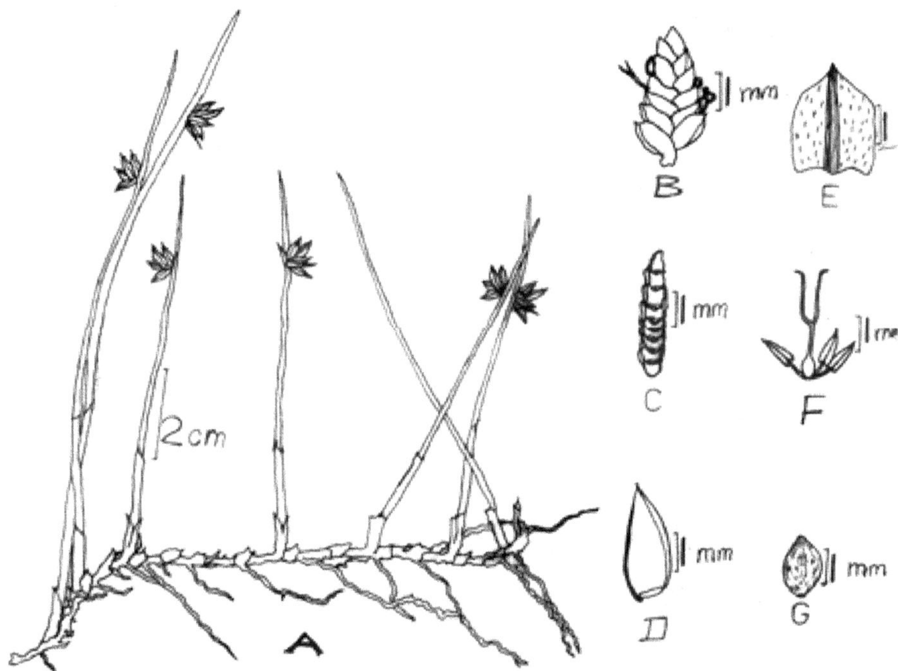

Figure 45. **Cyperus laevigatus** L. [Syn.: *Juncellus laevigatus* (L.) C.B.Clarke]
A. Habit, B. Spikelet, C. Rhachilla, D. Closed glume, E. Opened glume,
F. Pistil with stamens, G. Achene.

Cyperus michelianus (L.) Delile, Egypte, Hist. Nat. 3: 50. 1813 subsp. **pygmaeus** (Rottb.) Asch. & Graebn., Syn. Fl. Mitteleur 2(2): 312. 1936. *Scirpus michelianus* L., Sp. Pl. 52. 1753; FBI 6: 662. 1893. *Juncellus pygmaeus* (Rottb.) C.B. Clarke in Hook. f., Fl. Brit. India 6: 596. 1893; Fischer 3: 1629. 1931. *Cyperus pygmaeus* Rottb., Descr. Icon. 20. t. 14, f. 5. 1773.

Annual, dwarf, tufted herb; culms to 5 cm high, triquetrous. Leaves several, canaliculated, to 3 cm long, flaccid, scaberulous, green, acuminate; sheaths reddish. Spikelets very many, sessile, densely packed in a terminal head; involucral bracts 6, unequal, to 4 cm long, base dilated, acuminate, leaflike. Spikelets broadly ovate, to 3 mm, pale-brown, 5-10-flowered, compressed. Glumes pellucid, distichous, but often seemingly placed irregularly because of the twisted rachilla, oblong-laceolate or lanceolate, *c.* 2 x 0.5 mm, keeled, acute or with a short mucro, 3-5-nerved, very closely imbricate, at first white, finally stramineous, keel smooth; stamens 2. Achenes ellipsoid, trigonous, orange-brown.

Frequent along the sandy or silty beds of rivers, in rice fields as tufted cushions. Fl. & Fr.: September – March.

Talupula RF (ATP), *DV* & *BR* 44868; Peccheruvu (KNL), *BR* & *BSS* 30151; Horsley hills (CTR), *RVR* 14960; Nelapattu bird sanctuary (NLR), *MHR* 13327 A; Dhavaleswaram barrage (EG), *KH* & *BR* 9726; Narabylu (CTR), *GVS* 46026 (MH); Alluru (KSN), *CAB* 7939 (MH); Katekalur (EG), *Bourne* 3339 (MH); Hogenakkal (SLM), *S.India Flora* 18012 (MH); Serango (GJM), *SX, MB* & *GP* 2420 (RRL-B).

INDIA: Assam, Uttar Pradesh, Madhya Pradesh, Karnataka, Goa, Andhra Pradesh, Tamil Nadu.

WORLD: Africa, Europe, Russia, SW Asia, Pakistan, Myanmar, Nepal, China, Philippines, Thailand, Vietnam, Australia.

Cyperus mitis Steud., Syn. Pl. Glumac. 2: 316. 1855. *C. stenostachyos* Benth. var. *indica* C.B.Clarke in Hook. f., Fl. Brit. Ind. 6: 614. 1893. *C. subcapitatus* C.B. Clarke in Hook. f., Fl. Brit. Ind. 6: 616. 1893.

Perennial, stoloniferous herb, stolons terminated by a globose tuber; culms to 60 cm high, solitary, triquetrous above, forming globose, corm like enlargement at base. Leaves few, shorter than the culms; blades flattish-plicate; sheaths to 15 cm long, stained with light reddish brown, the basal sheaths dusky, disintegrating into dark brown fibres. Inflorescence terminal with the lowest involucral bract upright, simple, to 7 cm long, rather lax. Involucral bracts leafy, 4, elongated, the lowest much surpassing the inflorescence, to 10 cm long; primary rays *c.* 6, erect-patent, filiform, unequal, the longer ones to 5 cm long. Spikes broadly obovate, bearing 5 spikelets. Spikelets obliquely patent, to 20 mm long, flattened, *c.* 30-flowered, yellow brown; rhachillia winged. Achenes oblong, trigonous.

Reported from Tirumala hills by Naidu and Prakasa Rao (1969). Fl. & Fr.: October – December.

INDIA: Peninsular India.

WORLD: Magagascar, Sri Lanka, Myanmar, China, Thailand, Indian Ocean Islands.

Note: No specimens in any herbaria from Eastern Ghats, doubtful occurrence in Eastern Ghats.

Cyperus niveus Retz., Observ. 5. 12. 1789; FBI 6: 601. 1893; Fischer 3: 1640. 1931.

Perennial, glaucescent, rhizomatous herb, 20-70 cm high; rootlets wiry; culms nodose at the base. Leaves much shorter than the stem, to 25 cm long, narrowly linear, setaceo-acuminate. Spikelets 6-16 in one head, 2 cm long, oblong-elliptic, compressed, pale or sometimes almost white, 20-30-flowered, rhachilla not winged. Involucral bracts 2, longest reaching 5 cm long, finely acuminate, leaflike; spikelets straw-coloured, compressed, lanceolate, ovate-lanceolatee or ovate-oblong; glumes ovate, keeled; stamens 3; ovary trigonous, style up to 3 mm long, stigmas 3, style and stigmas much accrescent in fruit. Achenes obovoid, triquetrous, dark brown.

Rare in sandy localities of river beds, in gravelly soils and in dry grass lands. Fl. & Fr.: May – November. Vern.: Ori.: *Balsurya mukhi.*

Gani RF (KNL), *RVR* 1506; Nagalooty (KNL), *VBH* 83910 (BSID); Nekkerikal (GNT), *S.India Flora* 17474 (MH); Devipatnam (EG), *GVS* 24443 (MH & CAL); Ongole (PKM), *CAB* 7952 (MH); Nagarjunakonda valley (GNT), *KT* 9713 (CAL); Karaka (VSKP), *sine coll. s.n.* (MH); Lakshmipuram (SKLM), *CAB* 1756 (MH); Similipahar (MBJ), *SX* 349 (RRL-B), *SX* & *MB* 4308 (RRL-B); Pampasar RF, Satkosia Tiger Reserve, *KCM* 5139 (BSID).

INDIA: Throughout India.

WORLD: Tropical Africa, SW Asia, Afghanistan, Pakistan, Myanmar, Nepal, China, Thailand, Vietnam.

Cyperus nutans Vahl, Enum. Pl. 2: 363. 1806; FBI 6: 697. 1893; Fischer 3: 1640. 1931; Matthew, Fl. Tamilnadu Carnatic 3: 1740. 1983.

Key to Subspecies

1. Spikes loose, sub erect, open; achenes ellipsoid, 1.5 mm long .... subsp. **nutans**
1. Spike dense, erect, close; achenes obovoid, 1.2 mm long .... subsp. **eleusinoides**

Cyperus nutans Vahl subsp. **nutans**

Perennial herb with short corm like rhizome. Culms solitary, to 40 cm high, trigonous. Leaves few, shorter than the culm, to 20 cm long, subcoriaceous, flattish, 3-costate, whitish beneath, scabrous in upper margins; sheaths subloosely surrounding the culm base, the lower ones tinged with reddish brown, the upper most pale green, to 10 cm long. Inflorescence compound, to 10 cm long. Involucral bracts 5, foliaceous, lower 2 much surpassing the inflorescence, to 25 cm long;

primary rays 6, unequal, to 10 cm long. Spikes loose, suberect, open, to 4 cm long, spreading. Spikelets erect, patent, linear, 4 mm long, light brown. Achenes ellipsoid, 1.5 mm long, maturing brown.

In wet low lands, in shallow ponds, and also as a weed in low land cultivations. Fl. & Fr.: June – October.

Gavi Siddeswaram (ATP), *CJP* & *VS* 14169; Tirumala (CTR), *MHR* 14906; Punyagiri hills (VZN), *MCK* 18813; Nulakamaddi (EG), *GVS* 24533 (CAL).

INDIA: NW. to NE India and Sotuhwards, C. Himalaya.

WORLD: Sri Lanka, S. China, Malesia.

Cyperus nutans Vahl subsp. **eleusinoides** (Kunth) T. Koyama, Gard. Bull. Singapore 30: 136. 1977. *Cyperus eleusinoides* Kunth, Enum. Pl. 2: 39. 1937; FBI 6: 608. 1893; Fischer 3: 1640. 1931.

Culms to 90 cm high. Leaves as long as the culms, thick. Inflorescence decompound, 15 cm long. Involucral bracts 7, longest to 30 cm long; primary rays 7. Spikes erect, oblong, close, densely spikeletted, bearing spikelets nearly to the base; glumes ovate, to 2 mm, cuspidate, sides pinkish-nerved, margin hyaline apex retuse. Achenes obovoid, trigonous 1.2 mm long, more suddenly narrowed to apiculate tip.

In swamps or wet areas, along banks of streams and rivers. Fl. & Fr.: September – December.

Pennahobilam (ATP), *TP* & *NY* 358; Thungabhadra river (KNL), *KH* 10946; Seethampeta (SKLM), *MCK* 25292; Rollapenta (KNL), JLE 42253 (MH); Vishnunandi – Mahanandi (KNL), *JLE* 25449 (MH); Horsleykonda (CTR), *JSG* 15062 (CAL); Bhupathipalem (EG), *VNS* 239 (CAL); Kiliyur water falls, Yercaud (SLM), *SKK* 28309 (MH); PennagaramRF (SLM), *EV* 22423 (MH); Inner Javadi RF (NA), *KS* 6050 (MH); Tulka, Satkosia Tiger Reserve, *KCM* 6773 (BSID); Tikarpada, Satkosia Tiger Reserve, *KCM* 7628 (BSID).

INDIA: Throughout India.

WORLD: Tropics of Old world.

Cyperus pangorei Rottb., Descr. Icon. Rar. 31. 7. f. 3. 1773; Fischer 3: 1641. 1931; Matthew, Fl. Tamilnadu Carnatic 3: 1740. 1983. *C. tegetum* Roxb., Fl. Ind. 1: 211. 1820; FBI 6: 613. 1893. *C. corymbosus* Rottb. var. *pangorei* (Rottb.) C.B.Clarke, J. Linn. Soc. 21: 292. 1884 & in FBI 6: 613. 1893.

Perennial, erect, glaucous green, rhizomatous, robust, marshy herb; clulms to 130 cm high, 7 mm wide, rigid, trigonous, ribbed, sulcate. Leaves flat, to 25 cm x 1 cm, acuminate, nerves prominent; sheath long, purplish, lax. Umbel compound, to 10 cm long; involucral bracts 3, unequal, overtopping, longest 15 cm long, erect-patent, with strong midrib and scaberulous margins which are usually recurved in dried specimens; primary rays 5-7, to 7.5 cm long; secondary rays 5, to 2 cm long. Spikes corymbose, 3-10-spikeletted. Spikelets spicate, narrow-oblong, to 3 cm long, purplish brown, acute, compressed; rhachilla with large, dark-brown, deciduous wings. Achenes ellipsoid, trigonous, tapering towards the base, yellowish-brown.

Occasional along the water streams in marshy and swampy areas. Fl.& Fr.: September – December.

Amagondapalem RF (ATP), NY 1029 (MH); Macherla RF (GNT), *VRK* 3283; Peddavasantha (VZN), *MCK* 18862; Diguvametta, Nallamalais (PKM), *JLE* 32466 (MH & CAL); Srisailam (KNL), *BSN* 4368 (BSID); Chitvel RF (KDP), *A.M.Rao* 3036 (BSID); Rajampet (KDP), *JSG* 10800 (DD & CAL); Penchalakona (NLR), *A.S.Rao* 4117 (BSID); S.N. Gollapalem (KSN), *PV* 5549 (MH); Polavaram agency (WG), *DCSR* 190 (CAL); Maredumilli (EG), *MM* 100827 (BSID); Palakonda RF (VSKP), KCJ 17262 MH); Parnasala RF (KMM), *RCS* 99051 (MH & BSID); Rathamhutta hills (KMM), *RCS* 98795 (MH); Hogainakkal (KGR), *KCJ* 18018, 18021 (MH); Puliyur RF, Chengam range, Javadi hills (NA), *MBV* 1057 (MH); Kambakkam hills (NA), *S. India Flora* 16438 (MH);

INDIA: Throughout India.

WORLD: Sri Lanka, Nepal, Mynmar.

Cyperus paniceus (Rottb.) Boeckeler, Linnaea 36: 381. 1870. *Kyllinga panicea* Rottb., Descr Ic. Rar. 15, t. 4, f. 1. 1773. *Mariscus paniceus* (Rottb.) Vahl, Enum. Pl. 2: 373. 1806; FBI 6: 620. 1893; Fischer 3: 1644. 1931; Matthew, Fl. Tamilnadu Carnatic 2: 1774. 1983.

Perennial, rhizomatous, stoloniferous herb; rhizome decumbent, knotty, 2 cm long, clothed with dark red brown fibres, emitting stolons; stolons covered with reddish-brown scales, soon disintegrating into brown fibres, culms to 40 cm high, trigonous. Leaves 4 to a culm; blades nearly equaling the culm, flattish-plicate; sheaths to 5 cm long, reddish-brown. Inflorescence simple. Umbel of spikes or digitate-contracted, to 3 cm long, light glaucous-greenish. Involucral bracts 7, patent, surpassing the anthela, the longest to 25 cm long. Spikes cylindrical, to 1.5 cm long, densely bearing up to 30 spikelets. Spikelets horizontally divergent, lanceolate, whitish green, slightly recurved, 3 mm long, bearing 4 glumes, and 1-flowered. Achenes elliptic-oblong, trigonous, yellow brown.

Rare along forest margins, in damp places preferably under shade. Fl. & Fr.: July – December.

Batrepalli (ATP), *KRKS* 39607; Rampa hill (EG), *KH* & *BR* 9734; Pulusumamella (KNL), *VBH* 83931 (BSID); Nagapatla RF (CTR), *KS* 6854 (MH & CAL); Mamandur (CTR), *GVM* 115194 (BSID); Tiger camp, Maredumilli (EG), *MM* 105031 (BSID); Sirumalai (DGL), *sine coll.* 9105 (MH); way to Komattiyur (NA), *KS* 7473 (MH); on the way to Puliyur from Kattathur (NA), *KS* 6527 (MH).

INDIA: Throughout India.

WORLD: Sri Lanka, Nepal, Malesia, Thailand, Indo-China.

Cyperus pilosus Vahl, Enum. Pl. 2: 354. 1806; FBI 6: 609. 1893; Fischer 3: 1641. 1931; Matthew, Fl. Tamilnadu Carnatic 3: 1741. 1983.

Perennial, slender, erect, rhizomatous herb; culms to 90 cm high, rigid, triquetorus. Leaves as long as stem, to 35 x 0.8 cm wide, flat or canaliculated, nerves

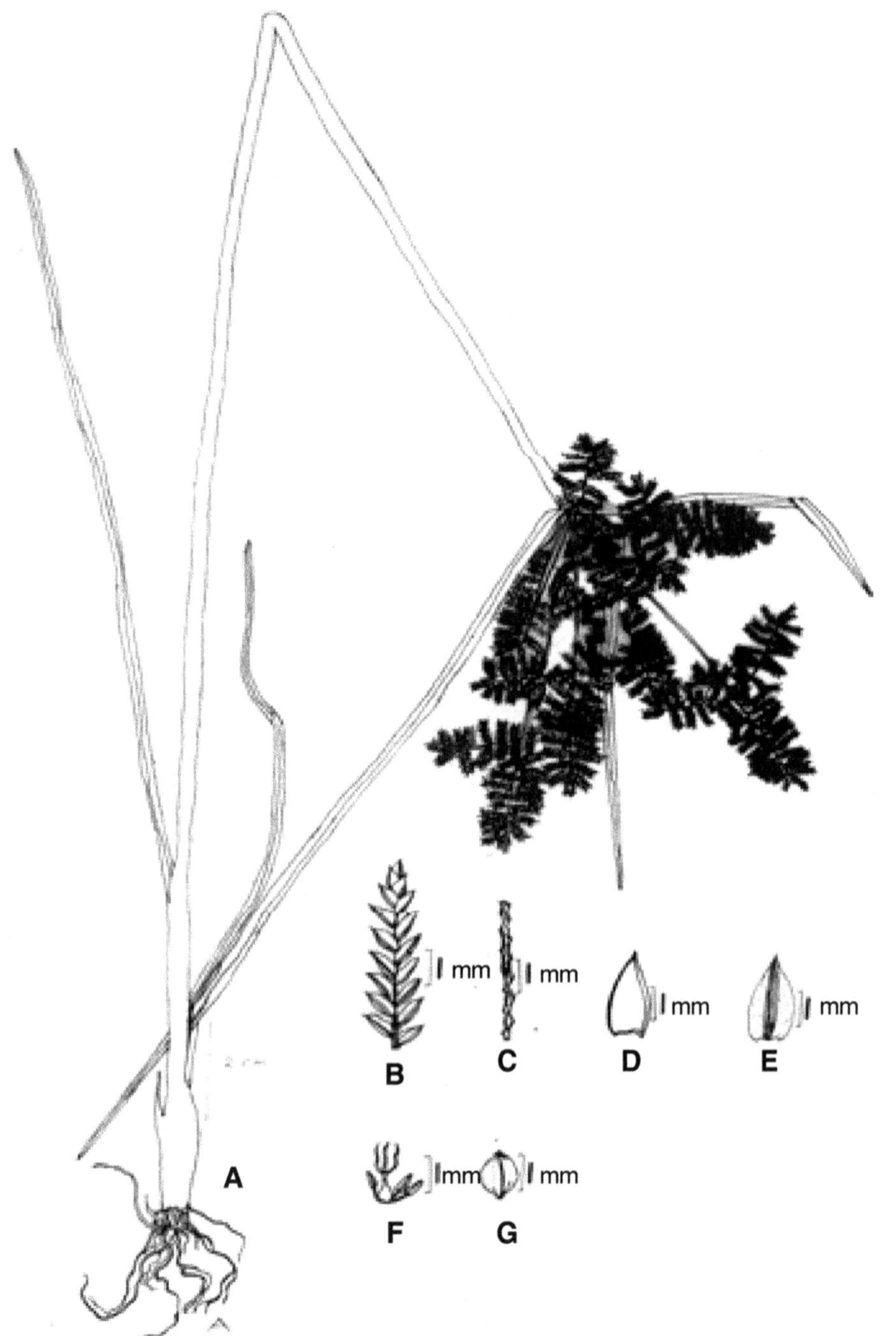

Figure 46. **Cyperus pilosus** Vahl
A. Habit, B. Spikelet, C. Rhachilla, D. Closed glume, E. Opened glume, F. Pistil with stamens, G. Achene

prominent; sheaths pale to purplish. Umbel compound, to 17 cm long; involucral bracts 3-7, overtopping the inflorescence, longest to 20 cm long, primary rays 5-7, to 15 cm long, trigonous; secondary rays 5, 1-5 cm long. Spikes 15-20-spikeletted, at right angles to the rhachis, rhachis densely hispid. Spikelet spicate, oblong, narrow, 7 mm long, compressed, pale to grey-purple, 5-10-flowered; rhachilla hispid. Achenes ellipsoid, triquetrous, dark grey.

A common weed, usually seen in rice fields, river banks, and open wet places. Fl. & Fr.: May – December.

Pennahobilam (ATP), *TP* & *NY* 358; Dharmagiri (CTR), *KI* 21569; Rampa hill (EG), *KH* & *BR* 9736; Maredumilli RF (EG), *KH* & *BR* 10907, *MM* 100810 (BSID); Seethampeta (SKLM), *MCK* 25284; Araku (VSKP), *GVS* 21575 (MH & CAL); Simalagud (VZN), *MV* 5221 (AU); Near Gummada (SKLM), *GVS* 62438 (MH & CAL); Kwadoli, Raigoda, Satkosia Tiger Reserve, *KCM* 6715 (BSID).

INDIA: Throughout India.

WORLD: Tropical Africa, through India to China, Japan and Malesia.

Cyperus platyphyllus Roem. & Schult., Syst. Veg. 2: 876. 1817; FBI 6: 618. 1893; Fischer 1642. 1931.

Perennial, densely tufted, rhizomatous herb; rhizome coarse, obliquely ascending, to 8 cm long; roots dusky brown, culms robust, to 200 cm high, 12 mm thick, sharply triquetrous with shallowly concave sides, scabrous on angles below the inflorescence. Leaves 7 to a culm, up to 2/3 as long as the culm; blades broadly linear, to 3 cm wide, to 100 cm long, scabrous lower midrib and margins toward apex; coriaceous, deep green; sheaths to 50 cm long, sub-cylindric, straw-coloured. Inflorescence compound, to 30 cm long. Involucral bracts 5, patent, broadly linear, two lower ones much surpassing the inflorescence, the lowest to 100 cm long and 3 cm wide; rays 10, the longer ones rather equal, robust, to 20 cm long. Spikes linear, cylindrical, to 6 cm long, yellow-brownish, the rhachis hidden by the densly crowded spikelets. Spikelets erect-patent, linear, to 8 mm long, subterete, to 14-flowered, rhachilla flattened, straightish, the internodes winged. Achenes elliptic, trigonous, apiculate at apex, maturing greyish-brown.

Occasional in marshy areas. Fl. & Fr.: August – December.

Kolleru lake (WG), *KH* & *BR* 10909.

INDIA: Andhra Pradesh, Tamil Nadu.

WORLD: Sri Lanka.

Cyperus platystylis R. Br., Prodr. 214. 1810; FBI 6: 598. 1893; Fischer 3: 1639. 1931.

Perennial, rhizomatous, aquatic, herb, to 90 cm high, culms solitary, triquetrous, stiff, scabrid on angles. Leaves few, equaling the culm; blades linear, to 12 mm wide, coriaceous, 1 costate, flattish-plicate, septate, nodulose, scabrous on margins; sheaths light brown. Inflorescence compound, depressed hemispherical, to 10 cm long, loose. Involucral bracts 10, slightly spaced, spreading, the longest to 70 cm long, to 12 mm wide, primary rays 15, spreading, to 10 cm long, slender and rigid;

secondary rays slightly spaced, hence anthelules truly corymbose, to 4 cm long. Spikelets digitate in groups of 8 at apices of raylets, each lance-oblong, flattened, to 2 cm long, densely to 50-flowered, straw-coloured and stained with brown. Achenes oblong-elliptic, trigonous with concave sides, glaucous brownish, the angles markedly spongy thickened, pale brownish.

Usually seen in marshy areas, swamps, and in the margins of ponds in large communities. Fl. & Fr.: September – December.

Seethampeta tank (SKLM), *MCK* 25287; Lingala (KSN), *PV* 5416 (AU); Lanjigarh (KHD), *MB* & *Rout* 7071 (RRL-B).

INDIA: Throughout India except NW. India.

WORLD: Sri Lanka, India to Taiwan and through Malesia eastwards to northern and eastern Australia.

Cyperus procerus Rottb., Descr. Icon. Rar. 29. t. 5. f. 3. 1773; FBI 6: 610. 1893; Fischer 3: 1641. 1931; Matthew, Fl. Tamilnadu Carnatic 3: 1740. 1983. *C. procerus* Rottb. var. *lasiorhachis* C.B.Clarke in Hook.f., Fl. Brit. India 6: 610. 1893; Saxena & Brahmam, Fl. Orissa 4: 2144. 1996.

Perenial, rhizomatous, stoloniferous herb; stolons slender, long, distantly covered with scales; culms to 100 cm high, triquetrous. Leaves several, shorter than the culms; blades to 10 mm wide, sub coriaceous, scabrid margined distally; sheaths long, pale greenish, the basal sheaths stained with dusty brown. Inflorescence simple, subloose, to 15 cm long. Involucral bracts 4, erect patent, the lowest one surpassing the anthela, the longest to 50 cm long; primary rays 7, very unequal, erect – patent, to 15 cm long; secondary rays to 2 cm long. Spikes broadly ovoid, to 4 cm long, loose, bearing 8 spikelets, rhachilla glabrous; spikelets oblong, often weakly curved, to 8 cm long, patent, subdensely to 40-flowered, pale and tinged with red brown. Achenes elliptical-trigonous, brownish.

Usually seen in marshy areas, rice fields, grows with *C. pilosus*. Fl. & Fr.: August – January.

Mahanandi (KNL), RVR 14550; Pamarru (KSN), *MCK* 22924: Maredumilli (EG), *KH* & *BR* 9738; Jonnavalasa (VZN), *MCK* 18835; Palagudem – Elugu (EG), *VNS* 4464 (MH); Pennankanimedu (SA), *VNS* 5137 (MH);

INDIA: East India southwards.

WORLD: From Sri Lanka and India to Cochinchina, E. China, Taiwan, southwards to Queensland, Malesia.

Cyperus pulchellus R.Br., Prodr. Fl. Nov. Holl. 213. 1810; Saxena & Brhmam, Fl. Orissa 4: 2145. 1996.

Pereennial with short rhizome, stolons 0; stems very slender, tufted, trigonous below, triquetrous above, smooth, 10-20 cm. Leaves 1-2, filiform, shorter than the stem, scaberulous at the top. Inflorescence a single globose, whitish or pale head, 5-10 mm across; involucral bracts usually 3, much overtopping the inflorescence; spikelets 8-50, ovate, 2-3 x 1.5-2 mm, strongly compressed, 8-12-flowered; rachilla

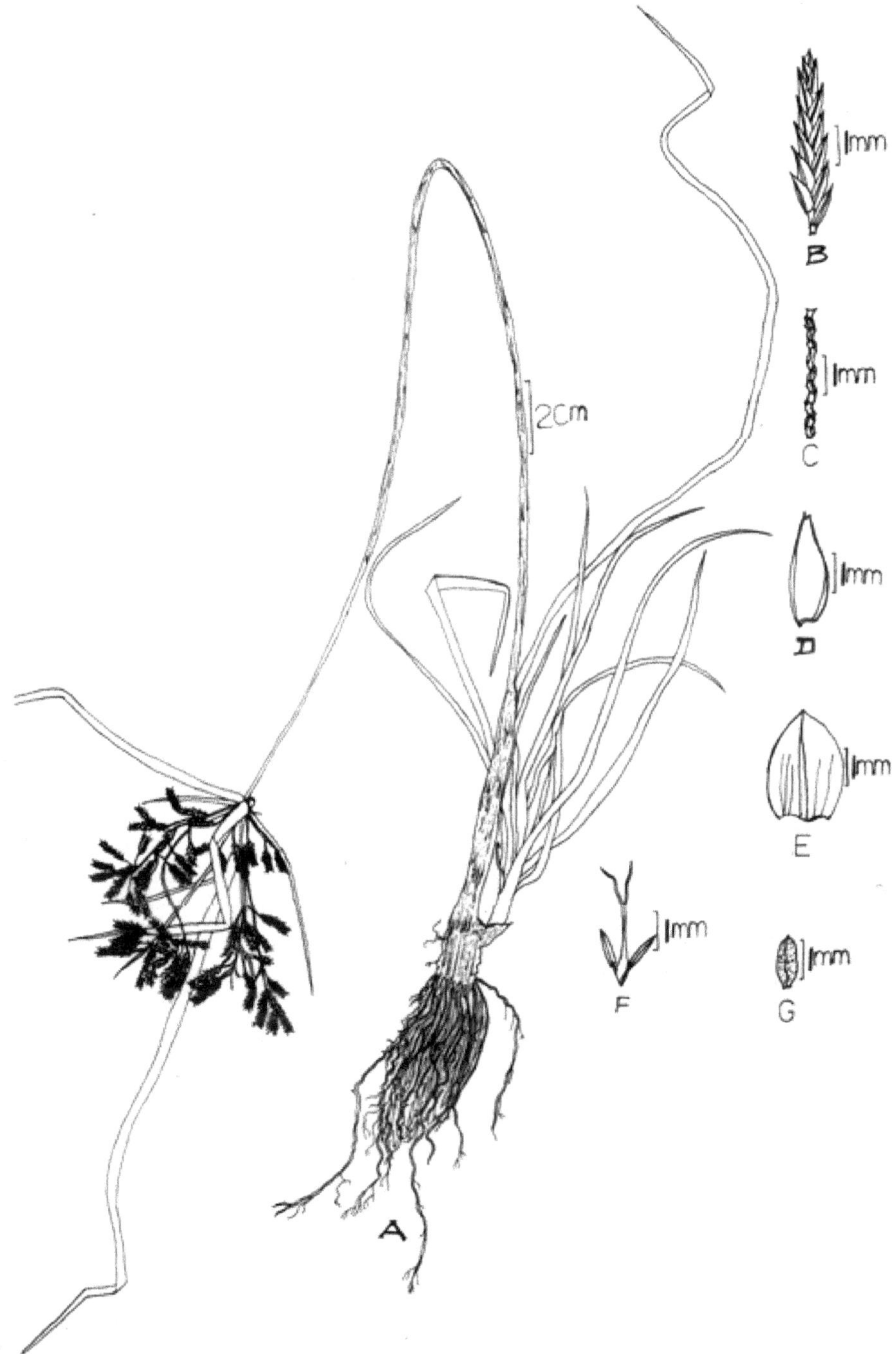

Figure 47. **Cyperus procerus** Rottb.
A. Habit, B. Spikelet, C. Rhachilla, D. Closed glume, E. Opened glume,
F. Pistil with stamens, G. Achene.

straight, wingless, persistent; glumes almost hyaline, elliptic-oblong or oblong-lanceolate, 1-1.75 mm long, slightly keeled, faintly 3-nerved, whitish with hyaline margins; stamen 1; stigmas 3. Nut trigonous, oblong-ellipsoid or obovoid, 0.5-0.9 mm long, shortly apiculate, yellowish to fuscous or black.

Rare in seasonlly wet places, swamps in Andhra Pradesh, Odisha (Kumar, 2013).

Motijharan (SBP), *HFM* 3451 (DD).

INDIA: Andra Pradesh, Bihar, Gujarat, Kerala, Karnataka, Maharashtra, Madhya Pradesh, Odisha, Rajasthan and Tamil Nadu

WORLD: Tropical Africa, Philippines, Australia,

Cyperus rotundus L., Sp. Pl. 45. 1753; FBI 6: 614. 1893; Fischer 3: 1641. 1931; Matthew, Fl. Tamilnadu Carnatic 3: 1742. 1983.

Perennial, glabrous, erect, stoloniferous herb; stolons bearing ovoid, tunicate, black, fragrant tubers; culms to 30 cm high, triquetrous at the top, tuberous at the base. Leaves several, narrowly linear, flat, 10-30 cm long, nerves prominent; sheaths brown. Umbel compound, 4-9 cm long. Involucral bracts 3-5, unequal, longest to 15 cm long; primary rays 4-6, to 6 cm long; secondary rays 3, bearing, short spikes of 4-10 slender, spreading, red-brown spikelets. Spikelets variable in length, linear, red-brown, compressed, spicate, to 3 cm long, pale or purplish, 15-30-flowered; rhachilla winged, pale. Achenes oblong, trigonous, greyish-black.

Very common weed in rice fields and cultivated lands. Fl. & Fr.: October – December. Vern.: Tel: *Tunga*; Ori.: *Mutha, Motha.*

Thummalapalli (KDP), *KRKS* 36979; Varigonda (NLR), *PMR* 22301; Achampeta (GNT), *VRK* 5865; Kinthali (SKLM), *MCK* 18881; Nandyal (KNL), *JSG* 10886 (CAL); Near Dongalacheruvu lake (CDP), *KS* 6305 (MH); Diguvametta (PKM), *JLE* 42191 (MH); Kondapalle (KSN), *CAB* 7980 (MH); Sitaramapuram (KSN), *JSG* 12658 (CAL); Polavaram (EG), *DCSR* 376 (CAL); Anantagiri (VSKP), *GVS* 19447; Salur (VZN), *NPBK* 1002 (CAL); Tiruvaiur (CPT), *JSG* 16312 (MH); Palur (SA), *CAB* 8345 (MH); Palakuppam (SA), VNS 4036 (MH); Hogainakkal (KGR), *EV* 24145 (MH); Woddapatti RF (SLM), *EV* 20703 (MH);

INDIA: Throughout India.

WORLD: Cosmopolitan.

Cyperus rubicundus Vahl, Enum. Pl. 2: 308. 1805; Matthew, Fl. Tamilnadu Carnatic 3: 1742. 1983. *C. teneriffae* Poir. in Lam., Encycl. 7: 245. 1806; FBI 6: 601. 1893; Fischer 3: 1639. 1931.

Annual, reddish-brown, glabrous tufted herb, to 18 cm high; culms trigonous. Leaves usually shorter than the stem or half (or less) as long as the stem, linear, acuminate; sheaths reddish, inflated. Head solitary, to 2 cm long. Involucral bracts 3, to 3 cm long, linear, leaf like; spikelets 3-12, sessile, much compressed, reddish-brown, oblong, digitate. Achenes obovoid, triquetrous, sides concave, apiculate, white, puncticulate, brown-stipitate.

Rare in rocky areas, in dry forests, at hight altitudes. Fl. & Fr.: November – January.

Penukonda (ATP), *BR* & *ANS* 35307; Madhavaram RF (KNL), *RVR* 1656; Guvvalacheruvu RF (KDP), *RVR* & *RV Reddy* 7905; Thummalapalle (KDP), *KRKS* 36987; Jayanthipuram (GNT), *VRK* 6969; Panapakam (CTR), *S.India Flora* 15763 (MH);

INDIA: Maharashtra, Karnataka, Andhra Pradesh, Kerala, Tamil Nadu.

WORLD: Africa, Madagascar, Malaysia, Australia.

Cyperus squarrosus L., Cent. Pl. 2: 6. 1756. *Mariscus squarrosus* (L.) C.B.Clarke in Hook. f., Fl. Brit. India 6: 623. 1893; Fischer 3: 1645. 1931; Matthew, Fl. Tamilnadu Carnatic 2: 1774. 1983. *Cyperus aristatus* Rottb., Descr. Icon. Rar. Pl. 22. 1773. *Illegitimate.*

Annual, densely tufted herb with reddish purplish fibrous roots, culms to 15 cm high, triquetrous with wing like angles. Leaves 3 to a culm, nearly equaling the culm; blades thinly herbaceous; sheaths to 2 cm long, membranous, pale-greenish and stained with reddish-purple. Inflorescence simple, capitate with numerous spikelets crowded into globose heads. Involucral bracts 3-5, patent, the lowest two much longer than the inflorescence, the lowest to 7 cm long. Spikes subglobose, light yellow-green. Spikelets densely spicately disposed on a short rhachis, oblong. Achenes broadly oblong, trigonous, greyish-green.

Common species usually seen in wet sandy soils, open grass lands. Fl. & Fr.: July – November.

Kalasamudram (ATP), *MHR* 14915; Ahobilam (KNL), *BR* & *BSS* 29330 (BSID); Palakonda hills (KDP), *CS* 8270; Seshachalam RF (KDP), *A.M.Rao* 328 (BSID); Near Komaticheruvu (CTR), *GVS* 45850, 46830 (MH); Gonupalle (NLR), *PMR* 19537; Allinagaram – Subbanabavi (PKM), *BR* & *BSS* 30812; Rampachodavaram (EG), *MM* 105077 (BSID); Near Anantagiri (VSKP), *NPBK* 11011 (MH); Anantagiri (VSKP), *GVS* 21737 (MH); Javadi hills (NA), *MBV* 907 (MH);

INDIA: Throughout India.

WORLD: Tropical Africa, extending to S. Africa, Tropical

Cyperus stoloniferus Retz., Observ. Bot. 4: 10. 1786; FBI 6: 615. 1893; Fischer 3: 1641. 1931; Matthew, Fl. Tamilnadu Carnatic 3: 1743. 1983.

Perennial, rhizomatous, stoloniferous herb; rhizome woody, creeping, forming ellipsoid or ovoid enlargement at base of short, clothed with dusky brown parallel fibres. Culms to 30 cm high, obtusely trignous, glaucous green, leaved at base. Leaves rigid, filiform, shorter than the culm; blades linear, to 4 mm wide, folded, recurved; sheaths pale, split into parallel brown fibres. Umbels simple and compact to 5 cm long; Involucral bracts 3, foliaceous, lowest to 30 cm long, second briefly surpassing the inflorescence; primary rays 4, unequal, rigid, to 5 cm long. Spikes ovate, to 2 cm long with 3-6 spikelets in short axis. Spikelets linear oblong, 10 mm long, to 18-flowered, slightly thickened with rather blunt edges, glaucous yellow and stained with sanguineous brown; rhachilla widely winged. Achenes ovate, oval, 3-sides, 1.5 mm long, maturing dark-brown.

Frequent in sandy river banks. Fl. & Fr.: May – December.

Pennahobilam (ATP), *TP* & *NY* 360; Kalasamudram (ATP), *KRKS* 39140; Tungabhadra river (KNL), *KH* 10944; 10949; S.Kota reservoir (VZN), *MCK* 18808; Nandyal (KNL), *JSG* 10886 (DD); Tulka, Satkosia Tiger Reserve, *KCM* 8316 (BSID).

INDIA: Madhya Pradesh, Maharashtra, Andhra Pradesh, Karnataka, Kerala, Tamil Nadu.

WORLD: Madagascar, Pakistan, China, Malaysia, Indonesia, Thailand, Vietnam, Australia.

Cyperus tenuiculmis Boeckeler, Linnaea 36: 286. 1870; Matthew, Fl. Tamilnadu Carnatic 3: 1744. 1983. *C. zollingeri* Steud.: sensu FBI 6: 613. 1893; Fischer 3: 1641. 1931.

Perennial herb with short, corm like, decumbent rhizome, emitting slender stolons, clothed with dark brown scales and their fibrous remnants; culms solitary, stiffy erect, 25 mm high, triquetrous. Leaves few, much shorter than the culm; blades narrowly linear, folded; sheaths light brown. Inflorescence simple. Involucral bracts 3, leafy, the lowest surpassing the anthela, the second one equaling the anthela; rays erect-patent, slender, unequal. Spikes ovoidal up to 4 cm long, bearing *c.* 12 spikelets. Spikelets linear, nearly quadrangular, straw-coloured, 1.5-2 mm wide; weakly flattened with acute edges, rhachilla dark brown, internodes winged, the wings hyaline, caducous, pale. Achenes elliptic, triquetrous with concave sides, 2 mm long, maturing black.

A very rare taxon seen only in swampy areas. Fl. & Fr.: September – December.

Kokkanti (ATP), *MHR* 13913; Lankamalai RF (KDP), *SRSR* 13164; Pottepalem (NLR), *PMR* 16488; Pulipadu tank – Darsi (PKM), *MCK* 22857; Kinthali (SKLM), *MCK* 18870; Dudurchampa, Similipahar (MBJ), *SX* & *MB* 4779 (RRL-B).

INDIA: Assam, Nagaland, Sikkim, West Bengal, Maharashtra, Odisha, Andhra Pradesh, Karnataka, Tamil Nadu, Kerala.

WORLD: Old World Tropics and Subtropics.

Cyperus tenuispica Steud., Synops. Pl. Glumac. 2: 11. 1855; Fischer 3: 1640. 1931; Matthew, Fl. Tamilnadu Carnatic 3: 1744. 1983. *C. flavidus* Retz: sensu FBI 6: 600. 1893.

Annual, slender, erect, tufted herb; culms to 20 cm high, triquetrous, finely ribbed. Leaves flat, linear, acute, to 15 cm long, Umbel compound, to 7 cm long; involucral bracts 3, longest to 13 cm long, overtopping; primary rays 7-10, to 6 cm long; secondary rays 5-7, to 1 cm long, bearing heads of 5-10, stellately spreading, minute spikelets. Spikelets narrowly linear, to 5 mm long, pale brown, *c.* 20-flowered; rhachilla flexuous. Achenes globosely obovoid, trigonous, round, at first pale-yellow becoming white when ripe.

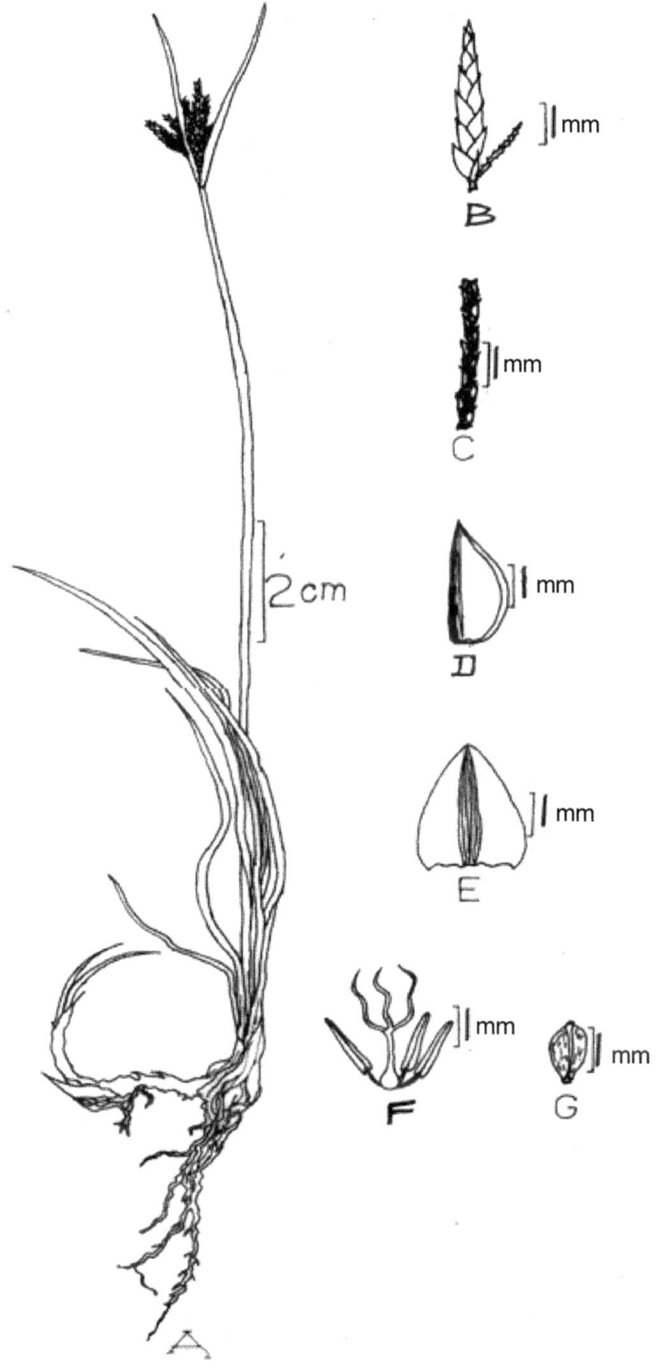

Figure 48. **Cyperus tenuiculmis** Boeckeler
A. Habit, B. Spikelet, C. Rhachilla, D. Closed glume, E. Opened glume, F. Pistil with stamens, G. Achene.

A very common weed in rice fields, and cultivated low lands. Common in marshy fields and margins of ponds. Fl. & Fr.: September – February.

Muniswarunikunta (ATP), *CJP* & *VS* 14154; Velugodu – Yoganandakonda (KNL), *BR* & *BSS* 29342; Kolleru lake (EG), *KH* & *BR* 10912; Near Padmapuram gardens (VSKP), *MHR* & *MCK* 14826; S. Kota (VZN), *KH* & *BR* 5599; Chelama (KNL), *GVS* 22492 (MH & CAL), *JLE* 18012 (MH); Near Komativaricheruvu (CTR), *GVS* 46834 (MH & CAL); Kollurupad (NLR), *CECF* 4232 (CAL); Ettukkuru (EG), *MM* 102611 (BSID); Mudurlanka, Maredumilli (EG), *MM* 101255 (BSID); Dornacheruvu (MBNR), *S.R.Srinivasan* 107460 (BSID); Hogainakkal (KGR), *KCJ* 18008 (MH); Anchetti (SLM), *KCJ* 18043 (MH); Vazhakadu (NA), *EV* 53438 (MH); Bhuvanagiri to Kurinjipadi (SA), *KRM* 60111 (MH); Pampasar RF, Satkosia Tiger Reserve, *KCM* 5136 (BSID).

INDIA: Rajasthan, Uttar Pradesh, West Bengal, Arunachal Pradesh, Assam, Nagaland, Bihar, Odisha, Maharashtra, Madhya Pradesh, Andhra Pradesh, Karnataka, Kerala, Tamil Nadu.

WORLD: Widespread in Tropical and Subtropical Africa, Asia.

DIPLACRUM R.Brown

1. Leaves oblong laceolate; sheaths not winged.................................**D. caricinum**
1. Leaves linear; sheaths slightly winged...**D. africanum**

Diplacrum africanum (Benth.) C.B.Clarke, Consp. Fl. Afric. 5: 668 1894. *Scleria africana* Benth., Gen. Pl. 3: 1071 1883; Jain, Indian Forester 95: 130. 1969; Saxena & Brahmam, Fl. Orissa 4: 2213. 1996.

Annual, slender, glabrous herb; stems tufted, few, to 12 cm high, leafy throughout. Leaves linear, 1.-4 x 0.2-0.4 cm, acuminate. Inflorescence of small, axillary or terminal cluster of spikelets, often continued nearly to the base of the stem; in very short-stemmed forms many spikelets crowded in lower axils; peduncled very short; 3-4 male spikelets at the base of the inflorescence; glumes 3-4, membranous or hyaline. Female spikelet terminal, 1-flowered; glumes 2. Ovate-lanceolate, 3-nerved, entire at tip, with small, hyaline, obtuse, lobes at about the middle of their margins, concave at base. Nut minute, subglobose or oblong, *c.* 1 mm long, with 10-14 longitudinal ribs from summit to base, ribs not anastomosing.

Near Laxmipur, Koraput, in a patch of ground overlying granite gneiss in association with small grasses and sedges *etc.* (Jain *loc. cit.*)

Laxmipur (KPT), *HFM* 4229 (K).

INDIA: Odisha, Goa, Karnataka.

WORLD: Africa and Madagascar.

Diplacrum caricinum R.Br., Prodr. Fl. Nov. Holl. 241. 1810. *Scleria caricina* (R.Br.) Benth., Fl. Austral. 7: 426. 1878; FBI 6: 688. 1893; Fischer 3: 1678. 1931; Saxena & Brahmam, Fl. Orissa 4: 2215. 1996.

Slender, small herbs with reddish-brown fibrous roots, stems up to 13 cm long, erect or decumbent, glabrous. Leaves oblong-lanceolate, up to 3.7 x 0.3 cm, apex acute, scaberulous on the margins towards apex; sheaths not winged. Flowers in axillary clusters; peduncles slightly exserted from the sheaths, primary bracts foliaceous; terminal spikelet of each cluster female with two lateral male flowers at base. Male spikelets: up to 0.2 cm long, linear-oblong; stamens 1. Female spikelet: up to 0.322 cm long, ovate-oblong; glumes 2, ovate-lanceolate, 3-lobed, lateral lobes shorter than the middle one, membranous, shortly apiculate, median lobe reticulate, long acuminate; style slender, stigmas 3, shorter than the style; disk obsolete. Nuts up to 0.1 cm across, globose, minutely apiculate, longitudinally and irregularly ribbed.

Common in wet places among grasses. Fl. & Fr.: September.

Kambakkam hills (CPT), *sine coll. s.n.* (MH); Paniganda (GJT), *SX* & *MB* 2002 (RRL-B); Manaskonda (KPT), *SX* & *MB* 6811 (RRL-B).

INDIA: Madhya Pradesh, Maharashtra, Andhra Pradesh, Karnataka, Tamil Nadu, Kerala.

WORLD: Sri Lanka, Myanmar, China, Laos, Cambodia, Indonesia, Malaysia, Philippines, Thailand, Vietnam, Australia.

ELEOCHARIS R. Brown

1. Spikelets cylindrical, as wide as the thick culms that range from 3-6 mm in width:
   2. Culms solid, not septate:
      3. Culms triquetrous:
         4. Glumes ovate; nut apically constricted into distinct neck with an expanded apex .... **E. acutangula**
         4. Glumes cuneate; nut without neck, apically annular ....**E. spiralis**
      3. Culms terete:
         5. Bristles shorter than the beak of nut......................................**E. swamyii**
         5. Bristles exceeding the beak of nut....................................... **E. paulestris**
   2. Culms hollow, transversely septate... **E. dulcis**
1. Spikelets ovoidal to ellipsoidal, much wider than slender culms that range from 0.3 – 2.5 mm in width;
   6. Stigmas 3; achenes olive-coloured to yellow green at maturity, trigonous:
      7. Stems 25-60 cm, striate; nut not pitted; style base conical........ **E. congesta**
      7. Stems less than 15 cm, punctulate; nut pitted; style base confluent.. **E. retroflexa**

6. Stigmas 2; achenes black when mature, biconvex:
   8. Hypogynous bristles 4-6, equallling or shorter than achene; stamens 2 .... **E. atropurpurea**
   8. Hypogynous bristles 5-7, longer than the achene; stamens 3 .... **E. geniculata**

Eleocharis acutangula (Roxb.) Schult. in Roem. & Schult., Mant. 2: 91. 1824; Matthew, Fl. Tamilnadu Carnatic 3: 1746. 1983. *Scirpus acutangulus* Roxb., Fl. Ind. 1: 216. 1820. *Eleocharis fistulosa* (Poir.) Schult. in Roem. & Schult., Mant. 2: 89. 1824; FBI 6: 626. 1893; Fischer 3: 1648. 1931. *Scirpus fistulosus* Poir., Encycl 6: 749. 1804.

Perennial, tufted, aquatic, partly submerged, rhizomatous and stoloniferous herb, to 70 cm high. Culms sharply triquetrous, light green, solid, not septate. Basal sheaths 4, the lower ones scale like, brownish, the uppermost ones to 10 cm long, pale green, obliquely truncate at orifice. Spikelets cylindrical, to 4 cm long, 4 mm wide, subacute at apex, obtusely angular, loosely many-flowered, pale green. Achenes broadly obovate, 2 mm long, compressed-triangular, the apex contracted to an annular neck, the sides yellowish brown, cancelloid with 15 rows of transversely oblong epidermal cells.

Rare in marshy areas near river sides, shallow stagnant waters in ponds. Fl.& Fr.: June – December.

Salur (VZN), *MCK* 25232; Seethalam river bed, Thandipandal (CTR), *MCB* 45094 (MH); way to Puliyur from Kuttathur (NA), *KS* 7454 (MH); Keonjhar garh (KJR), *HFM* 1372 (DD); Barapahar (SBP), *HFM* 3633 (DD).

INDIA: Assam, Odisha, Andhra Pradesh, Western Peninsula.

WORLD: America, Tropical Africa, Madagascar, Sri Lanka, Myanmar, China, Japan, Laos, Malaysia, Indonesia, Philippines, Thailand, Vietnam, Australia.

Eleocharis atropurpurea (Retz.) Presl., Rel. Haenk. 1: 196. 1828; FBI 6: 627. 1893; Fischer 3: 1648. 1931; Matthew, Fl. Tamilnadu Carnatic 3: 1746. 1983. *Scirpus atropurpurea* Retz., Observ. Bot. 5: 14. 1788.

Annual, erect, tufted, non rhizomatous herb, to 10 cm high; culms terete, capillary, smooth, sulcate. Leaves reduced to bladeless sheaths; sheaths membranous, purple-coloured. Inflorescence with a single spikelet; spikelets globose or ovoid, 3-5 x 1.5-2 mm, angular, densely many-flowered; glumes broadly oblong, 1.2-1.5 mm long, obtuse or rounded at tip, membranous, 3-nerved, keeled. Bristles 3-4, shorter than to about as long as the nut, whitish, transluscent, some times 0; stamens 1-2; style 2-fid. Achenes obovoid, biconvex, to 1 mm long, glossy.

Common in wet fields, marshy areas *etc.* Fl. & Fr.: November – December.

Palasamudram (ATP), *BR* & *ANS* 36593; Dhavaleswaram barrage (EG), *KH* & *BR* 9727; way to Guvvalacheruvu (KDP), *KS* 7787(MH); Balapalle (KDP), *JLE* 14317 (MH); Near Komativari cheruvu (CTR), *GVS* 46828 (MH & CAL); hill top of

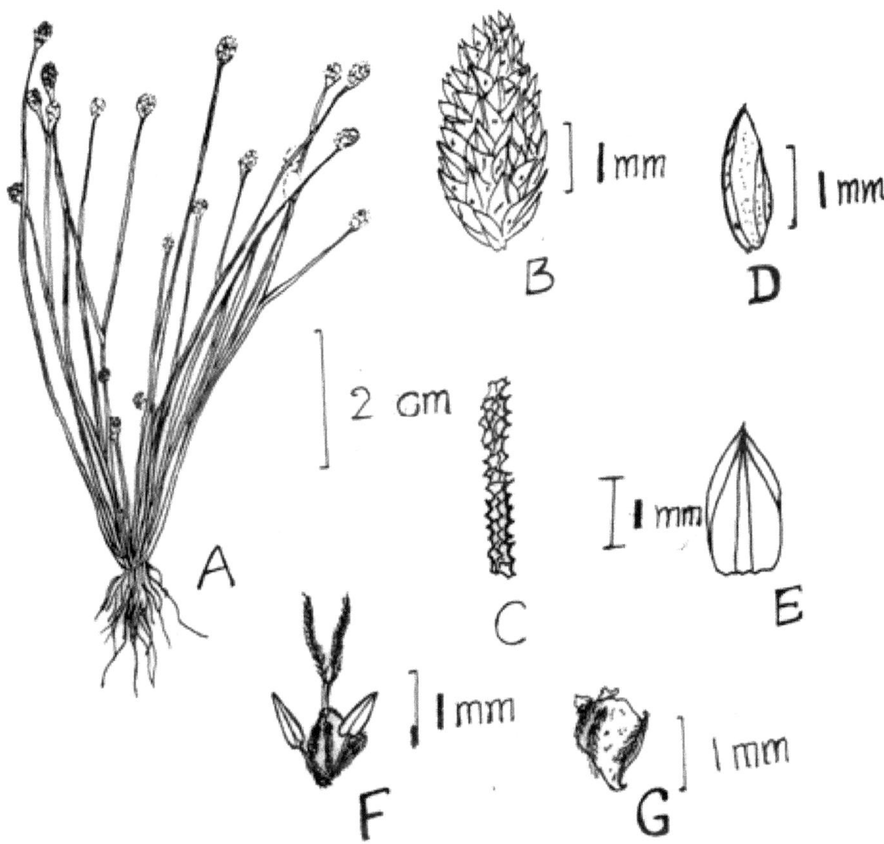

Figure 49. **Eleocharis atropurpurea** (Retz.) Presl.
A. Habit, B. Spikelet, C. Rhachilla, D. Closed glume, E. Opened glume, F. Pistil with stamens, G. Achene.

Chandrgiri fort (CTR), *NRR* & *TRS* 86161 (MH), *DRC* 2278 (MH); Craignoor swamps, Yercaus (SLM), *EV* 77726 (MH);

INDIA: Throughout India.

WORLD: Central, North and Sourth America, Europe, Pakistan, Bhutan, Nepal, China, Indonesia, Philippines, Indian Ocean Islands, Australia.

Eleocharis congesta D. Don, Prodr. Fl. Nepal 41. 1825; FBI 6: 630, 1893; Matthew, Fl. Tamilnadu Carnatic 3: 1747. 1983; Fischer 3: 1648. 1931 subsp. **congesta**

Annual, densely tufted herb. Culms thick, rigid, to 25 cm high, terete, deeply green, clothed at base with 3 sheaths. Basal sheaths membranous, reddish below, pale above, the uppermost one to 6 cm long, the orifice brown-tinged on margin. Spikelets ovate-oblong, to 10 mm long, terete brownish, densely many–flowered; glumes ovate-oblong, reddish brown at apex, margins scarious, slightly keeled; stamens 1 or 2; ovary oblong, style slender, stigmas 3. Achenes obovate, trigonous, 1.2 mm long, obtuse, edged, contracted to apex, the surfaces lightly greenish yellow, shiny, smooth.

Usually found in high elevations in hill sides. Fl. & Fr.: April – October.

Near Padmapuram gardens (VSKP), *MHR* & *MCK* 14825; Galikonda (VSKP), *GVS* 19616 (MH); Smilipahar (MBJ), *SX* 393 (RRL-B); Pampasar RF, Satkosia Tiger Reserve, *KCM* 5126 (BSID).

INDIA: Arunachal Pradesh, Manipur, West Bengal, Uttar Pradesh, Rajasthan, Gujarat, Madhya Pradesh, Karnataka, Andhra Pradesh, Tamil Nadu, Kerala.

WORLD: Pakistan, Sri Lanka, Myanmar, Bhutan, Nepal, China, Malaysia, Indonesia, Philippines, Thailand, Vietnam, Pacific Islands.

Eleocharis dulcis (Burm. f.) Trin. ex Hensch. in Vita Rumph. 186. 1833; Matthew, Fl. Tamilnadu Carnatic 3: 1747. 1983. *Andropogon dulce* Burm. f., Fl. Ind. 219. 1768. *Eleocharis plantaginea* (Retz.) Roem. & Schult., Syst. Veg. 2: 150. 1817; FBI 6: 625. 1893; Fischer 3: 1647. 1931. *Scirpus plantagineus* Retz., Observ. Bot. 5, 14. 1799.

Perennial, tufted, partly submerged, stoloniferous herb; stolons terminated by a small tuber. Culms to 90 cm high, 7 mm thick, hollow, transversely septate, the inter septa to 12 mm long, empty, deeply green, somewhat shiny. Basal sheaths to 20 cm long, coloured with red brown, truncate at orifice. Spikelet cylindrical, to 5 cm long, 4 mm wide, subobtuse at pex, terete, whitish to pale stramineous; glumes ovate-oblong, red gland-dotted, margins hyaline, keel distinctly raised below; stamens 3; ovary obovate, style slender, 4 mm long, stigmas 3; hypogynous bristles 6, retrorsely scaberulous. Achenes obovate, to 3 mm long, biconvexed with obtuse edges, yellow, shiny.

Occasional in marshy areas, stagnant or slowly moving water. Fl. & Fr.: September – December.

Jonnavalasa tank (VZN), *MCK* 18850; Siepenaidupeta (SKLM), *MCK* 18889; Thimmapuram (EG), *CAB* 8283 (MH); Araku Valley (VSKP), *NPBK* 10791 (MH & CAL); Atapaka (KSN), *PV* 5345 (AU); Rukeekonda (GJM), *VNS* 82958 (MH);

INDIA: NW India, Assam, Western Peninsula, Andhra Pradesh.

WORLD: Tropical Africa, Madagascar, Pakistan, Sri Lanka, Myanmar, China, Korea, Malaysia, Philippines, Thailand, Vietnam, N. Australia, Pacific Islands.

Eleocharis geniculata (L.) Roem. & Schult., Syst. Veg. 2: 150. 1817; Matthew, Fl. Tamilnadu Carnatic 3: 1748. 1983. *Scirpus geniculatus* L., Sp. Pl. 48. 1753. *Eleocharis capitata* R.Br., Prodr. 225. 1810; FBI 6: 627. 1893; Fischer 3: 1648. 1931. non *Scirpus capitatus* L. 1753.

Annual, erect, caespitose herb, to 30 cm high; culms slender, angular, sulcate, slightly glaucous green. Sheaths herbaceous, pale green, often tinged with red brown on lower part, the mouth oblique, acute. Spikelets ovoid-globose to ovoid-ellipsoidal, 5 x 3 mm, rounded at base, very obtuse at apex, reddish-ferrugineous, densely many-flowered. Achenes globosely obovoid, brown, shining, apiculate with broad style base.

A very common species usually on wet fields, marshy areas, including rice fields. Fl. & Fr.: June – October.

Yadiki RF (ATP), *TP* 498 (SKU & MH); Garugudu kona (ATP), *TP* 949 (SKU & MH); Madhavaram RF (KNL), *TP* & *RVR* 1353; Satyavedu (CTR), *BR* & *V.S.Rao* 32158; Kurichedu (PKM), *MCK* 22812; Macherla (GNT), *VRK* 5881; Rampa hill (EG), *KH* & *BR* 9732; Punyagiri hill, S. Kota (VZN), *KH* & *BR* 5595; Bughanka river (KDP), *KS* 6448 (MH); way to Guvvalacheruvu (CDP), *KS* 7787 (MH); Diguvametta (PKM), *JLE* 42185 (MH); Ramavaram (KMM), *R.Rajan* 112575 (BSID); Hoggenakkal water falls (DMP), *P. Venu* 111617 (BSID); Puliyur (NA), *KS* 6521 (MH); Takarani RF East (SA), *KRM* 52843 (MH); Gingee RF (SA), *KMS* 12322 (MH); Hogainakkal (KGR), *EV* 23524 (MH), *KCJ* 18015 (MH); Manbhang, Gandhamardan (Bargarh district), *SX* & *MB* 5983 (RRL-B).

INDIA: West Bengal, Bihar, Odisha, Rajasthan, Madhya Pradesh, Gujarat, Andhra Pradesh, Karnataka, Tamil Nadu, Kerala.

WORLD: Pantropic.

Eleocharis palustris (L.) Roem. & Schult., Syst. Veg. 2: 151 1817; Mooney, Suppl. Bot. Bihar & Orissa 147. 1950; Saxena & Brahmam, Fl. Orissa, 4: 2163. 1996. *Scirpus palustris* L., Sp. Pl. 70. 1753.

Slender, caespitose herb with creeping rhizome; stems 10-50 cm high and up to 3 mm diam. Leaf-sheaths loose, often reddish, uppermost sometimes produced on one side into a triangular point, otheriwise truncate. Spikelet ellipsoid or cylindric, 0.75-2.5 cm x 3-7.5mm, dense-flowered, yellow or brownish, with a supporting glume-like bract, oblong, green with broad scarious margin. Glumes much imbricate, boat-shaped with green keel with hyaline or coloured margins, lanceolate when unfolded, 4.5-5.5 mm long, obtuse. Bristles 6, as long as or exceeding the nut. Nut

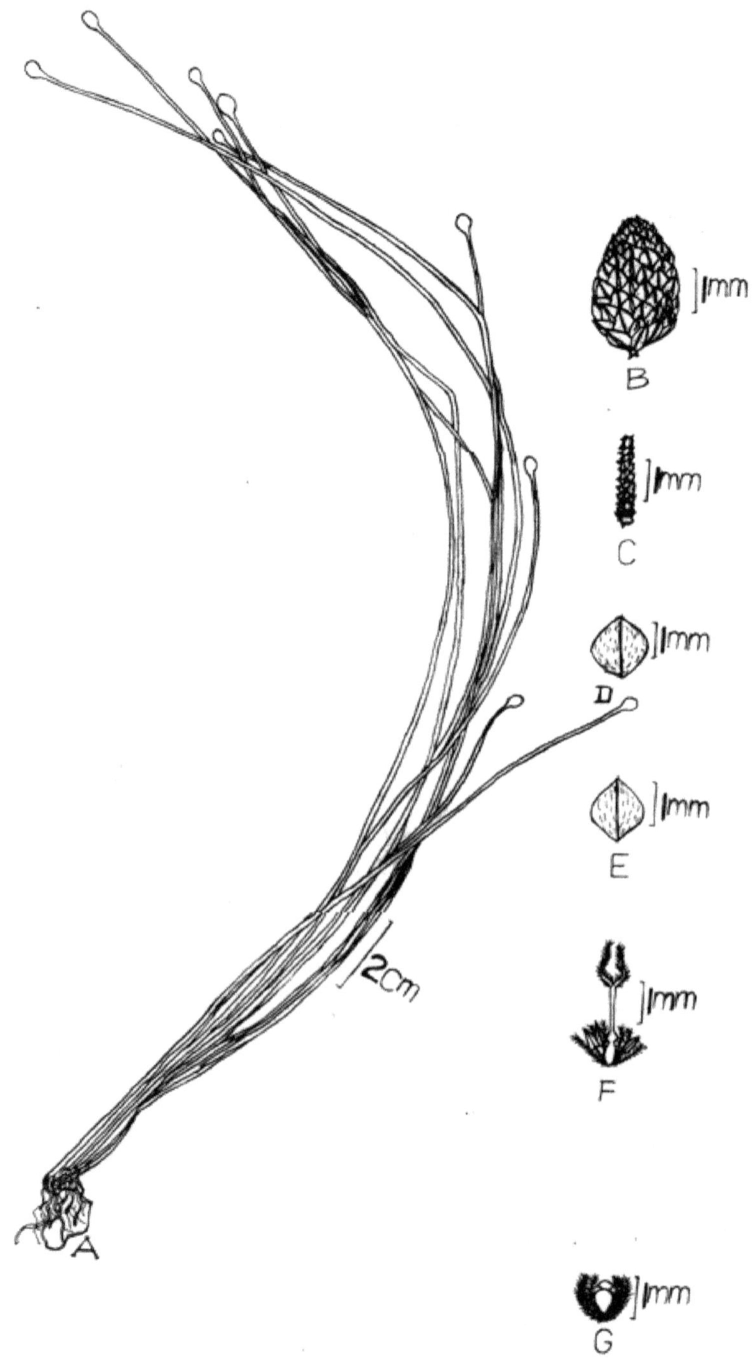

Figure 50. **Eleocharis geniculata** (L.) Roem. & Schult.
A. Habit, B. Spikelet, C. Rhachilla, D & E. Glumes,
F. Pistil with stamens and hypogynous bristle, G. Achene with hypogynous bristle.

obovoid or rounded, including the large style base, 1.5-2 mm long, biconvex, style base ovoid or broadly conical, nearly half as long as the nut.

In stagnant pools in Athmalik, Dhenkanal pool (Mooney, *loc.cit.*). Fl. & Fr.: December – April.

INDIA: From Western Himalays to West Bengal, Odisha.

WORLD: Cosmopolitan.

Eleocharis retroflexa (Poir.) Urb., Symb. Antill. 2: 165 1900; Matthew, Fl. Tamilnadu Carnatic 3: 1748. 1983; Saxena & Brahmam, Fl. Orissa 4: 2163. 1996. *Scirpus retroflexus* Poir., Encycl. 753. 1805.

Tufted, annual herb, 2.5-12 cm high; stems slender, capillary, deeply 5-ribbed, punctulate. Uppermost sheaths often with 1 or sometimes 2 scarious elliptic blades, 1-1.2 mm long. Spikelets ovoid, 1.2-4 mm long, compressed, few-flowered; glumes *c.* 10, basally distichous, oblong-lanceolate, 2-3 mm, thin, sides purplish, many-nerved, obtuse, margin, hyaline, keel strong; hypogynous bristles 8, equal, to 1 mm, yellow; stamens 3, style 3-fid. Nut obovoid, sharply triquetrous, to 1 mm, stramineous, pitted, cancellate, apically truncate.

Frequent in wet places close to ponds, rice fields, river *etc.* Fl. & Fr.: September-December.

Thakurmunda (MBJ), *SX & MB* 5390(RRL-B); Athmalik (Angul district), *HFM* 2864 (DD).

Eleocharis spiralis (Rottb.) Roem. & Schult., Syst. Veg. 2: 155. 1817; FBI 6: 627. 1893; Fischer 3: 1647. 1931; Matthew, Fl. Tamilnadu Carnatic 3: 1749. 1983. *Scirpus spiralis* Rottb., Descr. Ic. 45. t. 15.f.1.

Perennial tufted herb; rhizome short with creeping stolons, 3-5 mm in diameter, reddish brown, clothed with membranous scales; culms solid, dark green shiny, glabrous. Leaves completely reduced to sheaths, membranous, pale green to purplish red, obliquely attenuate with a scale like appendage. Spikelet yellowish green, cylindrical, 10-20 x 3-5 mm, densely many-flowered, apex obtuse. Glumes tightly imbricated, sub-orbicular, broadly obovate, 2.5 x 2 mm, keel 1-nerved, obscure, margin entire, hyaline, apex broadly obtuse. Hypogynous bristles 4, 1 mm long, equal; stamens 3, filaments 1.5 mm long, anthers yellow, linear, 1.2 mm long; pistil 3 mm long, ovary obovate, 1.5 mm long, style 1 mm long, base pyramidal, stigma 3-fid, rarely 2-fid. Achene yellowish brown, obovate, 1.5 x 1 mm, biconvex, each side obscurely with 18 rows of oblong cells.

Rare in swamps and marshes in Nellore district. Fl. & Fr.: October -December.

Near Kanigiri reservoir (NLR), *MCK* 22887.

INDIA: West Bengal, Madhya Pradesh, Andhra Pradesh, Karnataka, Tamil Nadu, Kerala.

WORLD: Sri Lanka, Thailand, Papua New Guinea, Australia.

Note: This is closely allied to *Eleocharis acutangula* (Roxb.) Schult., but differs in trigonous stem, dark green, cylindrical spikelets, 10-20 mm long, sub-orbicular to cuneate glumes and equal hypogynous bristles.

Eleocharis swamyii Govind., Proc. Indian Acad. Sci. 94: 13. f. 2. 1985.

Perennial, rhizomatous herb; rhizome elongated, obliquely creeping; culms erect, tufted, terete throughout, smooth, not transversely septate, to 50 cm high; sheaths membranous, dirty black, shining, tightly clasping, obliquely truncate, to 15 cm long. Spikelets cylindric, terete, later becoming angular due to squarrose glumes, broader than culms, obtuse, many-flowered, pale brown, to 20 mm long. Achenes turgid, symmetric, biconvex, distinctly cancellate, broadly orbicular or suborbicular with short neck, prominently rimmed above the neck, shining, yellowish-brown, 1.7 mm long.

Commonly found from mean sea level to 500-600 m. Fl. & Fr.: September – December.

Kambakkam (CTR), *E. Govindarajulu* 8016 (CAL, MH, BLAT, DD).

INDIA: Andhra Pradesh, Tamil Nadu. Endemic.

FIMBRISTYLIS Vahl *nom.cons.*

1. Achenes sub-orbicular, obovate; stigmas 3 or 2:
    2. Stigmas 3; achenes trigonous:
        3. Leaves surrounding culm bases bladed:
            4. Leaves with a ligule of a series of pubescence:
                5. Plant stoloniferous; involucral bracts subulate or glume like .... **F. pierotii**
                5. Plant not stoloniferous; involucral bracts leaf-like:
                    6. Culm to 35 cm high, glabrous; leaf sheath densely hairy; rachilla winged .... **F. naikii**
                    6. Culm to 40 cm high, pubescent; leaf sheath glabrous; rachilla not winged .... **F. complanata**
            4. Leaves without ligule:
                7. Nut verrucose, tuberculate:
                    8. Rhizome thick, oblique, to 7 cm long; leaves rosette like, recurved:
                        9. Leaves not distichous, glabrous or scabrid beneath, falcate .... **F. falcata**
                        9. Leaves distichous, pubescent beneath, not falcate .... **F. fusca**
                    8. Rhizome inconspicuous; leaves erect or rosette, not recurved:

10. Leaves erect, sheath 25 mm long; stem irregularly angled.. **F. tenera**

10. Leaves rosette, sheath minute; stem trigonous..**F. fimbristyloides**

7. Nut other than verrucose, smooth:

11.Stamen 1; involucral bracts regid.................................**F. littoralis**

11.Stamens 3; involucral bracts not rigid:

12. Lower involucral bract exceeding inflorescence...**F. paupercula**

12. Lower involucral bract shorter than inflorescence:

13. Spikelets 4-7-rays; leaves glabrous.................. **F. uliginosa**

13. Spikeletes 7-15-rays; leaves scabrid................ **F. salbundia**

3. All or upper few leaves surrounding culm bases blade less:

14. Leaf blades dorsiventral; culms pentagonous.......**F. quinquangularis**

14. Leaf blades laterally flattened, culms compressed, tetragonal:

15. Glumes spiral...**F. quinquangularis**

15. Glumes distichous at base:

16. Spikelets narrow-linear, less than 1 mm across..**F. cinnamometorum**

16. Spikeletes ovoid, more than 2 mm across ....**F. eragrostis**

2. Stigmas 2; achenes biconvex: (Occasionally 3 in *F. cymosa*)

17. Inflorescence corymbose with more than 2 rays and 5 spikelets:

18. Styles hardly flattened, glabrous...**F. cymosa**

18. Styles dorsiventrally flattened, fimbriate or long ciliate at least on upper part:

19. Leaves with a ligule, usually a fringe of dense pubescence:

20. Glumes pubescent on upper half:

21. Lowest involucral bract shorter than inflorescence; leaves less than 10 cm long......... **F. ferruginea**

21. Lowest involucral bract exeeding the inflorescence, leaves 15-40 cm long:

22. Perennials with decumbent rhizome; involucral bracts 3-5; stamens 2....................**F. hookeriana**

22. Annuals without rhizome; involucral bracts 1-5; stamens 3.................................... **F. pubisquama**

20. Glumes glabrous:

23. Ligule present:

24. Spikelets 1-1.8 mm wide, oblong .... **F. bisumbellata**

24. Spikelets 2-4 mm wide, ovate......................... **F. dichotoma**

23. Ligule absent .... **F. squarrosa**

19. Leaves without a ligule:

25. Achenes strongly trabeculate .... **F. albicans**

25. Achenes faintly cancellated or nearly smooth......... **F. aestivalis**

17. Inflorescence head like or consisting of 1-4 solitary spikes only:

26. Inflorescence head like with many spikelets; style glabrous .... **F. argentea**

26. Inflorescence with 1-4 solitary spikes; styles fimbriate (except in *F. acuminata*):

27. Inflorescence with 1-3 spikelets; leaf blades well elongated .... **F. schoenoides**

27. Inflorescence with a single spikelet; leaves bladeless:

28. Spikelets erect:

29. Spikelets obtuse at apex; leaf blades occasionally remaining as a setaceous appendages of sheath:

30. Style bifid; nut biconvex:

31. Rhizome absent; nutlet white, verruculose .... **F. hookeriana**

31. Rhizome very short; nutlet grayish black.. **F. polytrichoides**

30. Style trifid; nut trigonous:

32. Leaves and stems densely hairy:

33. Culm to 25 cm high; spikelets to 12 mm long; achenes 1 mm across............. **F. kingii**

33. Culm to 10 cm high; spikelets to 5 mm long; glumes to 2 mm across .... **F. monospicula**

32. Leaves and stems sparsely scabrid.................... **F. ovata**

29. Spikelets acute at apex; leaf blades always none .... **F. acuminata**

28. Spikelets obliquely patent.. **F. natans**

1. Achenes oblong-cylindrical; stigmas 2:
    34. Inflorescence corymbose with few to several small spikelets; spikelets 6 mm long .... **F. dipsacea**
    34. Inflorescence of a single terminal spikelet; spikelets 12 mm long .... **F. tetragona**

Fimbristylis acuminata Vahl, Enum. Pl. 2: 285. 1806; Fischer 3: 1658. 1931. *F. setacea* C.B.Clarke in Hook. f., Fl. Brit. India 6: 631. 1893; *p.p.*, non Benth.

Annual, densely tufted, rhizomatous herb; culms slender, to 35 cm high, obtusely trigonous, glaucous green, clothed at base with 3 bladeless sheaths only. Basal sheaths 1 cm long, cylindrical, membranous, pale green, the upper most sheath to 4 cm. Inflorescence a single terminal spikelet with a scale like bract. Spikelet ovate, terete, acute, to 12 mm long, greenish white, tightly *c.* 15-flowered. Achenes obovate-orbicular, biconvexed, maturing whitish or cream-coloured, transversely coarsely wrinkled with 8 ridges.

Usually seen in marshy places at low altitudes. It is common weed in paddy fields. Fl. & Fr.: October – November.

Seethalam river bed (CTR), *MCB* 45102 (MH); Rathamhutta hills (KMM), *RCS* 104228 (MH); Depisahi, Satkosia Tiger Reserve, *KCM* 9409 (BSID).

INDIA: Assam, Sikkim, West Bengal, Odisha, Himachal Pradesh, Gujarat, Madhya Pradesh, Maharashtra, Andhra Pradesh, Karnataka, Kerala.

WORLD: Sri Lanka, Myanmar, Philippines.

Fimbristylis aestivalis (Retz.) Vahl, Enum. Pl. 2: 288. 1806; FBI 6: 637. 1893; Matthew, Fl. Tamilnadu Carnatic 3: 1751. 1983. *Scirpus aestivalis* Retz., Observ. Bot. 4. 12. 1786.

1. Culm to 25 cm high; leaves 4-6; corymb simple.................................... var. **major**
1. Culm to 10 cm high; leaves a few; corymb compound. ....var. **aestivalis**

Fimbristylis aestivalis (Retz.) Vahl, var. **aestivalis**

Annual, densely tufted, small, erect herb, to 10 cm high; culms trigonous, light green, few leaves at base. Leaf blades much shorter than the culm, flattish with incurved margins, light green, subdensely to sparsely pilose on both surfaces with spreading hairs; sheaths to 15 mm long, pale greenish to yellow-brownish, subdensely pubescent with spreading hairs, the ventral side hyaline, light brown; orifice obliquely truncate. Corymbs compound, to 4 cm long; rays 7, filiform, patent; involucral bracts 3, filiform. Spikelets solitary, lance-ovate, subterete, to 6 mm long, densely to 40-flowered. Achenes obovate, thickly biconvexed, rounded at apex, short-stipitate cuneate base, yellowish at maturity and smooth.

Occasional in wet or moist sandy low lands and as a weed in rice fields. Fl. & Fr.: January – April.

SEDS – Penukonda road (ATP), *BR* & *ANS* 35355; Rangapuram RF (KNL), *RVR* 1496; Omkaram (KNL), *SS* & *AMR* 21352; Papanasanam (CTR), *MHR* 13357; Krishna river (GNT), *VRK* 6743; Koruturu (WG), *DN* 84291 (BSID); Simhachalam hill (VSKP), *KH* & *BR* 5592; Sunkarimetta (VSKP), *GVS* 19649 (MH); Satkosia Wild Life Sanctuary, *D.Hazra* & *D.Das* 19627 (BSID).

INDIA: Throughout India, except N.W.region.

WORLD: Sri Lanka, Nepal, China, Japan, Laos, Indonesia, Thailand, Vietnam, Papua New Guinea, Australia.

Fimbristylis aestivalis (Retz.) Vahl var. **major** Trimen, Cat. Pl. Ceylon 101. 1855; Veeranjaneyulu & Rao, Indian J. Forest. 34: 487. 2011.

Annual, densely tufted herb, culms 5-25 cm high, trigonous, smooth. Leaves 4-6 per culm, as long as or longer than the culm, slenderly linear, sub-acute at apex, densely pilose on both surfaces with spreading hairs; sheaths to 2 cm long, yellow-brownish, pubescent; orifice truncate. Corymb more frequently simple, each ray terminated by mostly one spikelet, rarely 2; bracts 3, filiform, lowest one longer than the corymb, pilose. Spikelets lanceolate-oblong, 5-7 x 1.5 mm, 10-30-flowered; glumes imbricated, ovate, boat-shaped keel, mucronate apex, membranous, slightly pubescent in upper half, 1.5-2.5 x 0.8-1 mm. Achenes obovate, biconvex, 1-2 x 0.6-1.5 mm, style slender, flattened, fimbriate above, stigmas 2-fid; stamens 2, rarely 1.

Occasional in moist localities in Southern Eastern Ghats. Fl. & Fr.: August – December.

Bheemunikolanu (KNL), *BR* & *BSS* 33109; Japalitheertham (CTR), *BR* & *DV* 39307; Bangarupalyam (CTR), *AL* 28570.

INDIA:Peninsular India.

WORLD: China, Japan

Fimbristylis albicans Nees in Wight Contrib. Bot. India 100. 1934; FBI 6: 641. 1893; Fischer 3: 1659. 1931.

Perennial herb, to 30 cm high. Culms few, slender, leafy at base, compressed. Leaves half as long, flat, narrowly ligulate, tip suddenly narrowed, acute. Spikelets 15 in the head; involucral bracts 3, to 2.5 cm long, suddenly acute. Spikelets ellipsoid, rusty grey, clustered, all sessile. Achenes obovate, stipitate, stramineous, shining, transversely trabeculate.

Occasional in marshy areas (Fischer). Fl. & Fr.: August – January.

INDIA: Peninsular India.

WORLD: Sri Lanka.

Fimbristylis argentea (Rottb.) Vahl, Enum. Pl. 2: 294. 1806; FBI 6: 640. 1893; Fischer 3: 1659. 1931; Matthew, Fl. Tamilnadu Carnatic 3: 1752. 1983. *Scirpus argenteus* Rottb., Descr. Pl. Rar. 27. 1777 & Descr. Icon. Rar. 51. t. 17. f. 6. 1773.

Annual, caespitose, glaucous herb, to 10 cm high; culms filiform, trigonous, striate. Leaves filiform, to 5 cm long; ligule 0; sheaths stramineous. Inflorescence capitate, 5-12 in a cluster. Involucral bracts 4, overtopping, longest to 3 cm long. Spikelets sessile, cylindric-oblong, terete, obtuse, many-flowered, 3 mm long; glumes spiral, ovate, subacute. Achenes obovoid, compressed, biconvex, fulvous.

Occasional on wet sandy grounds of open grasslands and rice fields. Fl. & Fr.: September – December.

Chintalapalli (ATP), *KH* 7468; Rudrakod (KNL), *BR* & *BSS* 30172; Veyinuthulakona (KDP), *MHR* 13310; Rollamadugu (KDP), *BR* & *BSS* 33102 (BSID); Gundlamotu reservoir (PKM), *BR* & *BSS* 33149; Guvvalacheruvu (KDP), *KS* 6387 (MH & CAL); Kambakam hill (CTR), *CAB* 45129 (MH); near Komaticheruvu (CTR), *GVS* 45855 (MH); Pernadu (NLR), *PV* 111218 (BSID); Narasimhulukonda (NLR), *A.S.Rao* 2102 (BSID); Alluru (KSN), *CAB* 7937 (MH); Hogainakkal RF (KGR), *EV* 20663 (MH); Gingee RF (SA), *KRM* 13015 (MH); Edaikal RF (SA), *KRM* 60390 (MH); Tulka RF, Satkosia Tiger Reserve, *KCM* 6771 (BSID).

INDIA: Peninsular, C, N. & E. India.

WORLD: S. and SE Asia.

Fimbristylis bisumbellata (Forssk.) Bubani, Dodecanthea 30. 1850; Fischer 3: 1658. 1931; Matthew, Fl. Tamilnadu Carnatic 3: 1752. 1983. *Scirpus bisumbellata* Forssk., Fl. Aegypt. – Arab. 1: 15. 1775. *Fimbristylis dichotoma* C.B.Clarke in Hook. f., Brit. India 6: 636. 1893.

Annual, erect, tufted herb; culms to 30 cm high, trigonous. Leaves flat, to 12 cm long, pubescent; ligule hairy; sheaths membranous. Umbels compound, terminal, 5 x 5 cm. Involucral bracts 3, longest to 5 cm long; primary rays 3-5, to 3 cm long; secondary rays 3-5. Spikelets solitary, oblong, 7 x 1.8 mm, many-flowered; glumes reddish-brown, ovate, apiculate, keel green; stamens 1; ovary obovate, style flattened, villous, dilated at base; stigmas 2, half as long as style. Achenes obovoid, biconvex, creamy-yellow.

This is common species usually found in rice fields, wet places, marshy areas. Fl. & Fr.: September – December.

Hussenapuram (KNL), *RVR* 3124; Palakonda hill (KDP), *CS* 7696; Kailasakona (CTR), *CPR* & *MHR* 14809; Gudur (NLR), *MCK* 23576; Darsi (PKM), *MCK* 22840; Rampa hill (EG), *KH* & *BR* 9733; Tyada Forest (VSKP), *KH* & *BR* 9721; Krishna river (GNT), *VRK* 6743; Koruturu (WG), *DN* 84292 (BSID); Rampachodavaram (EG), *GVS* 27245 (MH); Sunkarimetta (VSKP), *GVS* 19648, 19649 (MH); on the way to Jvapuram peak from Mdugula (VSKP), *GVS* 47267 (MH); Punyagiri village (VZN), GVS 19410 (MH); Bhadrachalam RF (KMM), *RCS* 99034 (MH & BSID); Rathamhutta hills (KMM), *RCS* 99056 (MH & BSID); Karkalpadu, Amangal RF (MBNR), *S.R.Srinivasan* 104547 (BSID); Kolli hills (NMK), *A.Mohan* 1040 (MH); Hogainakkal (KGR), *EV* 20553 (MH), Ohnakal (SLM), *KCJ* 17986 (MH); Tikarpada RF, Satkosia Tiger Reserve, *KCM* 6184 (BSID); Raigoda, Satkosia Tiger Reserve, *KCM* 7727 (BSID).

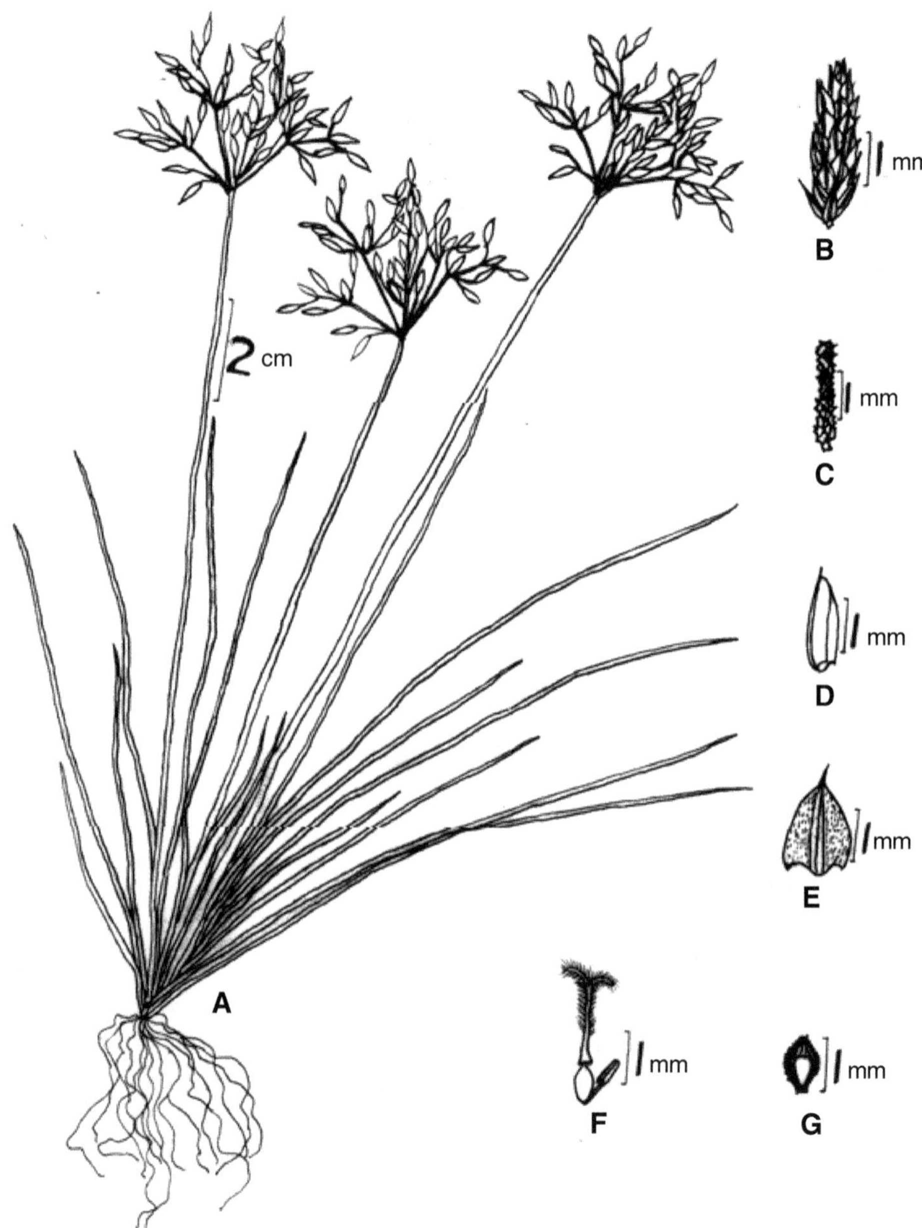

Figure 51. **Fimbristylis bisumbellata** (Forssk.) Bubani
A. Habit, B. Spikelet, C. Rhachilla, D. Closed glume, E. Opened glume, F. Pistil with stamens, G. Achene.

INDIA: Through out India

WORLD: N. Africa, W. Asia, Malesia, Taiwan.

Fimbristylis cinnamometorum (Vahl) Kunth, Enum. Pl. 2: 229. 1837; Matthew, Fl. Tamilnadu Carnatic 3: 1753. 1983. *Abildgaardia cinnamometorum* (Vahl) Thw., Enum. Pl. Zeyl. 347. 1864. *Scirpus cinamometorum* Vahl, Enum. Pl. 2: 278. 1806. *Fimbristylis cyperoides* R. Br., Prodr. Fl. Nov. Holl. 228. 1810. *F. cyperoides* R. Br. var. *cinnamometorum* (Vahl) C.B. Clarke in Hook. f., Brit. India 6: 650. 1893; Fischer 3: 1659. 1931.

Perennial, rhizomatous, caespitose herb, to 40 cm high, culms very slender, 3-sided, striate, leaved at base. Leaves many ranked; blades filiform, *c.* ½ the length of the culm, to 20 cm long, folded, glaucous green, distantly hispid on distal margins, subdensely ciliate at base with long white hairs; ligule a fringe of dense white hairs; sheaths folded with keel, the ventral side hyaline, pale brownish, the orifice obliquely truncate, ciliate. Corymbs compound, to 7 cm long, subdensely to loosely bearing 80 spikelets; involucral bracts 3, elongated, the lowest leaf like; rays 7, filiform, unequal, to 5 cm long, patent; secondary corymbs bearing 9 spikelets. Spikelets solitary, lance-ovate, to 8 mm long, bearing 10 glumes; glumes reddish-brown, densely red gland-dotted, lanceolatee, keeled; stamens 3; ovary obovate, style glabrous, base dilated, stigmas 3 Achenes narrowly obovate, very obtusely trigonous, maturing cream-white, transversely ridged.

Occasional in wet low lands and along grassy slopes. Fl. & Fr.: August – December.

Mogilipenta (KDP), *JSG* 21334 (CAL & DD); Seethalam river bed, Tandipandal (CTR), *MCB* 45098 (MH); Araku valley (VSKP), *NPBK* 10774 (MH); Kambakkam hills (CPT), *sine coll. s.n.* (MH); Champabandhmali, Karlapat (KHD), *HFM* 3505 (DD).

INDIA: Peninsular, C. & N.E.India.

WORLD: From SE Asia to S. China and Tropical Australia.

Fimbristylis complanata (Retz.) Link, Hort. Berol. 1: 292. 1827; FBI 6: 646. 1893; Fischer 3: 1659. 1931; Matthew, Fl. Tamilnadu Carnatic 3: 1753. 1983. *Scirpus complanatus* Retz., Observ. Bot. 5; 14. 1791.

Perennial, tufted, rhizomatous herb, to 40 cm high; culms densely caespitose, apex compressed, base trigonous. Leaves crowded at the base, 20 cm long, coriaceous, glaucous, ligule shortly hairy; sheaths compressed, keeled. Umbels decompound, effuse, 6 cm long; involucral bracts 2, not overtopping, to 6 cm long, largest leaf like; primary rays 3-5, to 3 cm long; secondary rays 3-4. Spikelet solitary, ovoid-oblong, to 8 mm long, acute, ca 12-flowered, brown with distichously arranged glumes. Achenes obovoid, trigonous, stipitate and stramineous.

Usually found in wet or moist places, on grassy slopes, river banks, open grass lands at low altitudes, in cultivated lands and in forest margins. Fl.& Fr.: August – April.

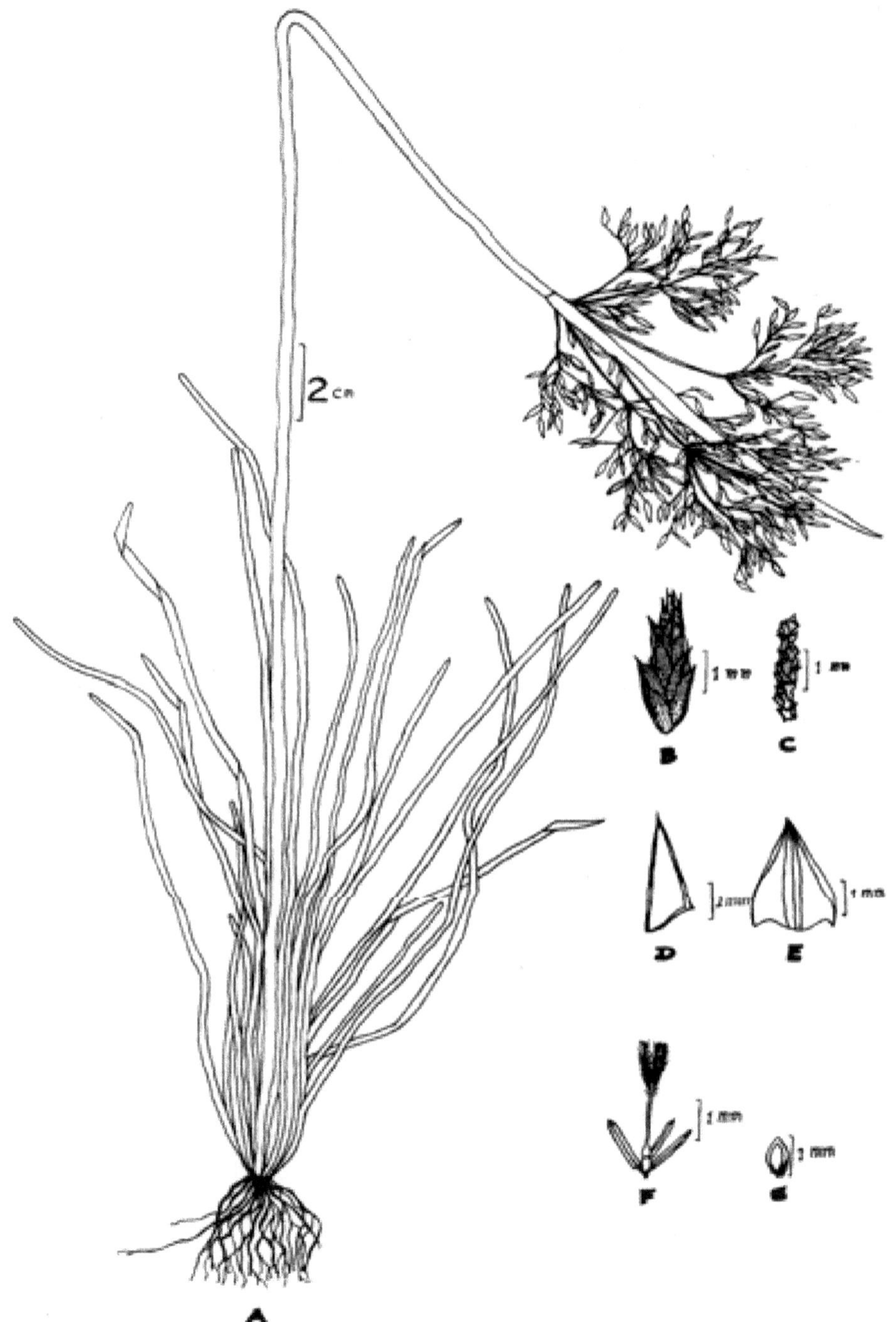

Figure 52. **Fimbristylis complanata** (Retz.) Link
A. Habit, B. Spikelet, C. Rhachilla, D. Closed glume, E. Opened glume, F. Pistil with stamens, G. Achene.

Madhavaram RF (KNL), *TP* & *RVR* 1337; Velgode (KNL), *BR* & *BSS* 29271 (BSID); Reddypalem (NLR), *PMR* 23069; Pamarru (KSN), *MCK* 22923; S.Kota reservoir (VZN) *MCK* 18809; Kannemitta (SKLM), *MCK* 18875; Nandyal (KNL), *JSG* 20360 (DD); Balapalle (KDP), *JLE* 14801 (MH & CAL), *A.M.Rao* 432 (BSID); Gajapathinagaram (VZN), *VNS* 4586 (MH); Annavaram (KSN), *CAB* 8190 (MH); Bainnachela (PKM), *RKM* 0854 (CAL); Rampur (KHD), *HFM* 1358 (DD); Katrang, Satkosia Tiger Reserve, *KCM* 6792 (BSID).

INDIA: Through out India.

WORLD: Pantropical extending to S.China and temperate Japan.

Fimbristylis cymosa R. Br., Prodr. Pl. Nov. Holl. 228. 1810; Matthew, Fl. Tamilnadu Carnatic 3: 1754. 1983. subsp. **cymosa.** *F. cymosa var. spathacea* (Roth) T. Koyama, Micronesica 1: 83. 1964 & in Dassan. & Fosb., Rov. Handb. Fl. Ceylon 5: 301. 1985. *F. spathacea* Roth, Nov. Sp. Pl. 24. 1821; FBI 6: 640. 1893; Fischer 3: 1659. 1931.

Perennial, erect, tufted, rhizomatous herb, to 40 cm high; culms densely leafy below, obtusely trigonous. Leaves densely crowded on the root stock, spreading and recurved, narrowly linear, to 15 cm long, rigidly coriaceous, margins recurved and scaberulous, sheaths short, coriaceous. Umbels compound, to 6 cm long; involucral bracts 3, filiform, to 2.5 cm long, leaf like, with broadly dilated base; primary rays 3-5, to 4 cm long; secondary rays 3, spikelet globose, to 3 mm long, pale brown, rhachilla winged; glumes ovate, keeled, obtuse, often notched, margins hyaline; stamens 1-3; ovary obovate, style glabrous, stigmas 2. Achenes obovoid, biconvex, turbinate, compressed.

Widely distributed species. Usually seen in sandy areas near sea shores, river beds and marshy low lands. Fl.& Fr.: March – October.

Gutturukona (ATP), *NY* 986; Bukkapatnam (ATP), *BR* & *ANS* 38110; Rajampeta (KDP), *AL* & *BSS* 28633; G.V.Palli (PKM), *BR* & *BSS* 33125; Guraja (KSN), *PV* 5538 (AU); Polavaram (WG), *CAB* 1507 (MH); Valamuru (EG), *MM* 101352 (MH); Nattavaram (VSKP), *CAB* 1901 (MH); S. Kota (VZN), *GVS* 21826 (MH); Singarapet on Tirupattur – Dharmapuri road (NA), KS 6516 (MH); Kalpadi (SA), *S.India Flora* 17861 (MH); Raigoda, Satkosia Tiger Reserve, *KCM* 6592 (BSID).

INDIA: Peninsular, N. & E. India.

WORLD: Pantropic.

Fimbristylis dichotoma (L.) Vahl, Enum. Pl. 2: 287. 1806; FBI 6: 635. 1893; Fischer 3: 1658. 1931; Matthew, Fl. Tamilnadu Carnatic 3: 1754. 1983. *Scirpus dichotomus* L., Sp. Pl. 50. 1753.

Key to Subspecies

1. Leaves glaucous to glabrescent .... subsp. **glauca**
1. Leaves other than:
   2. Leaves and bracts densely pubescent.................................subsp. **podocarpa**
   2. Leaves and bracts glabrous..subsp. **dichotoma**

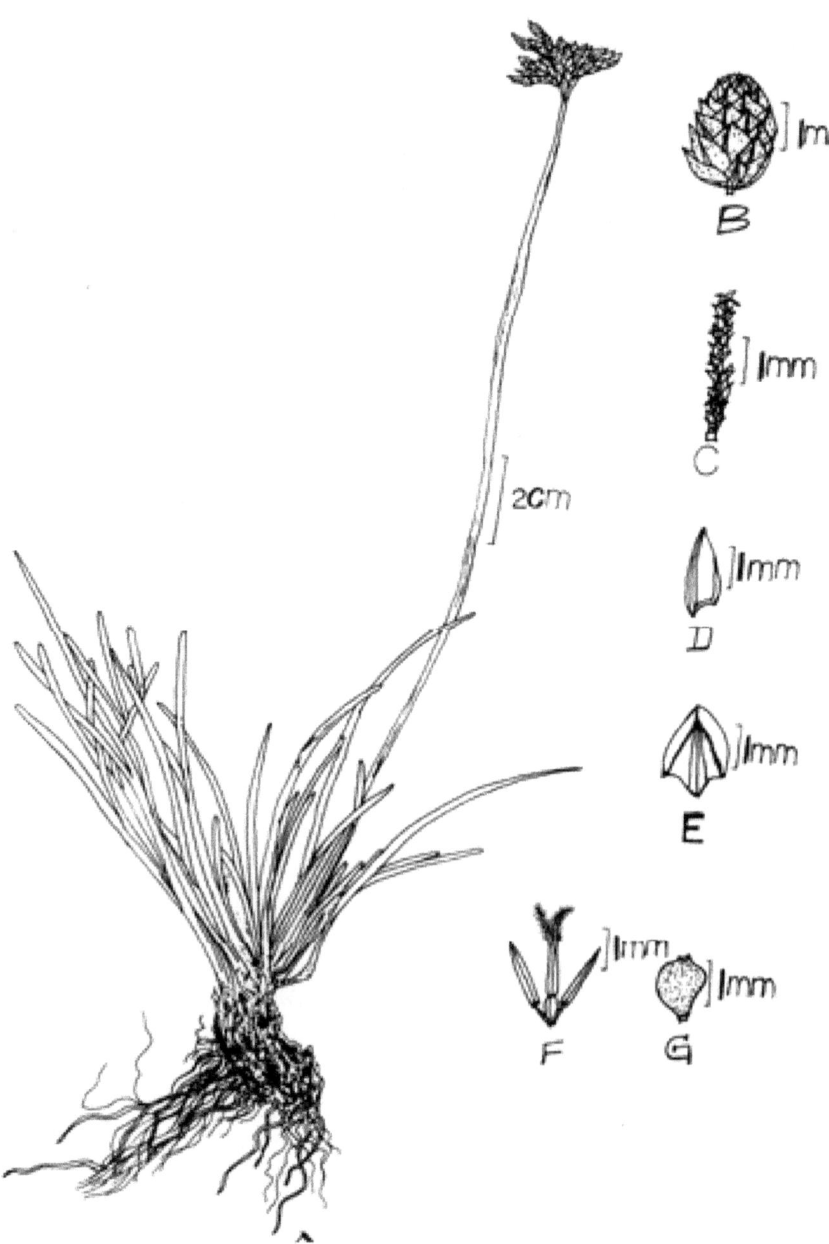

Figure 53. **Fimbristylis cymosa** R.Br.
A. Habit, B. Spikelet, C. Rhachilla, D. Closed glume, E. Opened glume, F. Pistil with stamens, G. Achene.

Fimbristylis dichotoma (L.) Vahl, subsp. **dichotoma**

Annual, tufted herb in a small clump; culms to 40 cm high, trigonous, leaves at base. Leaves to 10, as long as the culms; blades erect, patent, linear, to 30 cm high, herbaceous, glaucous green beneath, whitish green on upper surface, several veined; ligule of a fringe of dense white hairs; sheaths cylindrical with rounded back, the dorsal side pale greenish, the hyaline ventral side lightly brown, Umbels compressed, bearing *c.* 6 spikelets, lax to subdense, to 7 cm long. Involucral bracts to 6, the lowest one foliaceous, remaining setaceous; rays to 5, unequal, to 3 cm long, compressed, patent, terminated by 3 spikelets. Spikelets solitary, ovoid, to 7 mm long, terete, chestnut brown, densely many-flowered; glumes spirally arranged, oblong-ovate, obtuse, margin hyaline, apex reddish; stamens 2; style flattened, stigma 2-fid. Achenes obovate, biconvex, 1.8 mm long, excluding gynophore, rounded-truncate at mucronate apex, each side cancellated with 7-13 rows of transversely oblong cells.

Common along the sides of water courses, open grassy places, sandy localities and as a weed in cultivated lands. Fl. & Fr.: March – December.

Muniswarunikunta (ATP), *CJP* & *VS* 14149; Veyinuthulakona (KDP), *MHR* 13308; Papavinasanam (CTR), *MHR* 13345; Diguvametta (PKM), *BSS* & *SKB* 33173 (SKU & BSID); Rampa hill (EG), *KH* & *BR* 9733; Tyada Forest (VSKP), *KH* & *BR* 9720; Mahanandi (KNL), *JLE* 25486 (CAL); Rollapenta (KNL), *JLE* 42251 (MH); Durgam, Udayagiri (NLR), *P.Venu* 110225 (BSID); Peyyuru (KSN), *PV* 5850 (MH); Palagudem (WG), *VNS* 4491 (MH); Forest near Anigeri village, near Addatigala (EG), *GVS* 68551 (MH); Muzumamidivalasa, Maredumilli (EG), *MM* 105006 (BSID); Araku towards Jolaput (VSKP), *GVS* 19689 (MH); Elwinpet (VZN), *MV* 2796 (AU); Rathamhutta hills (KMM), *RCS* 104229 (MH & BSID); Kayapakam (CPT), *S.India Flora* 11231 (MH); Kongampallam – Yercaud (SLM), *SKK* 26847 (MH); Kindiripdr, Kasipur, (KHD), *HFM* 2528 (MH); Satkosia Wild Life sanctuary, *D.Hazra* & *D.Das s.n.* (BSID); Pampasar RF, Satkosia Tiger Reserve, *KCM* 5156 (BSID).

INDIA: Through out India.

WORLD: All warmer parts of the world.

Fimbristylis dichotoma (L.) Vahl subsp. **glauca** (Vahl) T. Koyama, Bot. Mag. Tokyo 87: 324. 1974. *F. glauca* Vahl, Enum. Pl. 2: 288. 1806.

Perennials, with short oblique rhizome, culms 25-75 cm high, slender, 0.5-0.8 mm thick. Leaves markedly glaucous, velvety to glabrescent, nearly as long as the culm; ligule absent. Corymbs lax, simple or compound, 2-5 cm, as long as broad. Spikelets obtuse-tipped, brownish-grey, dull; glumes ovate-elliptic, 1.5-2.5 x 1-1.5 mm, membranous-scarious, broadly hyaline-margined; keel 3-5-nerved, hardly reaching the apex. Achenes obovate, 0.6-0.7 x 0.4-0.5 mm, light brown, deeply trabeculate with 6-8 rows of transversely oblong cells, vertically 6-8-ridged; style fimbriate, stigma 2-fid; stamens 3.

Occasional in moist fields and grasslands. Fl. & Fr.: August-December.

Kalasamudram (ATP), *KRKS* 39932; Sundipenta (KNL), *BR* & *DV* 36376; Arey village (CTR), *BR* & *DV* 36364.

INDIA: S. India.

WORLD: Sri Lanka.

Fimbristylis dichotoma (L.) Vahl subsp. **podocarpa** (Nees & Meyen) T. Koyama, Micronesica 1: 87. 1964 & in Dassan & Fosb., Rev. Handb. Fl. Ceylon 5: 310. 1985. *F. podocarpa* Nees & Meyen ex Nees in Wight, Contr. Bot. Inda 98. 1834. *F. diphylla* Vahl var. *pluristriata* C.B.Clarke in Hook. f., Fl. Brit. India 6: 637. 1893. *F. annua* Roem. & Schult. var. *podocarpa* (Nees & Meyen) Kukenth., Bot. Jahrb. Syst. 59: 48. 1924. *F. diphylla* Vahl var. *podocarpa* (Nees & Meyen) Kukenth., Bot. Jahrb. Syst. 69: 257. 1938. *F. tomentosa* Vahl, Enum. Pl. 2: 290. 1805; Saxena & Brahmam, Fl. Orissa 4: 2188. 1996.

Leaves and involucral bracts densely pubescent. Achenes orbicular – obovate, thickly biconvexed, 1.5 mm long excluding gynophore, 1 mm wide, mucronate at apex, contracted at base; tuberculate on shoulder, cream-coloured, each side cancellated by 25 series of trasversely oblong hexagonal cells, gynophore obdeltoid.

Wet places, moist fields and in grass lands. Fl.& Fr.: March – November.

Mahanandi (KNL), *AL* 28800; Palakonda hills (KDP), *CS* 7693; Rapur (NLR), *BR* & *BSS* 33104; Gundlamotu reservoir (PKM), *BSS* & *SKB* 33151; Boddam (VZN), *MCK* 18825; Chelama (KNL), *JLE* 22079 (MH); D.M. Bhavi (KSN), *PV* 5033 (MH); Borapahar (SBP), *HFM* 3080 (DD); Jantiabandh (KHD), *HFM* 3529 (DD).

INDIA: Western Himalayas, Assam, Bihar, Odisha, Andhra Pradesh.

WORLD: India to Malesia, Micronesia, Polynesia.

Fimbristylis dipsacea (Rottb.) C.B.Clarke in Hook.f., Fl. Brit. India 6: 635. 1893; Fischer 3: 1658. 1931; Matthew, Fl. Tamil Nadu Carnatic 2: 1755. 1983; Saxena & Brahmam, Fl. Orissa 4: 2176. 1996. *Scirpus dipsaceus* Rottb., Descr. Icon. Rar. Pl. 56. 1773.

Small, densely tufted, annual herb, 5-10 cm; stems slender, capillary. Leaves filiform, 2-5 mm, scaberulous, apex acute; ligule 0, sheaths membranous, obliquely truncate, pinkish-grey. Inflorescence simple or compound, spiklet solitary, globose or ovoid, squarrose, 3-5 mm, obtuse, rachilla terete, wingled. Glumes lanceolate, to 2 mm, membranous, fulvous-stramineous, awn recurved, to 1 mm, scabrid, keel green, 1-nerved; stamen 1, style dilated at base, stigmas 2. Nut subterete, oblong-linear, 0.5-0.75 mm long, ferruginous or brown.

Occasionally in foot hills and plains. Fl. & Fr.: December – May.

Hogainakkal (KGR), *KCJ* 18006 (MH).

INDIA: Central India, West Bengal, Bihar, Odisha, Assam, Tamil Nadu, Karnataka.

WORLD: Tropical Africa, S. and E. Asia, Malesia.

Fimbristylis eragrostis (Nees & Meyen ex Nees) Hance, J. Linn. Soc. Bot. 13: 132. 1873. *Abildgaardia eragrostis* Nees & Meyen ex Nees in Wight, Contr. Bot. India 95. 1834. *Fimbristylis nigrobrunnea* Thw., Enum. Pl. Zeyl. 434. 1864; Fischer 3: 1669. 1931.

Perennial, tufted, rhizomatous herb, rhizome clothed with dark brown old sheaths and their fibrous remnants. Culms stiffly erect, to 40 cm high, triquetrous, weakly compressed, light green, leaved at very base. Leaves many; blades flattish with incurved margins, sheaths to 3 cm long, yellow brown. Corymbs open, simple, subdense, to 8 cm long, final corymb bearing 12 spikelets; rays 6, unequal, to 6 cm long; involucral bracts 5, setaceous, to 2 cm long. Spikelets solitary, oblong, to 12 mm long, yellow brown and tinged with purple, distichously bearing 20 flowers. Achenes broadly obovate, rounded at apex, cuneate at base, trigonous with convexed sides, maturing in straw colour.

Occasional on stable sandy grounds, grassy hill sides, crevices of rocks. Fl. & Fr.: September – October.

Gosala, (CTR), *GVS* 4589 (MH); Valamuru (EG), *MM* 10351 A (BSID).

INDIA: Peninsular, C. & N.E.India.

WORLD: Sri Lanka, China, Indonesia, Laos, Malaysia, Papua New Guinea, Thailand, Vietnam, Australia.

Fimbristylis falcata (Vahl) Kunth, Enum. Pl. 2: 239. 1837; Matthew, Fl. Tamilnadu Carnatic 3: 1750. 1983. *Scirpus falcatus* Vahl, Enum. Pl. 2: 275. 1806. *Trichelostylis junciformis* Nees in Wight, Contrib. Bot. India 106. 1834. *Fimbristylis junciformis* (Nees) Kunth, Enum. Pl. 2: 239. 1837; FBI 6: 647. 1893; Fischer 1660. 1931.

Perennial, erect, rhizomatous herb; rhizome oblique, to 7 cm long, thickly clothed with brown leaf sheaths, culms solitary or sparsely tufted, rigid, 4-angled, to 40 cm high, deeply striate. Leaves crowded round the base of the stems, shorter than stems, spreading and recurved with triangular acute tip, to 20 cm long; sheaths short, ligule 0. Umbel compound, at times capitate, to 5 cm long, bearing distant or clustered small chestnut-brown spikelets. Involucral bracts 4, shorter than umbel, to 3.5 cm long; primary rays 3, slender, to 2 cm long; secondary rays 2. Spikelets ellipsoid, solitary, 5 mm long, purplish, rhachilla winged; glumes ovate, acute, margins hyaline, keeled; stamens 3; ovary trigonous, obovate, style slender, fimbriate towards apex, base much dilated, stigmas 3, as long as the style. Achenes obovoid, trigonous, stramineous.

Usually seen in open grass lands along the lateritic slopes. The plants which grow on rocky surfaces the inflorescence is rather compact, whereas in open grassy places, it is open and lax. Fl. & Fr.: June – November.

Peddayapalle (KNL), *BR* & *BSS* 33131; Nagalooty (KNL), *VBH* 83919 (BSID); Venkatagiri hills (NLR), *BSN* 3803 (VV); Stambalakona, Devudu Vellampalli (NLR), *J.Swamy* & *S.Nagaraju* 14582 (BSID); Karaka (VSKP), *CAB* 1638 (MH); Towards Punyagiri hills (VZN), *GVS* 44227 (MH); Korada RF (SKLM), *GVS* 62390 (MH & CAL); Vattavarlapalli (MBNR), *VBH* 84967 (BSID).

INDIA: Through out India.

WORLD: Pakistan, Sri Lanka, Nepal, Thiland, Indo-China, Malesia.

Fimbristylis ferruginea (L.) Vahl, Enum. Pl. 2: 291. 1806; FBI 6: 638, 1893; Fischer 3: 1658. 1931; Matthew, Fl. Tamilnadu Carnatic 3: 1757. 1983. *Scirpus ferrugineus* L., Sp. Pl. 50. 1753.

1. Culm to 140 cm high; leaves 10 cm long; umbel compound.......... subsp. **ferruginea**
1. Culm to 45 cm high; leaves 40 cm long; umber simple.......... subsp. **sieberiana**

Fimbristylis ferruginea (L.) Vahl subsp. **ferruginea**

Perennial, tufted, erect, rhizomatous herb; to 40 cm high; culms sub trigonous, rigid, to 7 mm wide. Leaves less than 10 cm long, scabrid, apex acuminate; ligule a dense fringe of short hairs; sheaths fuscous, margin hyaline. Umbel sub-compound of few spikelets, to 5 cm long. Involucral bracts 2, lowest involucral bract shorter than inflorescence. Spikelets solitary, ovoid or oblong, pale reddish brown, pubescent, to 2 cm long, lowest spikelet bractiform; rhachilla angular and winged. Achenes obovoid, biconvex, stipitate, stramineous.

Common in wet places, open swamps, marshy areas. Fl. & Fr.: August – December.

Kolleru lake (WG), *KH* & *BR* 10910; S. Kota (VZN), *KH* & *BR* 5596; Siepenaidupeta (SKLM), *MCK* 18887; Guvvalacheruvu (KDP), *KS* 7793 (CAL); Varadayapalem to Kambakkam (CTR), *KCJ* 45154 (MH); Gudur (NLR), *KCJ* 18529 (MH); Bhavanipuram (KSN), *PV* 6039 (MH); Hogainakkal (KGR), *KCJ* 17982 (MH); Palakuppam (SA), *VNS* 4052 (MH);

INDIA: Throughout India.

WORLD: Pantropic.

Fimbristylis ferruginea (L.) Vahl subsp. **sieberiana** (Kunth) Lye, Nordic J. Bot. 2: 335. 1982. *Fimbristylis sieberiana* Kunth, Enum. Pl. 2: 237. 1837.

Perennial, tufted, rhizomatous herb, rhizome small, woody; culms to 45 cm high, basally scabrid. Leaves flat, to 40 cm long, margins inrolled, scabrid; ligule short, hairy; sheaths purple brown, hairy. Inflorescence simple, to 3 cm long. Involucral bracts 3-5, lowest overtopping, to 5 cm long; rays 2, to 2 cm long. Spikelets solitary, terete, to 1 cm long, obtuse. Achenes obovoid, biconvex, stramineous, smooth, stipitate.

Occasional in open marshy places and in cultivated fields. Fl. & Fr.: July – December.

Gani RF (KNL), *RVR* 1531; Madhavaram RF (KNL), *TP* & *RVR* 1339; Sangam (NLR), *MCK* 22948; Pulipadu tank – Darsi (PKM), *MCK* 22860; Kolleru lake (WG), *KH* & *BR* 10910; way to Guvvalacheruvu (CDP), *KS* 7793 (MH).

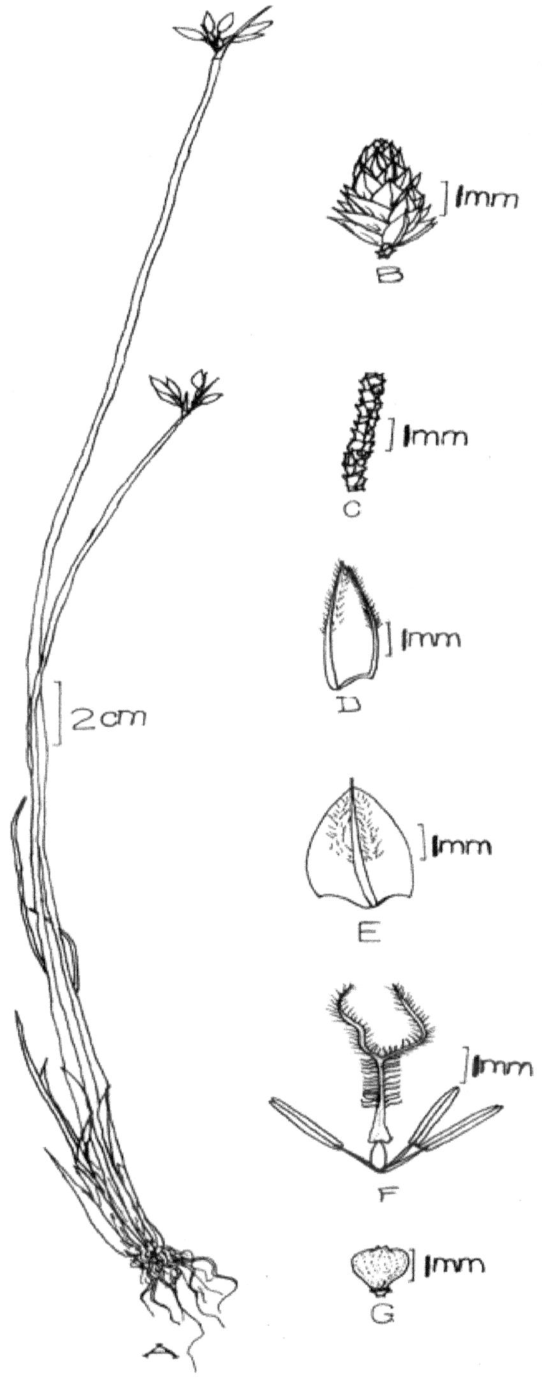

Figure 54. **Fimbristylis ferruginea** (L.) Vahl subsp. **ferruginea**
A. Habit, B. Spikelet, C. Rhachilla, D. Closed glume, E. Opened glume, F. Pistil with stamens, G. Achene.

INDIA: Delhi, Gujarat, Goa, Kerala, Karnataka, Lakshadweep, Maharashtra, Madhya Pradesh, Punjab, Rajasthan, Tamil Nadu, Tripura, Uttar Pradesh and West Bengal.

WORLD: Pantropic.

Fimbristylis fimbristyloides (F.Muell.) Druce, Bot. Soc. Exch. Club Brit. Isles 1916: 623. 1917; Chandramohan *et al.*, Indian J. Forestry 41: 327. 2018.

Tiny, dwarf, tufted, annual herb, up to 10 cm long, often scabrid all over; stems trigonous, capillary, subglabrous. Leaves eligulate, basal; sheath short, pale brown to reddish; blade flat, strongly falcate, 25-40 x 1-1.5 mm, apex obtuse, margins incurved. Involucral bracts 2-3, 5-25 mm long, foliaceous, shorter than inflorescence axis. Inflorescence simple to compound; primary branches 3-5, 0.5 – 2.5 cm long, densely hairy. Spikelets 3-6 per inflorescence, ovate to lanceolate, strongly compressed, 3-4 x 2-2.8 mm, acute, red brown; glumes 5-14 per spikelet, red brown, distichous, ovate to lanceolate, *c.* 2 x 1 mm, acute, keeled at the back, with nerveless sides and broadly hyaline margins, pubescent on the back; stamens 2, anthers oblong, *c.* 0.4 mm long; style triquetrous, pyramidally thickened at the base, glabrous; stigmas 3, shorter than style. Nut trigonous, with acute angles and convex faces, obovoid, 0.4-0.8 mm, with broadly tirangular base, whitish, verruculose, stipe distinct, epidermal cells isodiametric.

Rare growing in wet lands, near ditches in Satkosia Tiger Reserve in Odisha. Fl. & Fr.: September – November.

Satkosia Tiger Reserve, Mahanadi WL division, Purunapani, *KCM* 9409 (BSID).

INDIA: Meghalaya, Maharashtra, Eastern Uttar Pradesh, Odisha.

WORLD: Myanmar, E.Thailand, S.China, Ryukyu Is., S.Korea, Tropical Australia, Malesia.

Note: The speices was collected from a single locality in Purunapani, Mahanadi Wildlife Division, Nayagarh with scaty population (Chandramohan *et al.*, 2018). The species can be distinguished in the field by the rosette looking like a wheel with curved inner bars. Morphologically it could be confused with *F. straminea* Turrill but can be distinguished by densely hairy inflorescence branches and acute glumes.

Fimbristylis fusca (Nees) C.B.Clarke in Hook.f., Fl. Brit.India 6: 649. 1893; Haines, Bot. Bihar & Orissa 3: 917. 1924; Saxena & Brahmam, Fl. Orissa 4: 2178. 1996. *Abildgaardia fusca* Nees in Wight, Contrib. Bot. India 95. 1834.

Perennial, rhizomatous, stems tufted, 20-45 cm high, with radical leaves. Leaves 5-15 cm x 1.5-2 mm, glabrous or pubescent beneath, scabrid at the top. Involucral bracts 2-4, leaflike, 0.8-1.5 cm, hairy. Inflorescence compound. Spikelet solitary, oblong or lanceolate, 6-10 mm long, compressed. Glumes distichous, obliquely erect, narrowly lanceolate, 4-5 mm long, prominently keeled, 1-few-nerved, scabrid, whitish hyaline margins, lower 3 glumes empty; stamens 3; style very slender, stigmas 3. Nut pale brown, trigonous, obovoid, 1 mm long, verrucose.

Occasional in Orissa Eastern Ghats. Fl. & Fr.: May – July.

Motijharan (SBP), *HFM* 3821, 3825 (DD); Karlapat (KHD), *HFM* 3473 (DD); Tulka RF, Satkosia Tiger Reserve, *KCM* 5950 (BSID).

Fimbristylis hookeriana Boeck., Linnaea 37: 22. 1871; FBI 6: 641. 1893; Haines, Bot. Bihar & Orissa 3: 922. 1924; Saxena & Brahmam, Fl. Orissa 4: 2179. 1996.

Tufted annual, 10-25 cm high; stems compressed; rhizome absent. Leaves narrow, flat, often as long as the stems. Spikelets narrow, linear, 1-1.5 cm long, acute, clustered and solitary on the erecto-patent branches of umbels, often 10-15 cm diam.; spikelets 4-6 in a cluster, 2.5 mm or less broad. Median glume *c.* 3.7 mm long, narrowly lanceolate, very acute, rusty green, keeled, brown scarious on the sides; stigmas 2. Nut obovoid, 0.75 mm long, biconvex, obtuse or truncate, very shortly stipitate, light brown.

In crevices in northern Eastern Ghats. Fl. & Fr.: September – November.

Sonabera (KPT), *HFM* 3688 (DD); Barapahar (SBP), *HFM* 3616 (DD); Tulka RF, Satkosia Tiger Reserve, *KCM* 6784 (BSID).

Fimbristylis kingii Gamble ex Boeck., Beitr. Cyper. 2: 40. 1890; FBI 6: 633.1893; Fischer 3: 1658.1931; Matthew, Fl. Tamilnadu Carnatic 2: 1757. 1983.

Culms densely tufted, obtusely angled, glabrous, or shortly hairy, 10-20 cm high. Leaves shorter than, or as long as, culms, flat or rarely channelled, acute, hairy, to 0.1 cm wide, eligulate; sheaths membranous. Inflorescence simple. Involucral bracts 2, overtopping or not. Spikelet solitary, ovoid, acute, many-flowered, 0.3-0.5 cm. Glumes ovate, to 0.2 cm, membranous, obtuse, 3-5-nerved, mucronate, reddish brown, margins hyaline; stamens 3; style apically ciliate; stigmas 3, ciliate. Nut 3-gonous, obovoid, stalked, stramineous, *c.* 0.1 cm.

Waterlogged places, along banks of streams in Chitteri hills in Tamil Nadu (Matthew, 1983). Fl. & Fr.: October – January.

INDIA: Kerala, Tamil Nadu, Karnataka, Maharashtra. Endemic.

Fimbristylis littoralis Gaud. in Frey. Voy. Bot. 413. 1826; Sunitha & B.R.P. Rao, J. Econ. Taxon. Bot. 25: 24. 2003. *F. miliacea* (L.) Vahl, Enum. Pl. 287. 1806, *p.p.*; FBI 6: 644. 1893.

Culms tufted, erect portion arising from the axils of basal leaves, 15-60 cm high, 2-3 mm thick below, flattened, 4-angled with 2 sharp edges just below the inflorescence, smooth. Leaves reduced to bladeless sheaths, 2-10 cm long, clasping the stem, the ventral side splitting as the culm grows, laterally flattened with a sharp dorsal edge, brownish, hyaline on the margins, the lowest scale like. Involucral bracts 2-4, rigid, the lowest 1.8 cm, not overtopping the inflorescence. Inflorescence compound, 3-7 x 2-3 cm. Spikelets ovoid, globular, 2-3 x 2-2.5 mm, rusty brown, densely many-flowered. Glumes broadly ovate, 1-1.5 x 0.5-0.7 mm, boat-shaped, obtuse at the apex, membranous, reddish brown, the margin with white hyaline margins, keel 1-nerved ending below the apex; stamens 1, 1.5 x 0.5 mm, anthers 0.5

x 0.1 mm; pedicel 1 mm; style flattened, 3-cleft at the apex, 1.8 mm. Nut trigonous, obovoid, 0.5-0.8 x 0.4 mm, creamish yellow, surface warty particularly at the ridges.

Rare in open, moist to swampy places in dry deciduous forests. Fl. & Fr.: August – December.

Pennahobilam (ATP), *TP* & *NY* 374; Kottalacheruvu Sacred Grove (KNL), *BR* & *SS* 21259; Chintaparti river (CTR), *MHR* 13380; N.S.Right canal (GNT), *VRK* 3545; Kadiyam (WG), *MHR* & *MCK* 17347; Boddam (VZN), *MCK* 18823; Chelama (KNL), *JLE* 18006 (MH); Varadayapalem to Kambakam (CTR), *MCB* 45161 (MH); Maredumilli (EG), *GVS* 27298 (MH & CAL); Forest near Anigeri (EG), *GVS* 68858 B (MH); Balugram near Parvatipuram (SKLM), *VNS* 4553; Near Rangamatlagedda (SKLM), *GVS* 62335 (CAL); Vandalur RF (CPT), *ANH* 45548 (MH); Ammapettai (SA), *VNS* 4163 (MH); Bhuvanagiri to Krishnagiri (SA), *KRM* 60108 (MH); Sonanchadi (SA), *KRM* 64168 (MH); Kuppam (CTR), *S.India Flora* 10317 (MH); Palakuppam (CTR), *VNS* 4028 (MH); Tippukadu RF (NA), *KRM* 17651 (MH); Ohanakal (SLM), *KCJ* 17985 (MH); Tulka RF, Satkosia Tiger Reserve, *KCM* 6504 (BSID); Aska (GJM), *sine coll. s.n.* (MH).

INDIA: Peninsular, C. & N.E. India.

WORLD: Pantropic.

Note: On enquiry about the nomenclature Dr. Kanchi N. Gandhi of Harvard University Herbarium clarified about the nomenclature which is given below.

"***Fimbristylis miliacea*** (L.) Vahl (1805)" and its basionym "*Scirpus miliaceus* L. (1759)" are rejected names and must NOT be used. Of course, these 2 names may be cited as synonyms. Synonymy citation problem: please, see

Strong, M.T. 2004. Proposal to reject the name *Scirpus miliaceus* (Cyperaceae). Taxon 53(4): 1069-1070..

As mentioned in the proposal, the original material of *Scirpus miliaceus* may have consisted of two sheets in the Linnaean herbarium: No. 71.40 (= *F. quinquangularis* (Vahl) Kunth) and No. 71.41 (= *F. littoralis* Gaud.)

Therefore, since "*Fimbristylis miliacea* (L.) Vahl (1805)" is a rejected name, it may be cited as a *p.p.* synonym under both *F. quinquangularis* and *F. littoralis.*

1) **F. quinquangularis** (Vahl) Kunth

   Synonym: *F. miliacea* (L.) Vahl, p.p. (LINN no. 7140)

2) **F. littoralis** Gaud.

   Synonym: *F. miliacea* (L.) Vahl, p.p. (LINN no. 7141)

Fimbristylis monospicula Govind., Proc. Indian Acad. Sci., B 79: 169. 1974; Karthik. *et al.*, Fl. Ind. Enum. Monocot.: 53. 1989; Mahendra Nath *et al.*, Rheedea 27: 104. 2017.

Annual. Culms slender, tufted, 8–10 cm high, angled, 0.4–0.6 mm thick, hispid-hairy throughout. Leaves, shorter than culm, slender *c.* 1 mm wide, apex acute to acuminate, hispid-hairy throughout; sheaths rather glabrous with prominent veins; ligule absent. Inflorescence a solitary, terminal spikelet. Bracts usually glume-like, 2–3 mm long, hispid on the keel towards apex, often fall off as the spikelet mature. Spikelets ovoid, terete, 4–5.3 x 3–3.5 mm, apex obtuse-acute, yellowish brown to dark brown, many-flowered. Glumes somewhat thick, spirally imbricate, broadly ovate to suborbicular, *c.* 2.5-3.2 mm, apex obtuse, muticous; keel 3-veined. Rachilla shortly winged, excavated; stamens 3; filaments elongate up to 2.8 mm; anthers linear-oblong, *c.* 0.8 mm long; style *c.*1 mm long, pyramidally thickened at base, ciliate towards apex; stigmas 3, somewhat shorter or almost equal to the style, ciliate. Achenes obovoid, trigonous, *c.* 1.3 - 0.6 mm, base cuneate, apex truncate-obtuse, prominently tubercled, yellowish brown; epidermal cells in 6–8 rows, transversely oblong.

Horsely hills of Andhra Pradesh. Fl. & Fr.: July.

Horsely hills (CTR), *Mahendra Nath* HH-02934 (CAL).

INDIA: Andhra Pradesh and Karnataka. Endemic.

Fimbristylis naikii W. Khan & Lakshmin., J. Bot. Res. Inst. Texas 2 (1): 381. 2008; Wad. Khan, Cyperaceae W. Ghats, W. Coast & Maharashtra: 218. 2015.

Annual. Culms slender, trigonous, compressed, 10-35 cm tall, and 0.3-1 mm thick, glabrous. Leaves ligulate, subbasal, 1-2 mm wide, as long as culms, flat, hairy with long white cilia or glabresecnt; sheaths densely hairy, dark brown. Inflorescence small, simple to subdecompound, 1.5 – 3.5 cm long and broad with usually 3-9 spikelets, rays 1-3. Involucral bracts up to 6, foliaceous. Spikelets solitary, ovoid or ovoid-oblong, 3-7 x 3.5 mm, acute, brown-white variegated, many-flowered, rachilla winged; glumes spiral, ovate to broadly ovate, 2-2.8 x 1.8 – 2.2 mm, reddish brown; stamens 1 or 2; anthers elliptic oblong, 0.4-0.5 mm long, apiculate; style flat. Nutlet biconvex, broadly obovoid, 1.5 – 1.6 x 0.8 -1.2 mm, subtrabeculate with 15-18 vertical striations in between transversely oblong, thin-walled epidermal cells.

Occasional in wet open grassland and marshy areas. Fl. & Fr.: October – November

Allipudi (EG), 17/10/2015, *K. Yarrayya* 119057 (MH).

INDIA: Maharashtra, Andhra Pradesh, Endemic.

Fimbristylis nutans (Rez.) Vahl, Enum. Pl. 2: 285. 1803; FBI 6: 632. 1893; Fischer 3: 1658. 1931. *Scirpus nutans* Retz., Observ. Bot. 4, 12. 1786.

Perennial, densely tufted, rhizomatous herb, to 70 m high. Culms subterete, obtusely trigonous above, striate, clothed at base with 3 bladeless sheaths only. Basal sheaths cylindrical, membranous, dorsally lightly green, ventrally thinly membranous, pale brownish, the orifice obliquely truncate with broad, brown hyaline margin, the uppermost sheath to 12 cm long, second to 4 cm long, the radical ones scale like, rusty brown. Inflorescence a single terminal spikelet; involucral bract 1, scale like, broadly pale green on back. Spikelets obliquely patent, ovoidal, to 16 mm long, terete, brown, densely many-flowered. Achenes broadly obovate, unequally biconvexed, maturing cream-coloured, roughly transversely wrinkled with 6 ridges.

Occasional in wet places in Chittoor district. Fl. & Fr.: September – December.

Thandipandal (CTR), *MCB* 45102 (MH).

INDIA: S., C., N. & N.E. India.

WORLD: China, Philippines, Malaysia, Thailand, Vietnam, Australia.

Fimbristylis ovata (Burm. f.) Kern, Blumea 15: 126. 1967. & in Steenis, Fl. Males. 1. 7: 595. 1974; Mattew, Fl. Tamilnadu Carnatic 3: 1758. 1983. *Carex ovata* Burm. f., Fl. Indica 194. 1768. *Abildgaardia ovata* (Burm. f.) Kral, Sida 4: 72. 1971. *Cyperus monostachyos* L., Mant. 2: 180. 1771. *Abildgaardia monostachyos* (L.) Vahl, Enum. 2: 296. 1805. *Fimbristylis monostachyos* (L.) Hassk., Fl. Jav. Rav. 61. 1848, (*monostachya*); FBI 6: 649. 1893; Fischer 3: 1660. 1931

Perennial, erect, caespitose and leafy herb, to 40 cm high; culms filiform, trigonous. Leaves crowded on the swollen base of the stem, filiform, to 20 cm long, scabrid; ligule 0; sheaths stramineous. Spikelets solitary on the stem. Involucral bracts 2, not overtopping, to 1 cm long. Spikelets compressed, ovate, acute, shining; glumes subcoriaceous, broadly ovate, keeled, 2 lower glumes empty and longer than the others, cuspidate, to 1 cm long; stamens 3; ovary trigonous, style thick, dilated at base, flattened; stigmas 3, slightly shorter than style; rhachilla winged. Achenes obovoid, trigonous, stipitate, stramenious.

Occasional on wet grass land, stream banks, in bunds of rice fields and in marshy areas. Fl. & Fr.: December – March.

Gundumala RF (ATP), *TP* & *NY* 717; Gundlabrahmeswaram (KNL), *BR, BSS* & *DVL* 33137 (BSID); Lankamalai RF (KDP), *SRSR* 13163; Kondaveedu fort (GNT), *S.India Flora* 17492 (MH); Pulusumamella (KNL), *VBH* 83935 (BSID); Kambakkam hills (CTR), MCB 45074 (MH); Nagapatla RF (CTR), *KS* 6859 (MH); Venkatagiri (NLR), *BSN* 3498 (VV); Stambalakona, Devudu Vellampalli (NLR), *J.Swamy* & *S.Nagaraju* 7876 (BSID); Malakondapenta (PKM), *VBH* 83977 (BSID); Mylavaram (KSN), *CAB* 8172 (MH); Addathigala (EG), *GVS* 68557 (MH & CAL); Forest near Araku (VSKP), *NPBK* 10803 (MH); Gajapathinagaram (VSKP), VNS 4589 (MH); Salur (VZN), *NPBK* 949 (CAL); Rathamhutta hills (KMM), *RCS* 104228 (BSID); Ungaru forest (DMP), *EV* 57942 (MH); Anchetty – Kunthukottai RF (DMP), *EV* 57917, 57898 (MH); Sirupaakkam RF (SA), *KRM* 50696 (MH); way to Komatiyur (NA), *KS* 7469 (MH);

INDIA: Through out India.

WORLD: Pantropical north to S. China and Japan.

Fimbristylis paupercula Boeckler, Linnaea 38: 396. 1874; FBI 6: 647. 1893; Fischer 3: 1660. 1931; Matthew, Fl. Tamilnadu Carnatic 2: 1759. 1983; Saxena & Brahmam, Fl. Orissa 4: 2182. 1996.

Stems to 30 cm high, slender, 4-5-angled. Leaves flat or folded, to 7 x 0.1 cm, scaberulous, eligulate, apex acute, sheaths with hyaline margin. Inflorescence

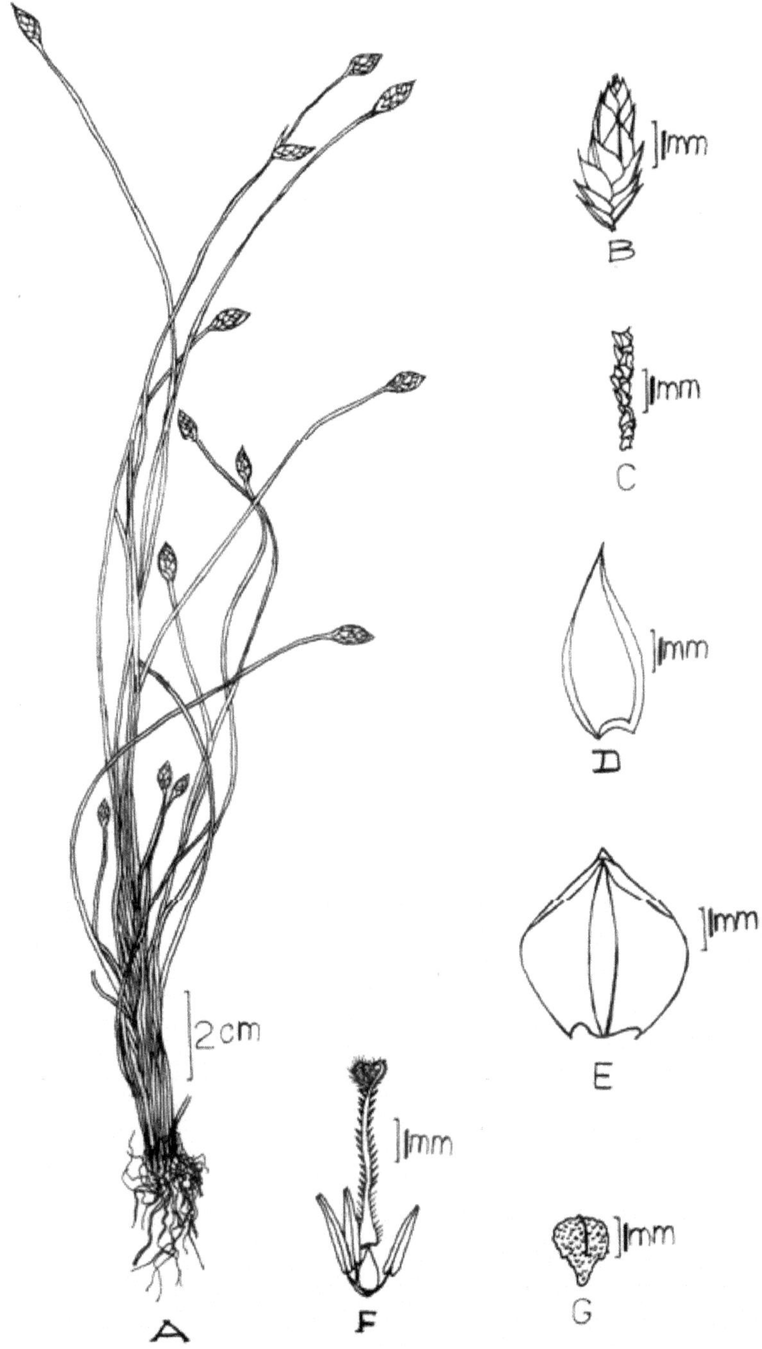

Figure 55. **Fimbristylis ovata** (Burm.f.) Kern [Syn.: *Abildgaardia ovata* (Burm.f.) Kral]
A. Habit, B. Spikelet, C. Rhachilla, D. Closed glume, E. Opened glume,
F. Pistil with stamens, G. Achene.

compound, to 3 cm; involucral bracts 3, longest to 2.5 cm, primary rays 3. Spikelets in clusters of 2-7, sessile, sometimes solitary, ovate or ellipsoid, 3.7-10 mm long. Glumes spiral, ovate, apical half glandular, keel or the whole glume often puberulous, minutely mucronate, dark, margins membranous and paler; stigmas 3; style 3-fid. Nut minutee, depressed obovid, trigonous, dark, trabeculate.

Hills bove 1000 m (Matthew, 1983). Fl. & Fr.: October – February.

INDIA: Peninsular India.

Fimbristylis pierotii Miq., Ann. Mus.Bot. Lugduno-Batavum 2: 145. 1865; FBI 6: 642. 1893; Saxena & Brahmam, Fl. Orissa 4: 2182. 1996.

Glabrous perennial with creeping rhizome; stems slender, solitary, compressed triquetrous, striate-sulcate, scabrid at the top, few-leaved at the base, 20-30 cm high. Leaves shorter than the stem, 1-2 mm wide, ligule 0. Inflorescence simple or subcompound, loose; involucral bracts very short. Spikelets solitary, ovoid-lanceolate, 7-15 x 3-4 mm, angular, acute, rather loosely flowered, rachilla broadly winged. Glumes spiral, ovate, 4.5-6 mm long, acute to obtuse, muticous, keel 3-nerved, margins hyaline, gland-dotted in the upper half; stamens 3; style pyramidally thickened at the base, 2.5-4 mm long, glabrous, stigmas 3. Nut trigonous, broadly obovoid, 1-1.2 mm long, shortly stipitate, verrucose, whitish.

Siramnda Parbat in Odisha (Saxena and Brahmam, 1996). Fl. & Fr.: ???

Sirimanda parbat, near Pottangi (KPT), *HFM* 3832 (DD).

INDIA: NW Himalaya, Odisha.

WORLD: India to Japan and Korea, Malesia.

Fimbristylis polytrichoides (Retz.) R. Br., Prodr. 226. 1810; FBI 6: 632. 1893; Fischer 3: 1658. 1931; Mattew, Fl. Tamilnadu Carnatic 3: 1759. 1983. *Scirpus polytrichoides* Retz., Observ. Bot. 4: 11. 1786/87.

Perennial, densely tufted herb, culms to 25 cm high, trigonous, shallowly striate, glaucous-green, clothed at base with few sheaths; sheaths to 3 cm long, herbaceous, yellowish-brown. Leaf blades to 10 cm long, canaliculated, remaining as setaceous appendages of sheath. Inflorescence a single, terminal spikelet. Involucral bracts 2, the longest up to 3 cm long, overtopping the spikelet. Spikelet ellipsoidal, to 6 mm long, terete, obtuse, brownish grey, densely many-flowered. Achenes obovoid, biconvex to 2 mm long, grey purple at maturity, minutely cancellated with transversely oblong cells.

Open wet grounds, sandy places, rock crevices. Fl. & Fr.: September – March.

Vankarakunta Ashramam (ATP), *CJP* & *VS* 14186; Punyagiri hill (VZN), *KH* & *BR* 9715; Varani (NLR), *JSG* 12681 (MH, DD, & CAL); Ettukkuru, Maredumilli (EG), *MM* 102616 (BSID); Parangipettai (SA), *KRM* 58179 (MH);

INDIA: S.C.& E. India

WORLD: Tropical regions of Old World from Tropical Africa to Tropical Australia.

Fimbristylis quinquangularis (Vahl) Kunth, Enum. Pl. 2: 229. 1807; FBI 6: 644. 1893; Fischer 3: 1659. 1931. *Scirpus quinquangularis* Vahl, Enum. Pl. 2: 279. 1806. *Fimbristylis miliacea* (L.) Vahl, Enum. Pl. 287. 1806, *p.p.*(LINN No. 7140); FBI 6: 644. 1893.

Annual, densely tufted, rhizomatous herb, to 70 cm high. Culms arising from axils of bladed leaves which first form a leaf shoot; pentagonous, clothed at base with 4 sheaths. Culm sheaths bladeless, subterete, pale green, the ventral side hyaline and lightly yellow brown, the orifice obliquely truncate, hyaline on margin, the uppermost sheath to 18 cm long, second to 10 cm. Leaf blades dorsiventrally flat, shorter than the culm, to 30 m long, herbaceous, whitish green and 5-nerved on upper surface, smooth except on upper margins where spinulose scabrid; sheaths to 7 cm long, pale green, yellow-brownish, ventral side split as culm grows. Corymbs compound, to 9 cm long; rays 7, patent, to 5 cm long, scabrous; secondary corymbs bearing to 30 spikelets. Involucral bracts 5, setaceous, much shorter than the rays. Spikelets solitary, ovoidal, to 5 cm long, rusty brown, *c.* 40-flowered.

Usually seen in wet places, grass lands, along canal banks and often in rice fields. Fl. & Fr.: August – November.

Guvvalacheruvu RF (KDP), *RVR* 7991; Jayanthipuram (GNT), *VRK* 6940; Sunkaravaripalli (NLR), *A.S.Rao* 2223; Rampa hill (EG), *VNS* 245 (CAL); Forest near Anigeri (EG), *GVS* 68858 A (MH); Rollapadu (KMM), *R.Rajan* 108081 (BSID).

Palur (SA), *VNS* 4439 (MH); Mukundapur (GJM), VNS 4675 (MH).

INDIA: Throughout India.

WORLD: Sri Lanka, S. China, Malesia, Troical Australia.

Note: See the clarification about thee nomenclatur under *F. littoralis.*

Fimbristylis salbundia (Nees) Kunth, Enum. Pl. 2: 230. 1837; Veeranjaneyulu *et al.*, Indian J. Forest. 36: 277. 2013. *Trichelostylis salbundia* Nees in Wight, Contrib. Bot. India 105. 1834. *T. pentaptera* Nees in Wight, Contr. Bot. India 105. 1834.

Perennials, culms erect, to 75 cm high, in tufts, clothed at base with 2-3 bladeless sheaths, glabrous and sparsely to densely hispid toward apex. Corymbs decompound bearing 7-15 spikelets, 3-8 x 3-6 cm; rays 7-14; bracts 3 or 4, scale-like, 3-7 cm long, membranous. Spikelets solitary, ovoid, 4.5-6.2 x 1.5-2.2 mm, sub-terete, sub-acute at apex, 8-60-flowered; glumes spirally imbricated, ovate, suddenly contracted at apex, 2.3 -3 x 1.5-2 mm, boat-shaped with convex keel, conspicuously 3-5-nerved; stamens 3; style slender, 1.5-2.2 mm long, glabrous; stigma 3-fid. Nuts orbicular-obovate, trigonous with convex sides, 1-1.2 x 0.6-0.8 mm, brownish-yellow at maturity, each side with 20 transverse rows of oblong cells.

Rare in muddy areas in Kadapa district. Fl. & Fr.: August – December.

Maddenayunipalli (KDP), *BR* & *DV* 39372.

INDIA: C, E., N.E. & N.W.India.

WORLD: Sri Lanka, Bangladesh, Myanmar, Nepal, China, Philippines, Papua New Guinea.

Fimbristylis schoenoides (Retz.) Vahl, Enum. Pl. 2: 286. 1806; FBI 6: 634. 1893; Fischer 3: 1658. 1931; Matthew, Fl. Tamilnadu Carnatic 3: 1759. 1983. *Scirpus schoenoides* Retz., Observ. Bot. 5: 14. 1788.

Annual, caespitose, glaucous herb, to 70 cm high; culms filiform, obtusely angled, striate. Leaves few, blades filiform, shorter than the culm, to 16 cm long, ligule shortly hairy; sheaths pinkish-brown. Involucral bracts 3, stiff, filiform, not overtopping, to 1.5 cm long. Spikelets 3, ovoid-subglobose, 1 x 0.4 cm long, rhachilla strongly winged; glumes broadly ovate or orbicular, mucronate, straw-coloured with reddish-tinge; stamens 2, filaments flattened; ovary obovate, style flattened, ciliate in the upper half, base dilated, stigmas much shorter than style. Achenes compressed, obovoid, biconvex to 2 mm long, stipitate, striate.

Usually seen in sandy areas along river banks and along the margins of rice fields. Fl. & Fr.: September – December.

Manchithirthalakona (CTR), *BR* & *AMR* 26064; Rollapenta (PKM) *BR* & *BSS* 30169; Rampa hills (EG), *KH* & *BR* 9729; Jonnavalasa (VZN), *MCK* 18840; Guvvalacheruvu (KDP), *KS* 6389 (MH & CAL); Near Tiger camp, Maredumilli (EG), *MM* 102584 (BSID); Araku Valley (VSKP), *NPBK* 10766 (MH & CAL); Bhadrchalam RF (KMM), *RCS* 104316 (BSID); Morappur (SLM), *S.India Flora* 11637 (MH); Hogainakkal (KGR), KCJ 18013 (MH); Ponnankanimedu (SA) *VNS* 5146 (MH); Mukundapur (GJM), *VNS* 4674 (MH); Buguda, Satkosia Tiger Reserve, *KCM* 8329 (KCM); Tulka, Satkosia Tiger Reserve, *KCM* 6751 (BSID).

INDIA: Throughout India.

WORLD: Sri Lanka, Nepal, Indo-China, S. China, Malesia, N.Australia.

Fimbristylis squarrosa Vahl, Enum. Pl.Obs. 2: 289. 1805; FBI 6: 635.1893; Matthew, Fl. Tamiladu Carnatic 2: 1760. 1983.

Stem tufted, slender, to 30 cm high, to 1 mm wide. Leaves flat, to 15 x 0.1 cm, scaberulous, apex obtuse, ligule 0, sheaths membranous, brownish-hairy. Inflorescence decompound or subcompound, 2-4 cm, involucral bracts 3-5, longest to 7 cm, primary rays *c.* 5; seondary rays *c.* 3. Spikelets solitary, oblong, subterete, to 7 mm, obtuse; rachilla persistent, winged. Glumes ovate, 1.5-2 mm, thin, purplish, mucronulate, keel 3-nerved; stamen 1, style 2-fid, ciliate, style-base pyramidal. Nut obovoid, biconvex, turgid, to 0.5 mm, stramineous, 7-ridged, striate.

Hills, *c.* 900 m in Tamilnadu Carnatic (Matthew, 1983). Fl. & Fr.: February – May.

INDIA: Andhra Pradesh, Tamil Nadu, Assam, Gujarat, Meghayala, Manipur, Punjab, Uttar Pradesh.

WORLD: South East Asia, Australia.

Fimbristylis tenera Schult., Mant. 2: 57. 1825; FBI 6: 642. 1893; Fischer 3: 1660. 193. *F. monticola* Hochst. ex Steud., Syn. Pl. Cyp. 3: 111. 1855.

Annual, slender, caespitose herb, to 30 cm high; culms filiform, obtusely trigonous, striate. Leaves erect, many at the base of the culm, shorter than the culm, filiform, glaucous-green, 15 cm long. Umbel subcompound, rays 4 with 2 spikelets

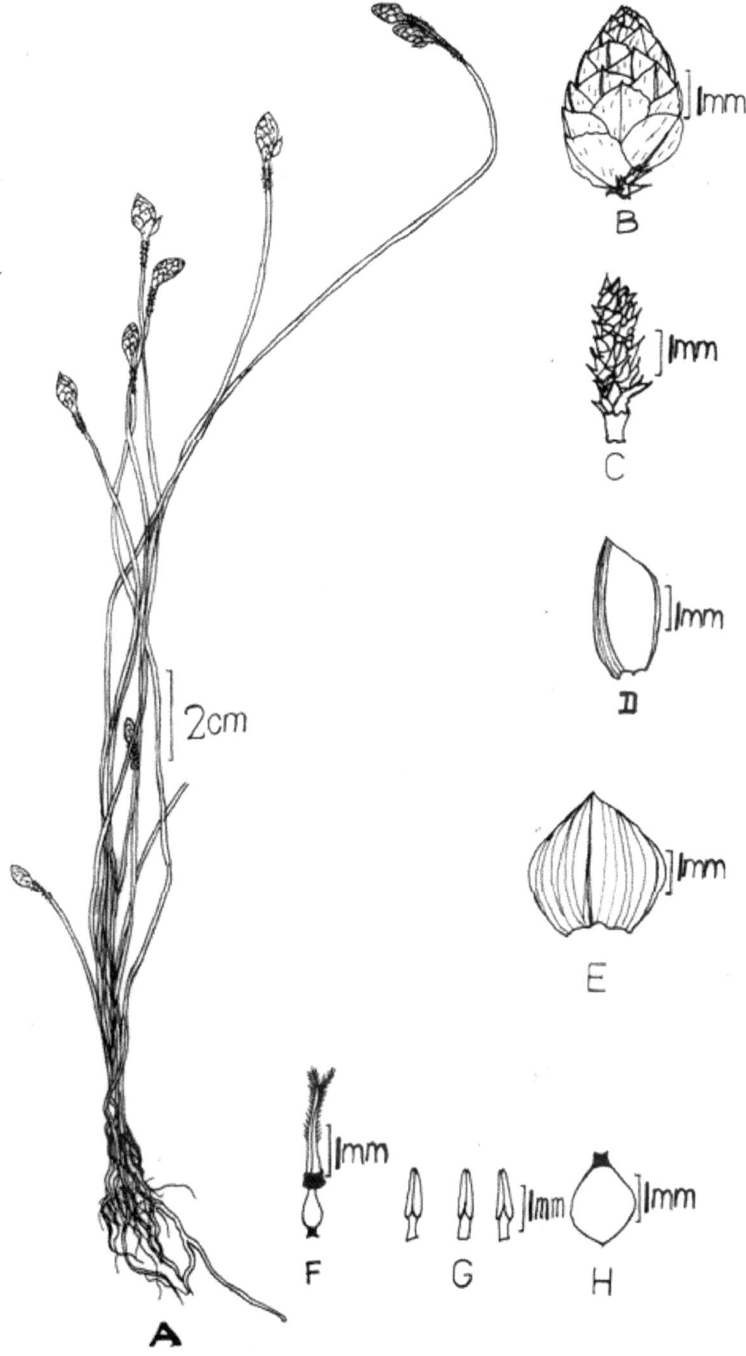

Figure 56. **Fimbristylis schoenoides** (Retz.) Vahl
A. Habit, B. Spikelet, C. Rhachilla, D. Closed glume, E. Opened glume, F. Pistil, G. Stamens, H. Achene.

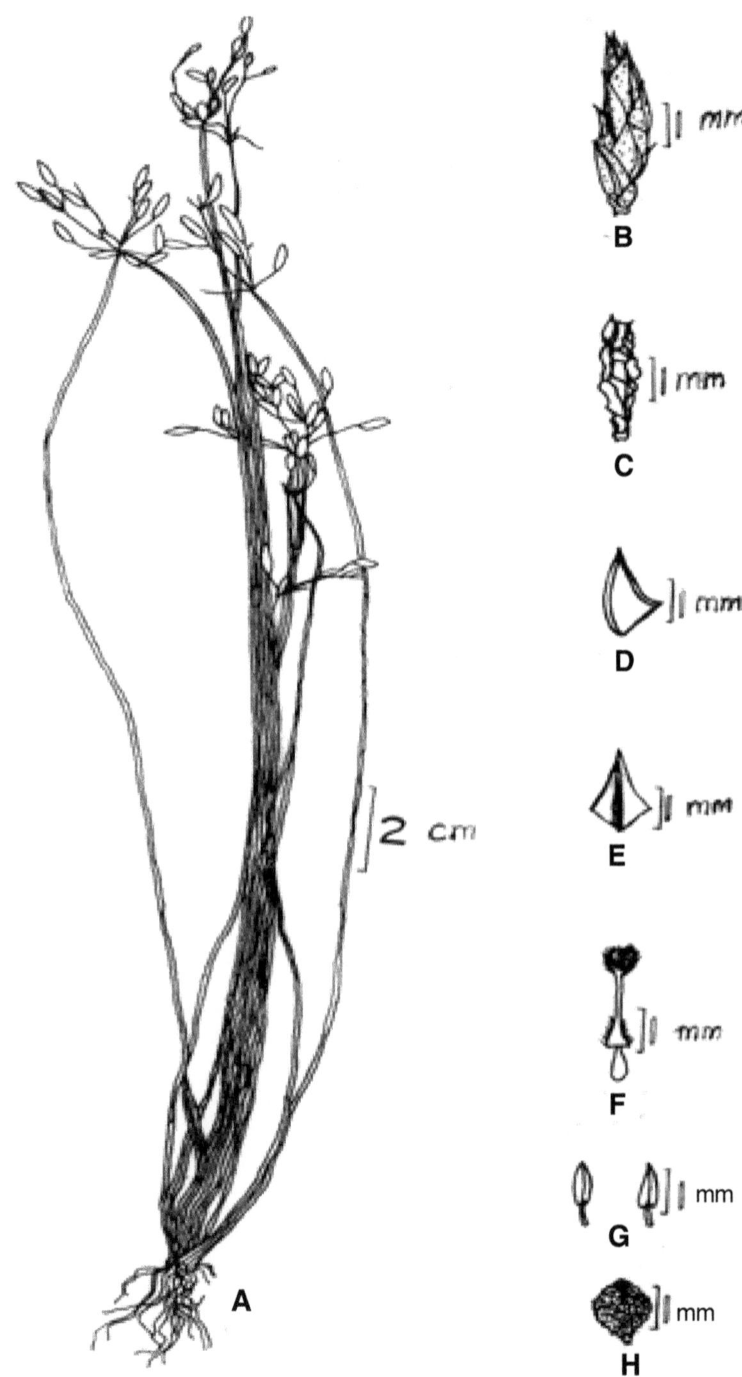

Figure 57. **Fimbristylis tenera** Schult.
A. Habit, B. Spikelet, C. Rhachilla, D. Closed glume, E. Opened glume, F. Pistil, G. Stamens, H. Achene.

on each. Involucral bracts very short, setiform. Spikelets to 1 cm long, oblong, acute, compressed, dark brown; rhachilla with deep pits. Achenes stipitate, globosely obovoid, obtusely trigonous, yellow or brown.

Usually seen in moist low lands, sandy soils, grass fields. Fl. & Fr.: August – November.

Kalasamudram RF (ATP), *TP* & *EC* 2856; Kuppagal RF (KNL), *KH* 10995; Guvvalacheruvu (KDP), *RVR* & *R.V.Reddy* 7906; *KS* 6388 (MH & CAL); Papavinasanam (CTR), *MHR* 13961; Venkatagiri hill (NLR), *BS* 3494 (VV); Redhakhol (SBP), *HFM* 2567 (DD); Sambalpur (SBP), *HFM* 2567 (DD).

INDIA: Peninsular, C. & N. India.

WORLD: Tropical Africa, Pakistan, Malaya.

Fimbristylis tetragona R. Br., Prodr. Fl. Nov. Holl. 226. 1810; FBI 6: 631. 1893; Fischer 3: 1658. 1931; Matthew, Fl. Tamilnadu Carnatic 3: 1761. 1983.

Perennial, caespitose herb, to 35 high; culms slender, ribbed, quadrangular. Leaves reduced to membranous sheaths, grey. Spikelet solitary, hemispherical, or sub-globose, lowest glume hardly brachiform, coriaceous, deciduous; rhachilla thickly studded with deep tetragonal pits with raised edges. Achenes oblong, cylindrical, biconvex, tapering towards the base, stipitate, fulvous grey, to 2 mm long.

Occasional in wet or moist open places, borders or rice fields, banks of ponds, edges of marshes and also in forest areas during monsoon. Fl. & Fr.: July – December.

Alurkona (ATP), *TP* 924; Chintalapalli Road (ATP), *KH* 7473; Rudrakod (KNL), *BR* & *BSS* 30172 (SKU & BSID); Kambakkam hills (CPT), *sine coll.* 8955 (MH); Purunakote RF, Satkosia Tiger Reserve, *KCM* 5186 (BSID).

INDIA: Throughout India except N.W.India.

WORLD: From Sri Lanka and India through Thailand and Indo-China, to S. China and Tropical Australia, Malesia.

Fimbristylis triflora (L.) K. Schum. ex Engl., Abh. Preuss. Akad. Wiss. 14. 1894. *F. tristachya* (Vahl) Thw., Enum. Pl. Zeyl. 434. 1854; FBI 6: 649; 1893; Fischer 3: 1660. 1931. *Abildgaardia triflora* (L.) Abeywick., Ceylon J. Sci., Biol. Sci. 2: 135. 1959. *Cyperus triflora* L., Mant. 2: 180. 1771.

Perennial, densely tufted herb often forming a large clump; rhizome knotty, woody, oblique, clothed with brown remnants of leaf sheaths; roots stout. Culms to 80 cm high, bulbous thickened at base, stiff, subterete, smooth, glaucous. Leaves many; some recurved; blades incurved, coriaceous, thickish, glaucous green, scabrous on margins; sheaths 7 cm long, cylindrical, dorsally pale, ventrally hyaline, lightly brown, transversely truncate at orifice. Corymbs simple, bearing 3 large spikelets. Involucral bracts 2, much shorter than the corymb; rays 2, to 4 cm long, each bearing 1 spikelet. Spikelets lanceolate, weakly laterally flattened, 2.5 cm long,

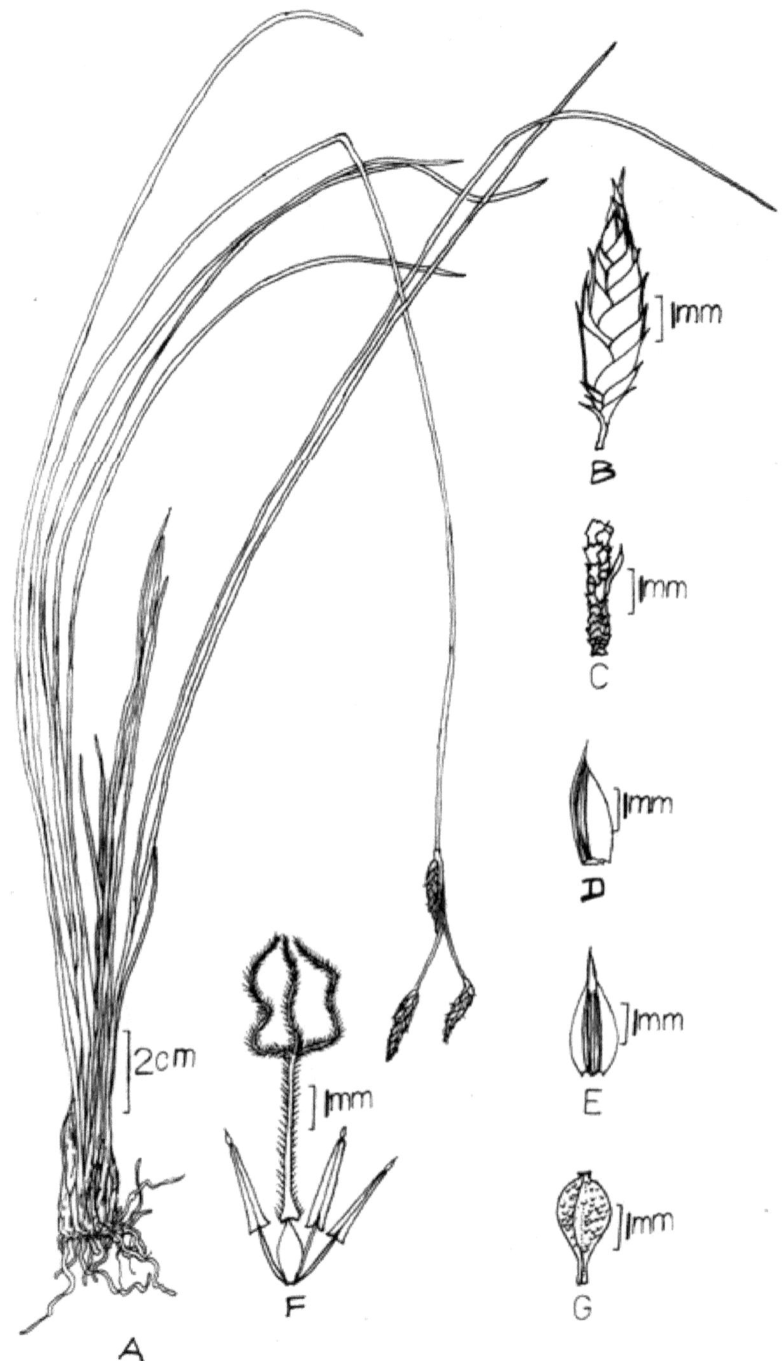

Figure 58. **Fimbristylis triflora** (L.) K. Schum. ex Engl.
[*Abildgaardia triflora* (L.) Abeywick.]
A. Habit, B. Spikelet, C. Rhachilla, D. Closed glume, E. Opened glume,
F. Pistil with stamens, G. Achene.

acute at both ends, gradually tapering to apex, straw brown. Achenes obovate, trigonous with convexed sides, 3 mm long, gradually attenuate below the middle to long strip like base, maturing brownish grey, shiny.

Rare in saline swamp area. Fl. & Fr.: July – December.

Guttur kona (ATP), *NY* 984; Old Chalama (KNL), *BR* & *BSS* 29359 (SKU & BSID).

INDIA: South India.

WORLD: Sri Lanka, S. Africa, Bangladesh.

Fimbristylis uliginosa Hochst. ex Steud., Syn. Pl. Glumac. 2: 109 1855; FBI 6: 648. 1893; Fischer 3: 1660. 1931; Matthew, Fl. Tamilnadu Carnatic 2: 1762. 1983.

Perennial herbs, up to 18 cm high; culms solitary, thickened at base. Leaves many, clustered, basal, to 15 x 0.2 cm, narrowed from a broad base, scabrid. Peduncle to 20 cm high, solitary; umbel simple with 3-7 spikelets. Spikelets 6-8 x 3 mm, ovoid, sessile; glumes spiral, 5 x 3 mm, ovate, mucronate. Nut obovoid, 1.5 x 1 mm, trigonous, tuberculate, dark brown.

In hills above 1200 m in Horsley hills and Tamilnadu Carnatic (Matthew, 1983). Fl. & Fr.: October – December.

Horsely hills (CTR), *Mahendra Nath* HH-02981 (CAL).

INDIA: Peninsula. Endemic

Fimbristylis woodrowii C.B.Clarke, Bull. Misc. Inform. 1898: 227.1898; Matthew, Fl. Tamilnadu Carnatic 2: 1762. 1983.

Stem sparsely tufted, 7-20 cm high, slender, rigid. Leaves flat, to 12 x 0,07 cm, flaccid, prominently nerved, scabrid, abruptly acute, sheaths membranous. Inflorescence a cluster of 3 spikelets, free, stalked; involucral bracts 2, not overtopping, to 6 mm. Spikelets oblong, to 5 mm, plicate, thin, fulvous, acute, mucro distinct; keel strongly 3-5-nerved; stamens 2; style 3-fid, apically ciliate, base pyramidal. Nut obovoid, trigonous, to 1 mm, glossy, white, vertically ridged, verruculose.

Hills above 1200 m in Tamil Nadu Carnatic (Matthew, 1983). Fl. & Fr.: September – November.

INDIA: Peninsula.

WORLD: Pakistan.

FUIRENA Rottboll

1. Perennials with decumbent rhizome; stem trigonous; leaves glabrous; hypogynous scales and bristles absent:
    2. Stem trigonous .... **F. cuspidata**
    2. Stem pentagonous .... **F. umbellata**
1. Annuals with fibrous roots; stem sulcate; leaves pilose; hypogynous scales and bristles present:

3. Scales quadrate or suborbicular:

  4. Inflorescence subcapitate; awn slender, pilose; bristles 4-6; claw of the petals (scales) as long as the blade ....**F. cilliaris**

  4. Inflorescence capitate; awn strong, thorn like, scabrid; bristles 3; claw of the petals shorter than the blade ....**F. uncinata**

3. Scales lunate, trilobate...**F. trilobites**

Fuirena ciliaris (L.) Roxb., Fl. Ind. 1: 184. 1820; Matthew, Fl. Tamilnadu Carnatic 3: 1763. 1983. *Scirpus ciliaris* L., Mant. Pl. 182. 1771. *Fuirena glomerata* Lam., Tabl. Encycl. 1: 150. 1791; FBI 6: 666. 1893; Fischer 3: 1669. 1931. *F. rotboelii* Nees in Wight, Contirb. Bot. India 94. 1834.

Annual, slender erect, leafy herb, to 30 cms high; culms sub-cylindric below and trigonous above, 2 mm wide, sparsely hairy above, glabrous below. Leaves linear-lanceolate, acuminate, to 9 x 0.4 cm; sheaths long, closed striate, slightly pubescent. Inflorescence umbellate, 3-10 in a cluster in terminal and axillary spikelets in subcapitate heads, 1.8 x 1.8 cm. Involucral bracts 0. Spikelets oblong, obtuse, brown, squarrose, lowest glumes longest, empty ciliate, 0.8 x 0.3 cm, many-flowered.

Common species usually seen in wet open groups, river banks, margins of ponds and rice fields. Fl.& Fr.: September – March.

Garugudukona (ATP), *TP* 954; Chintalapally Road (ATP), *KH* 7472; Gajulapalli (KNL), *BSS* & *SKB* 32251 (SKU & BSID); Guvvalacheruvu RF (KDP), *R.V.Reddy* 8116; Uppalamadugu (CTR), *RVR* & *CPR* 17721; Vinukonda (GNT), *VRK* 6867; Near Rayachoty (KDP), *KS* 7813 (MH & CAL); Polavaram agency Karagapadu (EG), *DCSR* 271 (CAL); Khammam to Dhaniyaigudem (KMM), *R.Rajan* 108512 (BSID); Hogainakkal (KGR), *KCJ* 18007 (MH); Madanancheri (NA), *MBV* 751 (MH); Tulka RF, Satkosia Tiger Reserve, *KCM* 5979 (BSID).

INDIA: Throughout India.

WORLD: Tropical and subtropical parts of Old World from Egypt south to South Africa, East to China, Japan and Korean Peninsula and south to Australia.

Fuirena cuspidata (Roth) Kunth, Enum. Pl. 2: 187. 1837. *F. wallichiana* Kunth, Enum, Pl. 2: 182. 1837; FBI 6: 665. 1893; Fischer 3: 1669. 1931; Matthew, Fl. Tamilnadu Carnatic 2: 1764. 1983.

Perennial, robust, rhizomatous herb, to 60 cm high; culms many, 3 mm wide, rigid, trigonous, hollow, slightly pubescent towards apex. Leaves linear-lanceolate, acute with strong midnerve, 10 cm long, glaucous; sheaths 3 cm long, trigonous, striate, with a short ligule. Spikelets in corymbose terminal clusters, oblong, terete, many-flowered, 1.1 x 0.4 cm. Achenes turbinate, with conical pubescent apex, trigonous, pale yellow, 1 mm long.

Found along the banks of streams and tanks, common in most pasture lands. Fl. & Fr.: Most of the year.

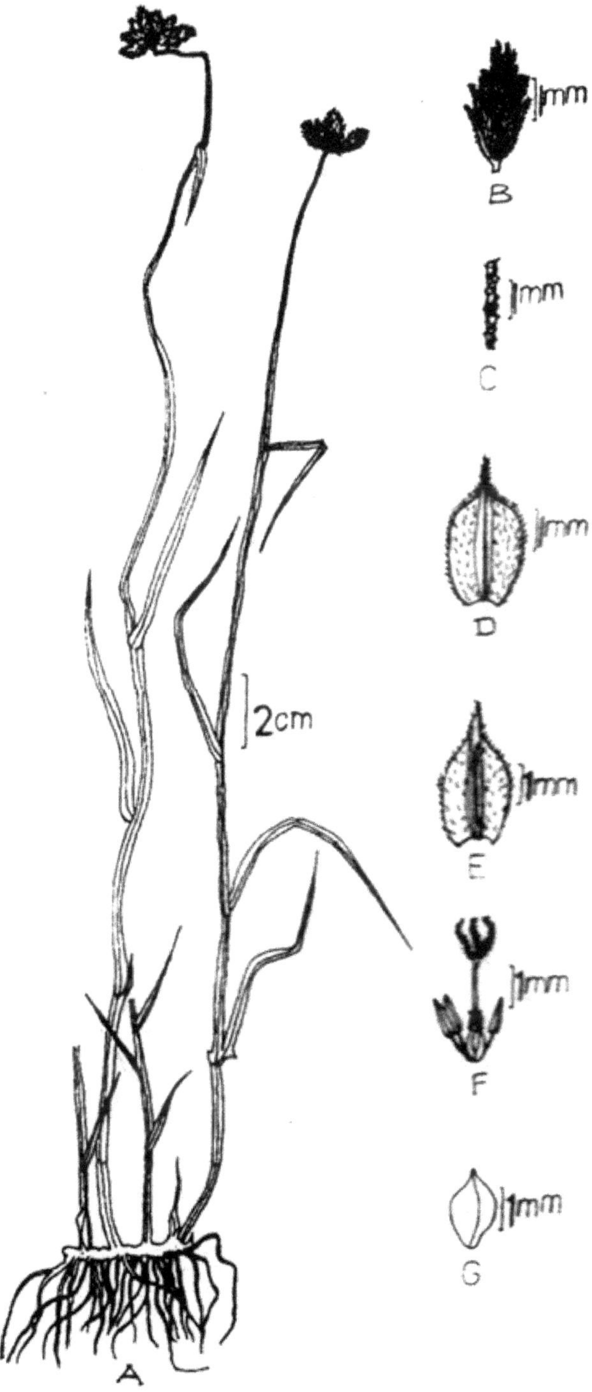

Figure 59. **Fuirena cuspidata** (Roth) Kunth (Syn.: *Fuirena wallichiana* Kunth)
A. Habit, B. Spikelet, C. Rhachilla, D & E. Glumes, F. Pistil with stamens, G. Achene.

Guttur Kona (ATP), *NY* 985; Amagondapalem (ATP), *BR* & *ANS* 36670; Kalasamudram (ATP), *KRKS* 39535; Peddathumabalam (KNL), *KH* 10957.

INDIA: Rajasthan, Uttar Pradesh, Maharashtra, Andhra Pradesh, Karnataka, Tamil Nadu. Endemic.

Fuirena trilobites C.B.Clarke in Hook. f. Fl. Brit. India 6: 666. 1893 & 111. Cyper. t. 5. 1909; Matthew, Fl. Tamilnadu Carnatic 3: 1764. 1983.

Annual, erect, caespitose herb, to 40 cm high; culms slender, striate, glaucous, leafy, sparsely hairy. Leaves to 11 cm long, linear-lanceolate, acuminate, 7-nerved, profusely hairy; sheaths 3 cm long, close, pubescent. Inflorescence cluster of 3-9 axillary and terminal spikelets in subcapitate heads; bracts absent but inflorescence invariably arising from leaf axil. Spikelets to 7 mm long, ovoid or oblong, obtuse, pale brown, squarrose, lowest glume empty, possess longest awn, densely ciliate. Achenes stipitate, triquetrous, acute, base trigonous, minutely tuberculate showing very faint horizontal striations.

Rare, found growing in rice fields and water logged grass lands. Fl. & Fr.: November – March.

Tada (NLR), *MCK* 23523; Darsi (PKM), *MCK* 22829; Buchireddypalem (NLR), *MCK* 22967; Melagiris (DMP), *KMM* 24023 (RHT, K).

INDIA: South India. Endemic.

Fuirena umbellata Rottb., Descr. Icon. Rar. 70. t. 19. f.3. 1773; FBI 6: 666. 1893; Fischer 3: 1669. 1931.

Perennial herb with long, creeping rhizome; stems tufted, erect, 30-50 cm high, acutely 4-5-angular, glabrous. Leaves linear, 7-8 cm long, acute or acuminate, glabrous or hirsute on the margins in the basal part, with 5 prominent nerves; cauline leaves distant. Inflorescence with a terminal, partial inflorescence and several axillary ones. Lower bracts similar to the leaves; spikelets in very dense clusters, ovoid to oblong-ovoid, 4-8 mm long, acute; glumes ovate or obovate, 2-2.5 mm long, membranous; stamens 3. Nut ellipsoid or obovate, 1-1.25 mm long, triquetrous, smooth, shining, stramineous to fuscous.

Rare in marshy localities. Fl. & Fr.: September - January.

Bodalanka (EG), *DN* 85581 (MH, BSID); Dasingabadi (GJM), *CAB* 1421 (MH).

INDIA: Throughout India except NW. India.

WORLD: South America, Tropical Africa, Sri Lanka, Laos, Thailand, Vietnam, Papua New Guinea, Australia.

Fuirena uncinata (Willd.) Kunth, Enum. Pl. 184. 1837; FBI 6: 666. 1893; Fischer 3: 1669. 1931; Matthew, Fl. Tamilnadu Carnatic 3: 1764. 1983. *Scirpus uncinatus* Willd., Sp. Pl. 1: 300. 1798. *Fuirena capitata* (Burm. f.) T. Koyama in Dssanayake & Fosberg, Rev. Handb. Fl. Ceylon 5: 151. 1985. *Scirpus capitatus* Burm. f., Fl. Ind. 21. 1768 illegitimate, non. L. 1753.

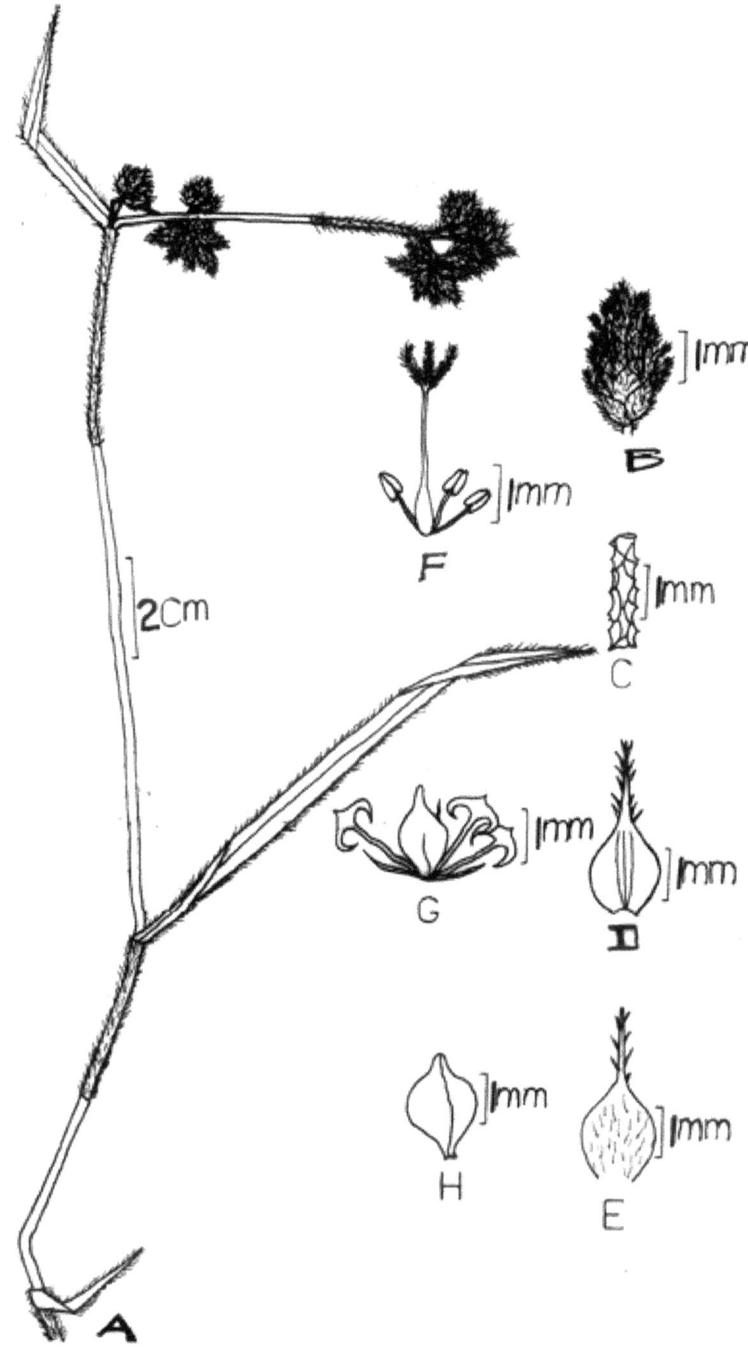

Figure 60. **Fuirena trilobites** Clarke
A. Habit, B. Spikelet, C. Rhachilla, D & E. Glumes, F. Pistil with stamens, G. Achene with hypogynous scales and bristles, H. Achene.

Annual, tufted, pubescent herb with reddish brown fibrous roots. Culms subdensely tufted, to 20 cm high, obtusely subpentagonal, ribbed, 4 nodes including inflorescence, the internodes pilose, to 5 cm long. Leaves basal and cauline; blades linear-lanceolate, to 8 cm long, several-costate, flattish, herbaceous, subdensely pilose on both sides with white, spreading, tubercle based hairs; sheaths to 5 cm long, laxly clothing the culm internode, pilose. Basal leaves reduced to sheaths, those rusty brown. Inflorescence capitate, bearing 5 glomerules of spikelets. Spikelets sessile, squarrose, ellipsoidal, polygonal, bearing up to 20 glumes. Achenes rhombic-oval, triquetrous with concave sides, stipitate at base, maturing light brown.

Usually seen in wet or moist sandy fields, along marshy places in sticky muddy soils of rice fields. Fl. & Fr.: December – March.

Hottabetta (ATP), *AL & KVS* 28691; Batrepallai (ATP), *KRKS* 39002; Sangam (NLR), *MCK* 22941; Balapalle (KDP), *JLE* 15783 (MH & CAL) & *CECF* 4278 (CAL); Kambakkam (NA), *sine coll. s.n.* (MH); Hogainakkal RF (KGR), *KCJ* 18009 (MH); Karai RF (SA), *KRM* 52861 (MH); Sunkaravaripalli (NLR), *A.S.Rao* 2228 (BSID); Salapa (Nabrangpur district), *MB* 2585 (RRL-B).

INDIA: Peninsular and W. India.

WORLD: Sri Lanka.

KYLLINGA Rottboel

1. Rhizome short or absent, stems tufted:
   2. Stem bulbous based; leaf sheaths disintegrate leaving a tuft of fibres; central spike globose. .... **K. bulbosa**
   2. Stem hardly thickned at base; leaf sheaths remain intact; spikes cylindrical .... **K. odorata**
1. Rhizome long, stems solitary in linear rows along the length of the rhizome:
   3. Glumes not winged:
      4. Involucral bract short, spreading, < 6 cm long; rhizome slender; culms <30 cm long, distant .... **K. brevifolia**
      4. Involucral bract elongate, deflexed, >15 cm long; rhizome thick; culms 30-50 cm, juxtaposed .... **K. melanosperma**
   3. Glumes winged .... **K. nemoralis**

Kyllinga brevifolia Rottb., Descr. Ic. Rar. Nov. Pl. 13, t. 4, f. 3. 1773; FBI 6: 588. 1893; Fischer 3: 1624, 1931; Matthew, Fl. Tamilnadu Carnatic 3: 1765. 1983. *Cyperus brevifolius* (Rottb.) Hassk., Cat. Hort. Bot. Bogor. 24. 1844; Saxena & Brahmam, Fl. Orissa 4: 2115. 1996.

Perennial, stoloniferous, rhizomatous herb; rhizome horizontal, creeping, clothed with brownish scales; culms many, spaced, arranged in a single row along the rhizome, to 25 cm high, slender, triquetrous, few-leaved at base. Leaf blades shorter than the culm, narrowly linear, flattish, plicate, herbaceous, scabrid margin and on the abaxial midvein; sheaths brownish. Inflorescence a terminal, single globose head. Involucral bracts 3, foliaceous, very unequal in length, lowest to 8 cm long; head globose, pale green, densely bearing numerous spikelets. Spikelets lance-oblong, laterally compressed; glumes 4, ovate-elliptic, folded with an acutee keel, 3-5-nerved, mucronate; stamens 1 or 2; style 2-fid. Achenes obovate, laterally lenticular, brownish.

A common ubiquitous taxon found growing almost every-where. Fl.& Fr.: July – December.

Kekati (ATP), *BR* & *ANS* 38330; Kuppagal RF (KNL), *KH* 10970; Veyithirthalakona (KDP), *MHR* 13305; Tirumala (CTR), *CPR* & *MHR* 14821; Vidavalur (NLR), *PMR* 16492; Simhachalam hill (VSKP), *KH* & *BR* 5588; Lakkavarapu reservoir (VZN), *MHR* & *MCK* 14867; Sunnipentaa vagu (KNL), *JLE* 22131 (MH); Chandragiri (CTR), *S. India Flora* 10048 (MH); Pedamantanala RF (PKM), *RKM* 0595 (CAL); P.Kona (NLR), *BSN* 5063 (BSID); Rampa hill (EG), *VNS* 424 (CAL); Devarapalli to Maredumilli (EG), *MM* 101311 (BSID); Muzumamidivalasa, Maredumilli (EG), *MM* 102514 (BSID); Krishnapuram (VSKP), *CAB* 1942 (MH); From Inspection Bungalow towards Galikonda (VSKP), *GVS* 19605 (MH); Pullajalamala (MBNR), *VBH* 84997 (BSID); Bhuvanagiri to Kurinjipadi (SA), *KRM* 60114 (MH); Baghmunda, Satkosia Tiger Reserve, *KCM* 7027 (BSID); Ganjam, *VNS* 5925 (MH).

INDIA: Throughout India.

WORLD: Tropical and subtropical regions of both hemispheres.

Kyllinga bulbosa Beauv., Fl. d' Oware & Benin 1: 11. t. 8. f. 1. 1804. *K. triceps* Rottb., Descr. Icon. Rar. 1 & t. 4. f. 6. 1773. *nom. illegit.*, C.B.Clarke in Hook. f., Fl. Brit. India 6: 587. 1893; Fischer, Fl. Madras 3: 1623. 1931.

Perennial, rhizomatous herb; culms to 15 cm high, densely tufted with short erect rhizome, obtusely trigonous, the base forming an ovoidal, corm-like enlargement clothed with brown fibres. Leaves few; blades flattish-plicate, sheaths stout, pale and basal ones tinged with light brown, sheaths disintegrate leaving a tuft of fibres. Inflorescence a head with 3 sessile spikes, pale greenish, central spike ovoid globose, to 8 mm long, obtuse, the lateral ones globose. Involucral bracts 3, eventually reflexed, the longest to 10 cm long. Spikelets oblong, pale greenish, 1-flowered; glume 4, up to 2 mm, ovate-oblong, 1-3-nerved, mucronate, strongly keeled; stamens 2; ovary oblong-obovate, style short, stigmas 2. Achenes oblong, laterally flattered, brownish.

Common species usually found in grass lands, lawns, and almost everwhere in the coastal plains. Fl. & Fr.: August – December.

Pennahobilam (ATP), *TP* & *NY* 332 (SKU & MH); Kalasamudram (ATP), *KRKS* 39698; Kuppagal RF (KNL), *KH* 10966; Guvvalacheruvu (KDP), *R.V.Reddy* 7970, KS 6381 (MH); Nagarjunakonda (GNT), *VRK* 6714; Tyada Forest (VSKP), *KH* & *BR* 9722; Balapalle (KDP), *JLE* 15785 (CAL); Mamandur valley (CTR), *KS* 6895 (MH & CAL); Satyavedu to Ambakkam (CTR), *MC* 45214 (MH); Narasimhulukonda (NLR), *BSN* 5224; Nuziveedu (KSN), *PV* 5217 (MH).

Note: *Kyllinga triceps* Rottb. is a superfluous illegitimate name.

Kyllinga melanosperma Nees in Wight, Contr. Bot. India 91. 1834; FBI 6: 588. 1893; Fischer 3: 1624. 1931; Matthew, Fl. Tamilnadu Carnatic 3: 1766. 1983. subsp. **melanosperma**. *Cyperus melanospermus* (Nees) Suringar, Het. Gest. Cyperus Mal. Arching. 50, t. 2, f. 8. 1898; Saxena & Brahmam, Fl. Orissa 4: 2137. 1996.

Perennial, rhizomatous herb; rhizome decumbent, to 8 cm long, knotty, clothed with lance-ovate, dark brown scales, very aromatic; culms to 50 cm high, disposed in a row along the rhizome, close together, triquetrous with almost winglike angles. Leaves all reduced to bladeless sheaths; sheaths 4, membranous, to 20 cm long, greenish, tinged with purplish brown, the orifice with hyaline margin. Inflorescence a single globose head, to 12 mm long, greenish, densely many spiculose. Involucral bracts 3, to 15 cm long, spikelets elliptic-oblong. Achenes elliptic-oblong, apiculate, laterally flattened.

Occasional on forest margins, in swampy areas. Fl. & Fr.: January – July.

Near Ondutla (KNL), *SS* & *AMS* 21293; Near Valamuru (EG), *KSK* 23448; Galikonda (VSKP), *MHR* & *MCK* 14854; Damakku (VSKP), *KSK* 23435; Jonnavalasa tank (VZN), *MCK* 18833; Nulakamaddi (EG), *GVS* 24535 (MH); Araku valley (VSKP), *NPBK* 10878 (MH & CAL); Sapparla (VSKP), *GVS* 29709 (MH); Kaveri peak, Yercaud (SLM), *DBD* 31303 (MH); Salur, Kollimalai (NMK), *S.India Flora* 12985 (MH); Similipahar (MBJ), *SX* & *MB* 6901(RRL-B); Karlapat, Kalahandi (KHD), *MB* & *N.K.Dhal* 7572 (RRL-B); Kalahandi (KHD), *HFM* 1235 (DD).

INDIA: Peninsular, C. & E. India.

WORLD: Tropical and subtropical Africa, S. Asia, Fiji, Malesia.

Kyllinga nemoralis (J.R. Forst. & G. Forst.) Dandy ex Hutch. & Dalziel, Fl. W. Trop. Africa 2; 486, 487. 1936; Matthew, Fl. Tamilnadu Carnatic 3: 1766. 1983. *Thryocephalon nemoralis* J.R.Forst. & G.Forst., Char. Gen. Pl. 130. 1776. *Kyllinga monocephala* Rottb., Descr. Ic. Rar. Nov. Pl. 13, t. 4, f. 4, 1773; FBI 6: 588. 1893; Fischer 3: 1624. 1931. *Cyperus kyllinga* Endl., Cat. Hort. Acad. Vindob. 1: 94. 1842; Saxena & Brahmam, Fl. Orissa 4: 2134. 1996.

Perennial, laxly tufted, rhizomatous herb; rhizome long-creeping, culms to 20 cm high, along the rhizome, triquetrous, not thickened at base. Leaves many, longer than the stem; sheaths brown. Inflorescence a single head, densely bearing many spikelets, whitish. Involucral bracts 4, spreading, the lowest to 15 cm long. Spikelets ovate, elliptic, 2-flowered. Achenes oblong-obovate, biconvexed, maturing brownish.

Common on road sides, waste places, open grass lands, near canals, as herbaceous undergrowth on moist soil. Fl. & Fr.: July – November.

Amarapuram (ATP), *BR* & *ANS* 35560; Guvvalacheruvu (KDP), *RVR* 8489, KS 6397 (MH); Penchalakona (NLR), *BR* & *BSS* 33106; Darsi (PKM), *MCK* 22837; Devarapalli (EG), *KSK* 23461; Tyada Forest (VSKP), *KH* & *BR* 9717; Near Nallamalais (KNL), *JLE* 22131 (MH); Papanasanam falls (CTR), *KS* 6956 (MH); Satyavedu to Ambakam (CTR), *MC* 45208 (MH); Durgam –Udayagiri (NLR), *P.Venu* 110223

(BSID); Forest Agiripalle (KSN), *PV* 5233 (MH); near Anigeri, Addatigala (EG), *GVS* 68554 (MH); Muzumamidivalasa, Maredumilli (EG), *MM* 102571 (BSID); Bison hill (EG), *CAB* 5178 (MH); Simhachalam (VSKP), *GVS* 1935 (MH); Araku (VSKP), *GVS* 21551 (MH); Kambukur (NA), *KS* 6094 (MH); Alagar hills, Nupura ganga (MDR), *KS* 3412 (MH); Lady's seat, Karadu – Yercaud (SLM), *SKK* 28296 (MH).

INDIA: Throughout India.

WORLD: Cambodia, China, Indonesia, Madagascar, Malaysia, Nepal, Pakistan, Philippines, Singapore, Sri Lanka.

Kyllinga odorata Vahl, Enum. Pl. 2: 382. 1906. subsp. **cylindrica** (Nees ex Wight) T. Koyama, Gard. Bull. Singapore 30. 161. 1977 & in Dassan & Fosb., Rev. Handb. Fl. Ceylon 5: 244. 1985. *K. cylindrica* Nees ex Wight, Contr. Bot. Ind. 91. 1834; FBI 6: 588. 1893; Fischer 3: 1624. 1931. *Cyperus sesquiflorus* (Torr.) Mattf. & Kuek., Pflanzenr. IV. 20(101). 19. 1935; Saxena & Brhmam, Fl. Orissa 4: 2151. 1996.

Slender, erect, tufted herbs with creeping rhizomes covered with brown scales; stems up to 18 cm high, rigid, glabrous, ribbed. Leaves shorter or longer than the stems, up to 3 mm broad, rigid, linear-lanceolate, acuminate, scaberulous towards apex, sheathing at base. Involucral bracts up to 3, foliaceous, rigid, unequal, linear-lanceolate. Heads sessile, cylindric or obovate, solitary or sometimes with 2 small subglobose, lateral heads at base; spikelets closely packed, ovate; empty glumes 2 or 3 at the base of the spikelets, small, hyaline, unequal; lower two empty glumes up to 1.5 mm long, third one when present up to 0.2 cm long; remaining 2 glumes up to 0.3 cm long, broadly ovate, apiculate, keeled; stamens 2; ovary obovate, stigmas 2. Nuts obovate, biconvex.

Usually grows in moist places and open grass lands along hill slopes. Fl. & Fr.: May – October.

Mallappakonda (ATP), *BR* & *ANS* 35321; Gundlabrahmeswaram (KNL), *MVS* & *MB* 38698; Guvvalacheruvu (KDP), *R.V.Reddy* 7970; Thummalapalle (KDP), *KRKS* 36991; Devarapalli, Rampa hills (EG), *KSK* 23459; Anantagiri (VSKP), *KSK* 23420, *GVS* 19489 (MH), 19574 (MH; Maredumilli (EG), *GVS* 68545 (MH); Way to Deomali (KPT), *SX* & *MB* 6626 (RRL-B); Dudurchampa, Similipahar (MBJ), *SX* & *MB* 4969 (RRL-B); Pampasar RF, Satkosia Tiger Reserve, *KCM* 5107, 5125 (BSID).

INDIA: Throughout India.

WORLD: Asia, Africa, Australia, S & C. America.

LIPOCARPA R. Brown *nom. cons*

1. Spikelet prophyll and glume absent:
    2. Inflorescence terminal; stigmata 2; achenes elliptic to narrowly rhombic in cross section, stamen 1 .... **L. kernii**
    2. Inflorescence pseudolateral, stigmata 3, achenes subtriquetrous in cross section; stamens 1-2 .... **L. squarrosa**

1. Spikelet prophyll and glume present:
   3. Plants robust, tufted perennials; spikelets 2-10; inflorescence white to creamy white with red dots, achenes trigonous in cross section, oblong to narrowly obovate .... **L. chinensis**
   3. Plants slender, tufted annuals; spikelets 2-4; inflorescence yellow to pale reddish brown; achenes subtrigonous to trigonous frontally obovate to subelliptical .... **L. gracilis**

Lipocarpha chinensis (Osbeck) J. Kern, Blumea, Suppl. 4: 167. 1958 & in Steenis, Fl. Males. 1, 7: 521. 1974; Matthew, Fl. Tamilnadu Carnatic 3: 1769. 1983. *Scirpus chinensis* Osbeck, Dagb. Osting. Rosa 220. 1757. *Lipocarpha argentea* (Vahl) R. Br. ex Nees, Linnaea 9: 287. 1835; FBI 6: 667. 1893; Fischer 3: 1670. 1931.

Annual, caespitose herb with reddish brown fibrous roots, culms to 50 cm high, obtusely trigonous, whitish green. Leaves 4 to a culm; blades linear, rather soft and slightly spongy, whitish green; sheaths to 15 cm long, subloosely surrounding the culm, pale, the lower ones stained with reddish purplish colour. Inflorescence a head with 7 spikes. Involucral bracts 3, unequal, dilated at base, much surpassing the head, the longest up to 20 cm long. Spikes oblong-ellipsoidal, terete, to 8 mm long, rounded at apex, whitish, densely bearing numerous bracteoles; glumes obovate-oblong, reddish, concave, obtuse at apex; hypogynous scales 2; stamens generally 1, sometimes 2; ovary obovate-oblong, style slender, stigmas 2 or 3, slightly shorter than the style. Achenes oblong, 1.5 mm long, trigonous, straight to weakly curved, straw-brown.

Common in marshy areas, swamps and rice fields. Fl. & Fr.: May – October.

Kalasamudram (ATP), *KRKS* 40107; Galikonda (VSKP), *MHR* & *MCK* 14856; Akkagari gudi (CTR), *GVS* 45931 (MH); Vettukkaru (EG), *MM* 100786 (MH); Valamuru (EG), *SS* 365 (AU); Araku valley (VSKP), *NPBK* 522, 10725 (CAL), *DDSR* 21326 (MH); Galikonda (VSKP), *GVS* 19606 (MH); Manuguru RF (KMM), *RCS* 104256 (MH); Kaveri peak – Yercaud (SLM), *DBD* 31295 (MH); Similipahar (MBJ), *SX* 392 (RRL-B), *SX* & *MB* 4372, 4686, 4941 (RRL-B); Dudhari (KPT), *SX* & *MB* 6574 (RRL-B).

INDIA: Throughout India.

WORLD: Tropical and subtropical regions of Africa, Asia south to Malesia and N. Australia.

Lipocarpha gracilis (Rich. ex Pers.) Nees, Linnaea 9: 287. 1834. *Hypolytrum gracile* Rich. ex Pers., Syn. Pl. 1: 70. 1805. *Lipocarpha sphacelata* (Vahl) Kunth, Enum. Pl. 2: 283. 1806; FBI 6: 667. 1893; Matthew, Fl. Tamilnadu Carnatic 2: 1769. 1983; Saxena & Brahmam, Fl. Orissa 4: 2196. 1996. *Hypaelyptum sphacelatum* Vahl, Enum. Pl. 2: 283. 1806. *Lipocarpha triceps* Nees in Wight, Contrib. Bot. Ind. 92. 1834; Fischer 3: 1670. 1931.

Annual, erect, caespitose herb, to 20 cm high; culms triquetrous. Leaves basal, canaliculated, shorter than stem, to 14 cm long; ligule 0. Involucral bracts 3, foliar, over-topping, to 7 cm long. Inflorescence terminal, capitate. Spikelets 3-5 in a cluster, globose or ovoid, many-flowered; glumes purple-striped, 5-nerved; hypogynous scales up to 1.8 mm, slightly exceeding achene; stamens 2, stigmas 3. Achenes ellipsoid, beaked, fuscous, stoutly stipitate and trigonous.

Occasional in damp grassy places and open wet fields and margins of swamps. Fl. & Fr.: August – January.

Batrepalli (ATP), *KRKS* 37857; Peccheruvu (KNL), *BR* & *BSS* 30157; Kambakkam hills (CTR), *RVR* & *CPR* 17727; S.Kota reservoir (VZN), *MCK* 18800; Balapalle (KDP), *JLE* 15785 (MH); Talakona RF (CTR), *GVS* 46960 (MH); Krishnagiri (KGR), *S.India Flora* 13890 (MH); Pakamalai RF, Gingee (SA), *KRM* 53537 (MH); Near Vellar bank (SA), *KRM* 60399 (MH); Tulka RF, Satkosia Tiger Reserve, *KCM* 5651 9BSID).

INDIA: Throughout India.

WORLD: Africa, Sri Lanka, Bangaldesh, Myanmar, Nepal, Thailand, Philippines, Laos, Malaysia, Indoneisa.

Lipocarpha kernii (Raymond) Goetgh., Wageningen Agric. Univ. Papers 89(1): 42. 1989; Ragan *et al.*, Rheedea 8: 99. 1998. *Rikliella kernii* (Raymond) J. Raynal, Adasonia ser. 2.13: 155. 1973. *Scirpus kernii* Raymond, Natur Canad 86: 230. 1959.

Small, tufted, annual herbs, 2-40 cm high. Leaves 1-3 to a culm, filiform, 16 cm x 2 mm, sub-acute at apex. Spikes 2-8, ovoid; involucral bracts 2-5, the largest up to 15 cm long; spikelet bract elliptical to obovate, long acuminate, scaberulous at the top, body yellowish green to pale brown; spikelet prophyll and glume absent; stamen 1; style 0.1 mm or shorter, with 2 branches. Achenes obovate with very small style base remnant, dorsiventrally flattened and elliptical to narrowly rhombic on cross section

Rare in moist places at low elevations in Nellore district. Fl. & Fr.: August – December.

INDIA: Peninsular & C. India.

WORLD: Ethiopia, Tanzania, Nigeria, and S. Africa.

Lipocarpha squarrosa (L.) Goetgh., Wageningen Agric. Univ. Papers 89(1); 71. 1989. *Scirpus squarrosus* L., Mant. Pl. 2: 181. 1771; FBI 6: 663; 1893; Fischer 3: 1666. 1931; Saxena & Brahmam, Fl. Orissa 4: 2210. 1996. *Rikliella squarrosa* (L.) Raynal in Adansonia ser. 2. 13: 154. 1973; Matthew, Fl. Tamilnadu Carnatic 2: 1782. 1983

Annual, slender, glabrous, tufted herb, to 25 cm long; culms filiform, 2 or 3 to a stem; sheaths pinkish-grey, short, tightly surrounding stem base. Inflorescence pseudolateral head with 1-3 spikes, deeply red brown. Involucral bracts 2, lower bract leaflike, erect, much surpassing the inflorescence, to 2.5 cm long, upper bract setaceous, slightly longer than the spikes. Spikes sessile, ovate-elliptic, squarrose, 6 mm long, deeply red brown with pale green awns, densely many-flowered; glumes obovate, 3-5-nerved, aristate, hypogynous bristles absent; stamen 1. Achenes obovoid, trigonous, yellow, obtuse at apex.

Occasional in wet sandy soil at the margin of ponds, ditches and paddy fields. Fl. & Fr.: August – December.

Guvvalacheruvu (KDP), *R.V.Reddy* 7969, 7972; Mamandur (CTR), *BSS* & *SKB* 33167; Penchalakona (NLR), *BR, BSS* & *SKB* 33105; Tyada forest (VSKP), *KH* & *BR* 9716; top of Gyanaiah gundlu (CTR), *GVS* 46820 (MH); Kollurupad (NLR), *CECF* 4227 (CAL); Valamuru (EG), *SS* 333 (AU); Araku (VSKP), *GVS* 19731 (MH); Salur (VZN), *NPBK* 952 (CAL); Tippukadu RF (NA), *KRM* 17669 (MH); Karai RF – Gingee (SA), *KRM* 52859 52859 (MH); near Kurinjipadi (SA), *KRM* 60130 (MH); Krishnagiri (KGR), *S.India Flora* 13893 (MH); Hogainakkal (KGR), *KCJ* 18014 (MH); Purunkote RF, Satkosia Tiger Reserve, *KCM* 5259 (BSID).

INDIA: Throughout India.

WORLD: S. Asia.

PYCREUS P. Beauvois

1. Achenes transversely wrinkled:
   2. Rhizomatous herb; inflorescence umbel, sometimes reduced to the head .... **P. diaphanus**
   2. Non rhizomatous herb; inflorescence spicate .... **P. stramineus**
1. Achenes smooth to puncticulate:
   3. Culms few to several nodose below the middle, the lower part decumbent or obliquely ascending, branching and rooting at lower nodes; culms clothed to considerably above the base .... **P. sanguinolentus**
   3. Culms not nodose above the base, erect from very base; culms with leaves only at the base:
      4. Glumes distinctly cuspidate, apex retuse .... **P. pumilus**
      4. Glumes not cuspidate nor retuse:
         5. Stems tufted:
            6. Achenes symmetric:
               7. Annuals; rhachilla straight, not at all winged; involucral bracts 2-4 .... **P. flavidus**
               7. Perennials; rhachilla winged; more or less zigzag; involucral bracts 3-4........ **P. polystachyos**
            6. Achenes asymmetric, concave .... **P. sulcinux**
         5. Stems robust, solitary........ **P. puncticulatus**

Pycreus diaphanus (Schrad. ex Roem. & Schult.) S.S. Hooper & T. Koyama, J. Jap. Bot. 51: 316. 1976. *Cyperus diaphanus* Schrad. ex Roem. & Schult., Mant. 2: 477. 1824 var. *latespicatus* (Boeck.) Kern in Steenis, Fl. Males. I. 7: 653. 1974; Saxena & Brahmam, Fl. Orissa 4: 2122. 1996. *C. latespicatus* Boeck., Flora 42: 433. 1859. *Pycreus latespicatus* (Boeck) C.B.Clarke in Hook. f., Fl. Brit. India 6: 590. 1893.

Annual, rhizomatous herb, to 30 cm high, culms slender, stiff, obscurely angled, striate. Leaves shorter than the stem, rigid. Umbel simple, sometimes reduced to one head; involucral bracts 3, unequal, the longest sometimes reaching 10 cm long. Spikelets straw-coloured, shaded with chestnut brown, oblong-lanceolate, compressed, to 6-flowered. Achenes obovoid, black, shortly apiculate, slightly compressed.

Occasional in wet places, marshy grounds, along water streams and banks. Fl. & Fr.: August – January.

Araku valley (VSKP), *NPBK* 10763 (MH); Paniganda (GJM), *SX* & *MB* 1984 (RRL-B).

INDIA: S., E. & N.E. India.

WORLD: Central Asia.

Pycreus flavescens (L.) P.Beauv. ex Rchb., Fl. Germ. Excurs. 1: 72. 1830; FBI 6: 589. 1893; Matthew, Fl. Tamilnadu Carnatic 2: 1776. 1983; R.R.V. Raju *et al.*, J. Econ. Taxon. Bot. 23: 1181. 2003. *Cyperus flavescens* L., Sp. Pl. 46. 1753.

Stem tufted, 15-50 cm high, 1.5 mm wide, erect, triquetrous. Leaves basal, flat, flaccid; stem scabrid; sheaths reddish-brown. Inflorescence subcompound or compound, brown, 2.5-7 cm; involucral bracts 3, overtopping, longest to 25 cm; spikes to 2 cm, fulvous brown, *c.* 15-40-spikeletted; spikelets spicate, oblong, to 1.5 cm long; glumes broadly ovate, to 2 mm; stamens 2. Nut obovoid, trigonous, to 1 mm, fuscous, obtuse, apiculate.

Rare in marshy lands of plains (R.R.V.Raju *et al.*, 2003). Fl. & Fr.: August – December.

INDIA: Peninsular and N.W.India.

WORLD: Europe, Russia, Africa, Lebanon, Syria, Iraq, Turkey, Iran, S. America, Vietnam, Australia and Pakistan

Pycreus flavidus (Retz.) T. Koyama, J. Jap. Bot. 51: 313. 1976; Matthew, Fl. Tamilnadu Carnatic 3: 1777. 1983. *Cyperus flavidus* Retz., Observ. Bot. 5: 13. 1788. *Pycreus globosus* (All.) Rchb., Fl. Germ. Excurs. 1: 140. 1830; Fischer 3: 1627. 1931. *Cyperus capillaris* Koenig ex Roxb., Fl. Ind. 1: 198. 1820. *Pycreus capillaris* (Koenig ex Roxb.) Nees ex C.B.Clarke in Hook. f., Fl. Brit. India 6: 951. 1893. *P.flavidus* (Retz.) T. Koyama var. *nilagiricus* (Hochst. ex Steud.) Karthik. in Kartik. *et al.*, Florae Indicae enumeratio Monocotyledonae 65. 1989. *Cyperus nilagiricus* Hochst. ex Steud., Syn. Pl. Glum 2:2. 1855. *Pycreus capillaris* var. *nilagiricus* (Hochst. ex Steud.) C.B. Clarke, J. Linn. Soc. 21: 49. 1884; Fischer 3: 1627. 1931. *P. globosus* var. *nilagiricus* (Hochst. ex Steud.) C.B. Clarke, J. Linn. Soc. 36: 204. 1903.

Annual, densely tufted herb, to 40 cm high, obtusely trigonous. Leaves few, setaceous, shorter than the culm; blades slenderly linear, plicate; sheaths elongated, tinged with reddish-brown. Inflorescence simple with 4 unequal rays, t time congested in a head like cluster of spikelets; rays slender, the longer ones

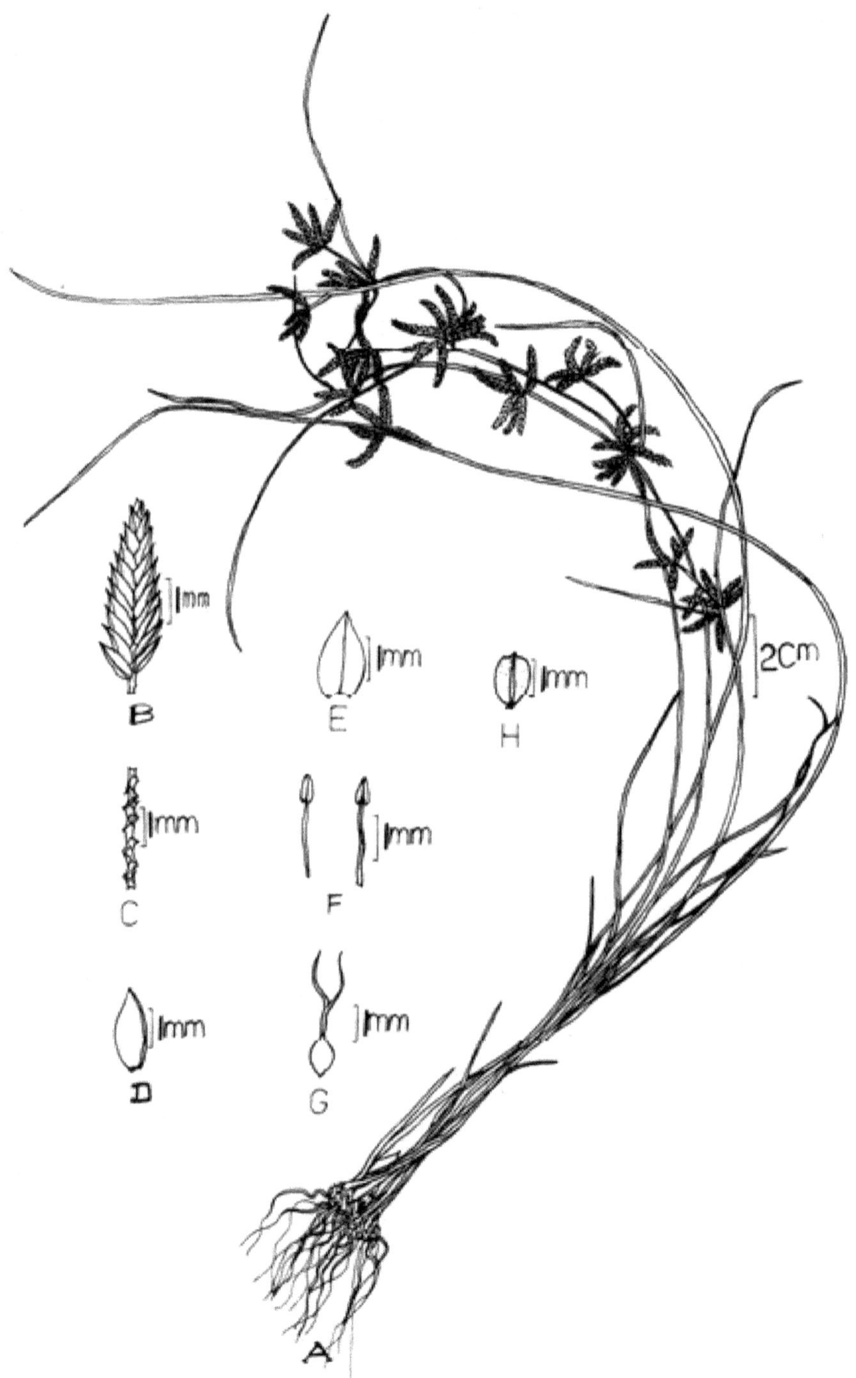

Figure 61. **Pycreus flavidus** (Retz.) Koyama.
A. Habit, B. Spikelet, C. Rhachilla, D. Closed glume, E. Opened glume, F. Stamens, G. Pistil, H. Achene.

up to 5 cm long. Spikes bearing *c.* 20 spikelets on short rhachis, ovoid, becoming subglobose cluster of spikelets in the upper part. Involucral bracts 3, erect-patent, the lowest one much surpassing the inflorescence, the longest up to 18 cm long. Spikelets spreading, linear, parallel-sided, strongly flattened, 20 mm long, densely 40-flowered, red brown; rhachilla straight, wingless, tetragonous. Achenes narrowly obovate, apiculate, lenticular and laterally flattened, 1 mm long, maturing dark brown.

Found in wet places in forests, near water falls and marshy localities. Fl. & Fr.: July – November.

Penakacherla (ATP), *NY* 593 (SKU & MH); Kuppagal (KNL), *KH* 10969; Krishnanandi (KNL), *VBH* 866 (BSID); Talakona (CTR), *BR* & *BSS* 33140; Darsi (PKM), *MCK* 22835; N.S. Right canal (GNT), *VRK* 3548; Sompalli reservoir (VSKP), *MHR* & *MCK* 14900; S. Kota-Punyagiri hill (VZN), *KH* & *BR* 5598; Kurnool district, *JSG* 10887 (K); Balapalle (KDP), *JLE* 14299 (MH & CAL); Bojeripakala, Maredumilli (EG), *MM* 101323 (BSID); Vettukkuru, Maredumilli (EG), *MM* 100785 (BSID); Araku valley (VSKP), *NPBK* 10757 (MH), *DDSR* 21408 (MH); Puliyur (NA), *KS* 6522 (MH); Guthirayan RF (SLM), *EV* 24126 (MH); Hogainakkal (KGR), *S.India Flora* 18010 (MH); Tulka RF, Satkosia Tiger Reserve, *KCM* 5977 (BSID).

INDIA: Throughout India.

WORLD: S. Europe, Africa, W. & C. Asia, China, Japan, Malesia.

Note: A detailed study of specimens in Kew from different regions of India made by Prasad and Simpson (2013) revealed continuous variation for a range of characters. Considering the high degree of variations within the collections of *Pycreus flavidus* in India Prasad and Simpson (2013) opined to treat the species *sensu lato* by merging all the infraspecific taxa described from India under the name *Pycreus flavidus* with relevant synonymy as given above.

Pycreus polystachyos (Rottb.) P.Beauv., Fl. De Oware & Benin 2: 48, t. 86, f. 2. 1807; FBI 6: 592. 1893; Matthew, Fl. Tamilnadu Carnatic 3: 1777. 1983. *Cyperus polystachyos* Rottb., Descr. Ic. Rar. Nov. Pl. 39. t. 11, f. 1. 1773; Saxena & Brahmam, Fl. Orissa 4: 2143. 1996. *Pycreus odoratus* Urban, Symb. Antill 2: 164. 1900; Fischer 3: 1627. 1931.

Key to Varieties

1. Inflorescence capitate, rays short .... var. **polystachyos**
1. Inflorescence anthelate, rays well developed .... var. **laxiflorus**

Pycreus polystachyos (Rottb.) P.Beauv. var. **polystachyos**

Perennial erect, tufted rhizomatous herb, to 50 cm high; rhizome covered with purple scales; culms trigonous, smooth. Leaves shorter than the culms, flat, linear, gradually acuminate, scabrid on margins in the upper part; sheaths purple brown, 5 cm long. Inflorescence simple, reduced to a single head, 3 cm long and wide, rays sometimes present with spikelets spicately arranged. Involucral bracts 8, lower 5 much exceeding the inflorescence, others short; spikelets strongly compressed, yellow to straw-coloured, 20 mm long, linear-lanceolate, rhachilla flexuous, narrowly winged; glumes distichously arranged, deciduous, elliptic-lanceolate, apex shortly stiff aristate; stamens 2, yellow; ovary oblong, styles 2. Achenes oblong, biconvex, 1.2 mm long, dark brown, sharply apiculate.

Common in marshy areas, rice fields, along water streams, river banks *etc.* Fl. & Fr.: September – December.

Batrepalli (ATP), *KRKS* 37951; Omkaram (KNL), *BSS* & *MSR* 29346; Lankamalai RF (KDP), *SRSR* 14659; Macherla (GNT), *VRK* 5880; Punyagiri hill-S. Kota (VZN), *KH* & *BR* 5600; Near Srisailam (KNL), *JLE* 16835 (MH); Kailaskona (CTR), *GVS* 32031 (CAL); Udayagiri (NLR), *BSN* 4771 (BSID); Pedayedlagadi (KSN), *PV* 116536; Maredumilli (EG), *GVS* 27296 (CAL); Maredumilli to Deverapalli (EG), *MM* 101308 (BSID); Araku (VSKP), *NPBK* 10863 (MH & CAL), *GVS* 19737 (MH); Rangamatiagedda (SKLM), *GVS* 62340 (MH & CAL); Parangipettai (SA), *KRM* 58176 (MH); Kaveri peak, Yercaud (SLM), *DBD* 31300 (MH); Balmadies estate- Yercaud (SLM), *AVNR* 26971 (MH); Surpanam Chanadi (SA), *VNS* 5062 (MH); Gollabandh (GJM), *SX* 997 (RRL-B); Dhanei (GJM), *SX* & *MB* 4047 (RRL-B); Brudhakol (GJM), *MB* 2843 (RRL-B); Tikarpada RF, Satkosia Tiger Reserve, *KCM* & *J.Swamy* 7619 (BSID).

INDIA: Assam, Meghalaya, Sikkim, West Bengal, Rajasthan, Uttar Pradesh, Madhya Pradesh, Odisha, Andhra Pradesh, Tamil Nadu.

WORLD: Cosmopolitan, Tropical and subtropical regions.

Pycreus polystachyos (Rottb.) P.Beauv. var. **laxiflorus** (Benth.) C.B.Clarke in Hook. f., Brit. India 6: 592. 1893; Matthew, Fl. Tamilnadu Carnatic 3: 1778. 1983. *Cyperus polystachyos* var. *laxiflorus* Benth., Fl. Austr. 7: 261. 1878.

Inflorescence compound with well developed rays, 7 cm long and wide, branches spicate.

Rare in a semi-evergreen forests. Fl. & Fr.: November – December.

Konapuram (ATP), *BR* & *ANS* 36614; Papavinasanam (CTR), *MHR* 13336; Venkatagiri hill (NLR), *BS* 3900 (VV); Devarapalli (EG), *MM* 100790 (MH); Maredumilli (EG), *GVS* 27296 (MH).

INDIA: Peninsular India.

WORLD: America, Argentina, China, Egypt, Costa Rica, Indonesia, Malaysia, Vietnam, Mexico, Myanmar, Namebia, South Africa, Sri Lanka.

Pycreus pumilus (L.) Nees, Linnaea 9: 283. 1834; FBI 6: 591. 1893. Fischer 3: 1627. 1931; Matthew, Fl. Tamilnadu Carnatic 3: 1778. 1983. *Cyperus pumilus* L., Cent. Pl. 2: 6. 1756; Saxena & Brahmam, Fl. Orissa 4: 2145. 1996. *Pycreus nitens* Nees, Nov. Act. Acad. Caes. Leopcarol. Nat. Cur. 19, Suppl. 1: 53. 1843; FBI 6: 591. 1893.

1. Achenes elliptic, lenticular, truncated with apex........................subsp. **pumilus**
1. Achenes obovoid, biconvex, apiculate with apex............subsp. **membranaceus**

Pycreus pumilus (L.) Nees subsp. **pumilus**

Annual, densely tufted herb, to 15 cm high, culms slender, triquetrous. Leaves few, surpassing the culm, blades narrowly linear, soft, flat, light green, sheaths pale. Inflorescence simple; involucral bracts 4, patent, lower 3 much surpassing the inflorescence, the longest 15 cm long; primary rays 4, patent, up to 3 cm long, slender; spikes subglobose, densely bearing many spikelets, the short rhachis up to 3 mm long. Spikelets spreading, oblong, 7 mm long, flattened, *c.* 30-flowered, light greenish, rhachilla straight, wingless. Achenes elliptic, laterally flattened, lenticular, maturing brownish, contracted to truncate apex.

Common along the banks, in the drying beds of canals of soft sticky soil, marshy fields, sandy river sides. Fl. & Fr.: August – November.

Chintalapalli Road (ATP), *KH* 7475; Madhavaram RF (KNL), *RVR* 1646; Ahobilam (KNL), *BR* & *BSS* 29331 (SKU & BSID); Palakonda hill (KDP), *CS* 7695; Gonupalli (NLR), *PMR* 19538; Nagarjunakonda (GNT), *VRK* 6715; Near Komativaricheruvu (CTR), *GVS* 46833 (MH & CAL); Chandragiri (CTR), *S. India Flora* 10049 (MH); Venkatagiri (NLR), *BSN* 5288 (BSID); Gaddada, Maredumilli (EG), *MM* 102644 (BSID); Salur (VZN), *NPBK* 947 (CAL); Parnasala RF (KMM), *RCS* 102474 (BSID); Lakshmipuram RF (KMM), *RCS* 102492 (MH); Kottaiyru RF, Yelagiri hills (NA), *MBV* 829 (MH); Poondi (CPT), *DN* 701 (MH); Berhampur (GJM), *VNS* 1717 (MH); Majhipada, Satkosia Tiger Reserve, *KCM* 7051 (BSID); Tulka RF, Satkosia Tiger Reserve, *KCM* 5958 (BSID).

INDIA: Throughout India.

WORLD: Sri Lanka, Nepal, Indo-China, SE. China, Malesia, Australia.

Pycreus pumilus (L.) Nees ex C.B. Clarke in Hook. f., Fl. Brit. India 6: 591. 1893. subsp. **membranaceus** (Vahl) T.Koyama, Gard. Bull. Singapore 30: 151. 1977. *Cyperus membranaceus* Vahl, Enum. Pl. 2: 330. 1806.

Culms densely tufted, 5-16 cm, smooth. Leaves shorter than culms or overtopping them, flat or channeled, acuminate, to 0.2 cm wide; sheaths pale brown. Inflorescence compound, to 5 cm, involucral bracts 3-5, unequal, overtopping, to 12 cm; spikes ovoid, 5-18-spikeletted, to 3 cm; spikelets oblong, compressed, subacute, 10-40-flowered, 0.4-1 cm; glumes ovate to elliptic, to 0.15 cm, membranaceous. Nut obovoid, biconvex, apiculate, stalked, brownish.

Rare in marshy lands of foot hills in Anantapuram, Kurnool, Prakasam and Vizianagaram districts. Fl. & Fr.: August – December.

Kalasamudram (ATP), *MHR* 680, 14919, *KRKS* 40076; SEDS – Penukonda Road (ATP), *BR* & *ANS* 35360; Pecheruvu (KNL), *BR* & *BSS* 30152; Perayapenta (PKM), *BR* & *BSS* 33156; Kolleru lake (WG), *KH* & *BR* 10913; Salur (VZN), *MCK* 25273.

INDIA: Throughout India.

WORLD: Widespread in Africa, Asia, Bangladesh, Australia, Malaysia, United Republic, China, Pakistan and Thailand.

Pycreus puncticulatus (Vahl) Nees in Fl. Brasil. 2 (1): 10, in note 1842; FBI 6: 593. 1893; Fischer 3: 1628. 1931; Matthew, Fl. Tamilnadu Carnatic 3: 1779. 1983. *Cyperus puncticulatus* Vahl, Enum, Pl. 2: 348. 1806; Saxena & Brahmam, Fl. Orissa 4: 2146. 1996.

Annual herb with brownish fibrous roots; culms robust, solitary, to 50 cm high; triquetrous, often some what enlarged at base. Leaves 3-5 to a culm, slightly overtopping the culm; leaf blades to 40 cm long, to 8 mm wide, subcoriaceous, glaucous green; sheaths to 20 cm long, broader than blade, pale brown. Inflorescence compound, lax, large, to 24 cm long and as broad. Involucral bracts 5, spreading, longer than the anthela, the lowest to 30 cm long; rays 10, unequal, patent, slender, the longer ones to 14 cm long. Spikes lax, to 6 cm long, distantly bearing 20 mm long, compressed, to 30-flowered, pale sanguineous. Achenes obovate, orbicular, laterally strongly flattened, rounded, emarginate at apex, stipitate at deltoid-contracted base, maturing orange-brown, densely puncticulate.

Found in the rice fields, swamps, marshy areas and tank margins (Fischer; V.S. Raju *et al.,* 2008). Fl. & Fr.: August – December.

Brudhakol (GJM), *MB* 2783 (RRL-B).

INDIA: Maharashtra, Karnataka, Andhra Pradesh, Tamil Nadu, Kerala, Odisha.

WORLD: Sri Lanka, Malay Peninsula, Cochinchina.

Pycreus sanguinolentus (Vahl) Nees ex C.B.Clarke in Hook. f., Fl. Brit. India 6: 590. 1893; Fischer 3: 1627. 1931; Matthew, Fl. Tamilnadu Carnatic 3: 1780. 1983. *Cyperus sanguinolentus* Vahl, Enum. Pl. 2: 351. 1806; Saxena & Brahmam, Fl. Orissa 4: 2150. 1996.

Perennial, rhizomatous herb, to 40 cm high; culms tufted, trigonous, clothed above the base, the lower part decumbent, 3 nodose, rooting and branching at nodes, the upper part erect. Leaves borne on the lower part of the culms, shorter than the culms; blades linear, plicate; sheaths loosely surrounding the culm, longer than the internode, pale green. Inflorescence simple with 3 short rays, 3 cm long; spikes ovoidal, densely bearing *c.* 15 spikelets on short rhachis; ivolucral bracts 5, spreading, the lower 3 much surpassing the corymb, the longest to 12 cm long, spikelets narrowly ovate, 10 mm long, flattened, 16-flowered, sanguineous brown; rhachilla tetragonous, straight, not winged. Achenes broadly obovate, laterally lenticular, 1.2 mm long, maturing black.

Usually found in marshy localities, rice fields, swamps and along water courses. Fl. & Fr.: July – December.

Darsi (PKM), *MCK* 22840; Maredumilli RF (EG), *KH* & *BR* 10903; Tyada Forest (VSKP), *KH* & *BR* 9718; Near Padmapuram gardens (VSKP), *MHR* & *MCK* 14831; Kalichedu (NLR), *A.S.Rao* 4264 (BSID); Maredumilli towards Kutravada (EG), *GVS* 68539 (MH); Minumuluru (VSKP), *GVS* 29748 (MH); Araku valley (VSKP), *NPBK* 10768 (MH & CAL), GVS 21553 (MH); Khotnighada (VSKP), *GVS* 19507 (MH); Dornacheruvu, Mannanur RF (MBNR), *S.R.Srinivasan* 107454 (BSID).

INDIA: Throughout India.

WORLD: Widespread from Central Asia and India through China to the temperate Far East, Malesia, Oceania, Africa.

Pycreus stramineus (Nees) C.B.Clarke in Hook. f., Fl. Brit. India 6: 59. 1893; Fischer 3: 1627. 1931. *Cyperus stramineus* Nees in Wight, Contr. Bot. Ind. 74: 1834. *C. substramineus* Kuek. in Engl., Pflanzenr. 101: 398. 1936; Saxena & Brahmam, Fl. Orissa 4: 2154. 1996.

Annual, slender, tufted herb, to 30 cm high; culms obtusely trigonous. Leaves 3 to a culm; blades nearly equaling the culm, slenderly linear, canaliculated, subrigid, gradually tapering to a long acuminate apex; sheaths to 4 cm long, stained reddish-purplish. Inflorescence congested in a single, spike like cluster of a *c.* 15 spikelets, to 40 cm long. Involucral bracts 2, leaflike, both erect-patent, surpassing the inflorescence, the lower up to 10 cm long. Spikelets erect-patent, lance-oblong, to 30 cm long, acute, flattened, straw yellowish, rhachilla straight, wingless. Achenes orbicular obovate, 1 mm long, laterally flattened, biconvex, maturing purplish-brown, finely transversely undulate with 15 discontinous, weak, slender wrinkles.

Usually found in marshy localities, rice fields and grass fields in Visakhapatnam district. Fl. & Fr.: August – December.

Araku valley (VSKP), *NPBK* 10763 (CAL).

INDIA: Throughout India.

WORLD: China, Vietnam, Thailand, Malaysia, Myanmar and Sri Lanka.

Pycreus sulcinux (C.B.Clarke) C.B.Clarke in Hook. f., Fl. Brit. India 6: 593. 1893. *Cyperus sulcinux* C.B.Clarke, J. Linn. Soc. 21: 56. 1884.

Perennial, slender, tufted herb. Culm slender, 3-angled, smooth, with a few leaves basally. Leaves shorter than culm; sheath purplish brown; leaf blade 0.5-2 mm wide, usually folded, apical margin scabrous. Involucral bracts 3-5, spreading, leaflike, usually basal 1 or 2 longer than inflorescence. Spikelets longer, up to 40-fid; glumes more remote; rays 3-8, mostly to 7 cm, slender, each with 3-15 spikelets arranged into a spike. Spikelets spreading, linear, 0.5-1.5cm x 1-1.5 mm, compressed, 8-20-flowered; rachilla flexuose, narrowly winged. Glumes straw-colored to brownish yellow on both surfaces, lax, ovate to oblong-ovate, 1.8 mm, medially veins 3 and green, laterally membranous and veinless, margin narrowly white hyaline, apex obtuse and mucronate; stamen 1; anther oblong, 0.3 mm; style of medium length; stigmas 2. Nutlet blackish brown, oblong, 1.2-1.3 mm, slightly compressed, both surfaces concave sulcate and puncticulate

Rare in marshy localities of Nallamalais. Fl. & Fr.: October – January.

Bheemunikolanu (KNL), *BSS* & *SKB* 33113 (BSID).

INDIA: S., E., N. & N.E.India.

WORLD: Bangladesh, Bhutan, Indonesia, Myanmar, Malaysia, Papua New Guinea, Thailand, Vietnam and northern Australia.

QUEENSLANDIELLA Domin.

Queenlansdiella hyalina (Vahl) Ballard, Hooker's Icon. Pl. 33: t. 3208. 1933. *Cyperus hyalinus* Vahl, Enum. Pl. 2: 320. 1805. *Kyllinga hyalina* (Vahl) T.Koyama, J. Jap. Bot. 51: 317. 1976; Matthew, Fl. Tamilnadu Carnatic 2: 1766. 1983.

Annuals with purplish fibrous roots, growing in small tufts, culms triquetrous, to 20 cm high, with 2-3 leaves, slightly overtopping the culm. Leaves narrowly linear, to 20 x 0.3 cm; sheaths 1-5 cm long. Inflorescence simple, open, spikes more or less congested into a head-like cluster, 2-7 cm, as long as wide; rays 2-6, to 6 cm long. Spikelets 5-20 in a spike, ovate-elliptic, strongly flattened, 4-8 x 2-3 mm, 7-10-flowered, rhachilla flexuous, widely winged; glumes ovate, folded with an acute keel, 2-3 x 2-2.5 mm, keel distinctly 3-nerved, sides hyaline, pale. Achenes elliptic, often asymmetric, bilaterally lenticular, 1.2-1.8 x 1-1.3 mm, shallowly emarginated at apex, puncticulate with isodiametrical cells; styles 2-fid, 2.5-4 mm; stamens 2.

Occasional in moist fields and grasslands in Sambalpur hills in Odisha and Chittoor hills in Andhra Pradesh. Fl. & Fr.: July – October.

Palamaneru (CTR), *AL* 27833.

INDIA: Peninsular & E. India.

WORLD: Tropical East Africa, Mascarene Islands, India to Tropical Australia,Malesia.

REMIREA Aublet

Remirea maritima Aubl., Hist. Pl. Guian. Franc. 1: 45. t. 16. 1775; FBI 6: 677. 1893. *R. pedunculata* R. Br., Prod. Fl. Nov. Holl. 236. 1810. *R. maritima* var. *pedunculata* Benth., Fl. Austr. 7: 347. 1878. *Mariscus pedunculatus* (R. Br.) T. Koyama, Gard. Bull. Singapore 30: 159. 1977 & in Dassan. & Fosb., Rev. Handb. Fl. Ceylon 5: 240. 1985. *Cyperus pedunculatus* (R. Br.) Kern, Act. Bot. Neerl. 7: 798. 1958 & in Steenis, Fl. Males. 1, 7: 644 1974.

Perennial, rhizomatous herb to 10 cm high; rhizome horizontally long creeping, branched, rooting at nodes, clothed with membranous, acute, brownish sheaths, the internodes to 5 cm long; culm tufted from the branched head of the rhizome, rigid, trigonous. Leaves crowded, rigid, canaliculate, scabrid on the margins in the upper part, very gradually narrowed into the triquetrous pungent tip. Inflorescence head like, consisting of digitately arranged sessile short spikes. Involucral bracts 3-5, patent, the longest up to 8 cm long, much overtopping the inflorescence. Spikes ovoid, to 12 mm long. Spikelets sessile, densely crowded, ovoid, acute, slightly compressed, 1-flowered, falling off as a whole; rhachilla disarticulating above the basal 2 glumes; uppermost internode strongly flattened finally thickned, corky, to 3 mm long. Achenes trigonous, oblong, slightly compressed, shining, tightly enclosed in the upper node of the rhachilla.

Along wet, sandy edges of rivers and ponds in Kurnool district. Fl. & Fr.: August – December.

Thungabhadra river – Kurnool town (KNL), *KH* 10953.

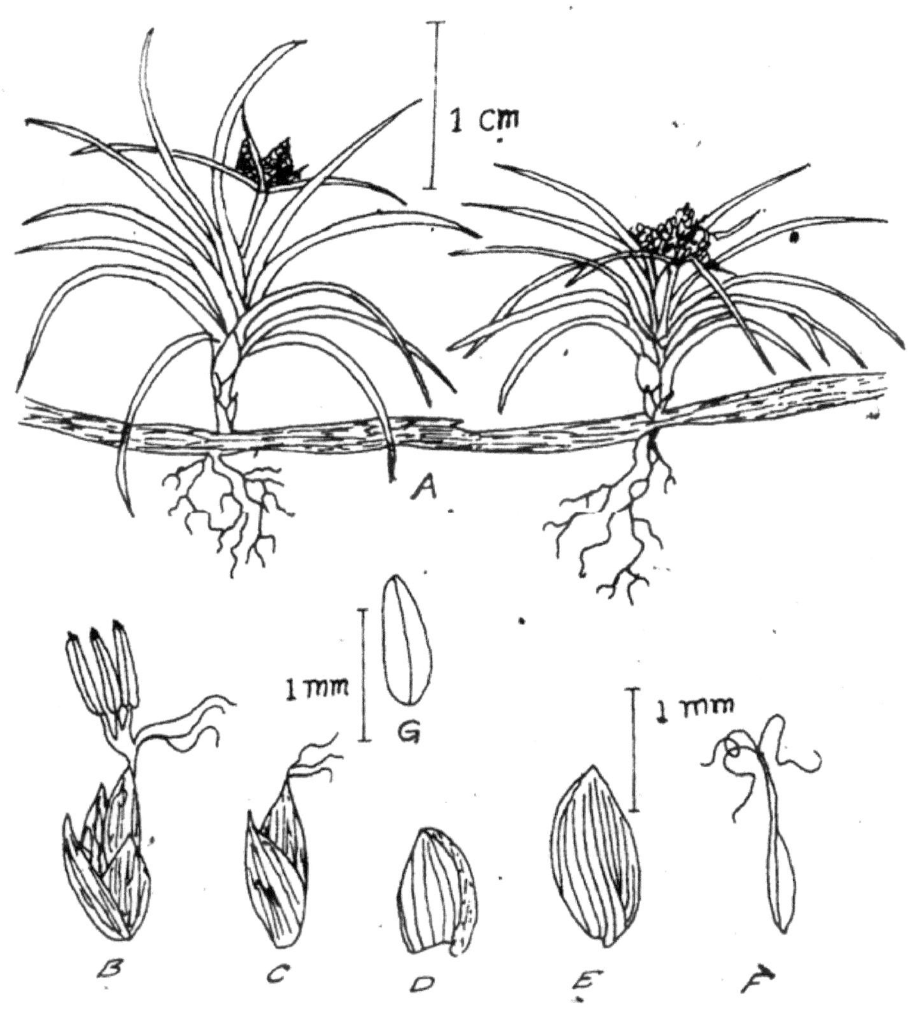

Figure 62. **Remirea maritima** Aubl.
A. Habit, B. Spikelet, C. Achene enclosed by glumes, D & E. Glumes, F. Pistil, G. Achene.

INDIA: Peninsular India, Andaman & Nicobar Islands.

WORLD: Tropical Africa, Madagascar, Sri Lanka, Myanmar, China, Malaysia, Indonesia, Philippines, Thailand, Vietnam, N. Australia, Pacific Islands.

RHYNCHOSPORA Vahl *nom. cons*

1. Style deeply bifid .... **R. gracillima**
1. Style sub entire or shallowly bifid:
   2. Inflorescence paniculate, lowest flower bisexual........ **R. corymbosa**
   2. Inflorescence a single, terminal globose or subglobose head, lower flower female:
      3. Nuts obovoid; bristles usually distinctly shorter than the nut, sometimes as long as the nut or absent .... **R. rubra**
      3. Nuts oblongoid; bristles always present, some of the bristles distinctly longer than the nut:
         4. Spikelets 10 – 12 mm long; bristles prominently antrorsely scabrous in the upper half and plumose in the lower; style-base prominently grooved on both sides .... **R. longisetis**
         4. Spikelets 5 – 7 mm long; bristles scabrous throughout; style-base terete........ **R. wightiana**

Rhynchospora corymbosa (L.) Britton, Trans. N.Y. Acad. Sci. 11: 84. 1892; Fischer 3: 1672. 1931; Matthew, Fl. Tamilnadu Carnatic 3: 1781. 1983; Dey & Prasanna, Rheedea 20: 5. 2010. *Scirpus corymbosus* L., Cent. Pl. 2: 7. 1756. *Rhynchospora aurea* Vahl, Enum. Pl. 2: 229. 1806; FBI 6: 670. 1893.

Perennial, rhizomatous herb, to 120 cm high; culms solitary, arising from short but rather thick rhizome, to 9 mm thick, several nodes, leaves up to the top, scaberulous on upper angles. Leaves many, aggregated at base of culm and several upper on the culm; blades broadly linear, to 60 cm long, to 20 mm wide, gradually long acuminate at apex, cauline sheaths shorter than the internode, bearing hyaline, contraligule at orifice; basal sheaths straw-brown. Inflorescence ample, to 35 cm long, bearing 4 inflorescence rather distantly borne on upper part of the culm; inflorescence corymbiform, to 14 cm long, diffuse, many-branched and bearing numerous spikelets. Involucral bracts leafy, to 25 cm long; primary rays to 12 cm long; secondary rays to 3 cm long. Spikelets lanceolate, acute, 9 mm long, rusty or chestnut brown, 3-flowered, the lowest bearing fruits. Achenes obdeltoid-obovate, compressed, to 4 mm long, yellow brown, finely transversely wrinkled in median portion, coarsely undulate-rugose towards margins, truncate at apex.

Occasional in moist hilly areas, along streams in open swamps and margins of rice fields (Fischer). Fl. & Fr.: August – January.

Pongal kovil shola, Kolli hills (NMK), *K.M.Matthew & V. Alphonse Amalraj* 2999 (RHT).

INDIA: Punjab, Rajasthan, Madhya Pradesh, Bihar, West Bengal, Assam, Meghalaya, Karnataka, Kerala, Tamil Nadu.

WORLD: Central and South America, Africa, Sri Lanka, Bangldesh, China, Indonesia, Malaysia, Philippines, Thailand, Vietnam, Australia.

Rhynchospora gracillima Thwaites & Hook., Enum. Pl. Zeyl. 435. 1864; FBI 6: 671. 1893; Fischer 3; 1672. 1931; Dey & Prasanna, Rheedea 20: 15. 2010.

Annual, densely tufted herb, with yellow brown fibrous roots, to 50 cm high; culms trigonous, leaves at and above the base. Leaves 4 to a culm, setaceous; sheaths to 3 cm long, light green. Inflorescence to 20 cm long, bearing 3 very lax anthelas distantly; terminal anthela larger than the lateral ones, to 3 cm long, bearing 4 capillary rays, to 4 cm long, rays terminated by 2 spikelets; lateral anthelas simple. Involucral bracts setaceous, to 3 cm long, the sheathing base to 15 mm long. Spikelets solitary, peduncled, lanceolate, acuminate, acute, to 7 mm long, 4-flowered, 2-fruited, brown. Achenes obovate-orbicular, rounded to short stipitate at base, truncate at apex, light grey, undulate rugose with 8 ridges.

Occasional in the opening of secondary forest and wet grass lands. Fl. & Fr.: August – December.

Araku valley (VSKP), *NPBK* 10773 (MH & CAL).

INDIA: S., E. & N.E.India.

WORLD: Tropical Africa, Madagascar, Sri Lanka, China, Indonesia, Thailand, NE Australia, Indian Ocean Islands.

Rhynchospora longisetis R.Br., Prodr. Fl. Nov. Holl. 230.1810; FBI 6: 669. 1996; Saxena & Brahmam, Fl. Orissa 4: 2198. 1996; Dey & Prasanna, Rheedea 20: 9. 2010.

Annual, erect herbs, with fibrous slender roots. Culms tufted, slender, 10 – 45 cm high, trigonous, striate, smooth. Leaves basal; blade linear, 6 – 20 x 0.1 – 0.3 cm, rigid, flat or conduplicate, scaberulous at margin, gradually tapering to a subacute apex; sheath-mouth truncate. Inflorescence a terminal, globose or semiglobose dense head, 1 – 2 x 2 – 2.5 cm; involucral bracts 4 or 5, patent or reflexed, 2 – 7 cm long (the lowest up to 7 cm long), densely ciliate at the dilated base. Spikelets numerous, sessile, linear-lanceolate, 10 – 12 mm long, acuminate, 2-flowered; lower flower female; upper one male; glumes 5 or 6, distichous, ovate-lanceolate, 2 – 10 mm long, acute, yellowish brown; bristles in the female flower 6, 5 of them rigid, 8 – 10 mm long, prominently antrorsely scabrid in the upper half and plumose in the lower, the remaining one slender, 4 – 5 mm long, antrorsely scabrid throughout; stamens 2; anthers linear, 2.5 – 3 mm long; style 5 – 6 mm long, shortly bilobed; style-base oblong conical, 2.5 – 3 x 2 – 2.5 mm, acute, grooved on both sides, smooth or weakly scabrous. Nut oblongoid, dorsiventrally compressed, 2.5 – 3 x 1 – 1.5 mm, hispidulous towards apex, brown; epidermal cells isodiametric.

Marshy places, in rock crevices, along road sides, near paddy fields and moist ditches in Eastern Ghats of Odisha. Fl. & Fr.: September – October.

Kalahandi (KHD), *HFM* 3675 (DD); Motijharan (SBP), *HFM* 2734 (DD); Sundargarh, Birmitrapur (Sundargarh Distr.), *D.Namhata* 3889 (CAL); Bargarh, Debrigarh Sanctuary, *C.Sudhakar Reddy s.n.* (CAL).

INDIA: Andhra Pradesh, Madhya Pradesh, Maharashtra, Odisha and Uttar Pradesh,

WORLD: Australia (Queensland), Cambodia, Laos, Myanmar, SW Pacific Islands (New Guinea), Thailand and Vietnam.

Rhynchospora rubra (Lour.) Makino, Bot. Mag. Tokyo 17: 180. 1903; Saxena & Brahmam, Fl. Orissa 4: 2199. 1996; Dey & Prasanna, Rheedea 20: 11. 2010.

Perennial. Culms tufted, erect, up to 1 m high, slender, trigonous below, triquetrous in the upper part, striate, smooth, glabrous. Leaves mainly basal, few cauline; blades linear, 5 – 60 x 0.3 – 0.4 cm, scaberulous at margin near apex, gradually tapering to a subacute apex, rigid, coriaceous; sheath-mouth truncate. Inflorescence capitate, subglobose to globose, 1–2 x 1–1.5 cm, dense, brown. Involucral bracts 4 – 8, patent, 2 – 8 cm long, rigid, densely ciliate at the dilated base. Spikelets sessile, many, ovate-lanceolate, 5 – 7 mm long, acute, compressed, 2-flowered; lowest flower female, the remainder male; glumes 6 – 8, distichous, ovate-lanceolate, acute, keeled; empty glumes 1.8 – 2.2 mm long; fertile glumes 4.2 – 6.5 mm long; bristles 0 – 3(– 6), 0.5 – 1.0 mm long, antrorsely scabrous; stamens 2 (or 3); anthers linear, *c.* 2 mm long; style 6 – 7 mm long, undivided; style-base pyramidal, *c.* 0.5 x 1 mm, obtuse, puncticulate. Nut obovate to oblong-obovate (young), 1.5 – 2 x 1 – 1.5 mm, cuneate at base, biconvex, minutely puncticulate, hispidulous-scabrous, brown to black; epidermal cells isodiametric.

Wet grasslands, hill slopes and along roadsides. Fl. & Fr.: June – November.

Rairakhol, Buromal (SBP), *HFM* 3995 (DD); Katrang, Satkosia Tiger Reserve, *KCM* 6790 (BSID).

INDIA: Andaman & Nicobar Islands, Assam, Bihar, Madhya Pradesh, Maharashtra,

Manipur, Meghalaya, Odisha, Sikkim, Tamil Nadu and Tripura.

WORLD: Australia, Borneo, Indonesia, Japan, Korea, Madagascan Region, Malaysia, Nepal, North West Pacific Islands, Philippines, SC and SE China, New Guinea, Sri Lanka, Taiwan, Thailand and Tropical Africa.

Rhynchospora wightiana (Nees) Steud., Cyper. 148. 1855; FBI 6: 669. 1893; Fischer 3: 1672. 1931. *Haplostylis wightiana* Nees, Nov. Act. Akad. Nat. Curr. 19 (Suppl. 1); 101. 1843.

Perennial, erect, caespitose herb, to 35 cm high, culms slender, trigonous, glabrous. Leaves many, at the base of the stem only, to 14 cm long, linear acuminate, grass like. Spikelets numerous in reddish-brown heads. Involucral bracts 6, unequal, the longest to 7 cm long, linear-lanceolate, acuminate, dilated and ciliate at the base. Spikelets to 1.5 cm long, lanceolate, acute, reddish-brown, smooth and shining. Achenes compressed, oblong, dark brown.

Occasional in rice fields, grass lands and in moist grounds in forests. Fl. & Fr.: September – January.

Jayanthipuram (GNT), *VRK* 6941; Motijharan hill (SBP), *HFM* 2743 (DD).

INDIA: Peninsular, C. & E. India.

WORLD: Sri Lanka, Vietnam, New Guinea.

SCHOENOPLECTIELLA Lye

1. Culms flowering in the lower half, septate when dry; glumes not mucronate:
   2. Culms robust; sheaths lax; achenes with transverse wavy lines .... **S. articulata**
   2. Culms slender; sheaths close; achenes with transverse undulate ridges:
      3. Perennial .... **S. senegalensis**
      3. Annual .... **S. praelongatus**
1. Culms flowering in the upper half, not septate (except *S. roylei*); glumes mucronate:
   4. Glumes multistriate, lax, inflated in fruit; stem septate when dry .... **S. roylei**
   4. Glumes not striate, imbricating; not inflated in fruit; stem not septate when dry:
      5. Culms trigonous; glumes ovate; achenes broadly elliptic .... **S. supinus**
      5. Culms terete; glumes elliptical; achenes broadly obovate .... **S. lateriflorus**

Schoenoplectiella articulata (L.) Lye, Lidia 6: 20: 2003. *Scirpus articulatus* L., Sp. Pl. 47. 1753; FBI 6: 656. 1893; Fischer 3: 1666. 1931; Saxena & Brahmam, Fl. Orissa 4: 2203. 1996. *Schoenoplectus articulatus* (L.) Palla, Bot. Jahrb. Syst. 10: 299. 1888; Matthew, Fl. Tamilnadu Carnatic 2: 1783. 1983.

Perennial, short living, robust herb, to 40 cm high; culms terete; 0.8 cm wide below, transversely septate at intervals of 1 cm, hollow, smooth, glabrous deeply green and slightly shiny when alive, clothed at base with 3 bladeless, stramineous-brown, lax sheaths; upper 2 basal sheaths cylindrical, to 12 cm long, herbaceous, pale green and later becoming yellow brownish, finely many-nerved, lower 1 sheath reduced and scale like. Inflorescence a pseudolateral head, globose or spherical, 4 cm in diameter, located at the midway portion of the (culm) stem or much lower down near the apex of upper-most sheath (depending upon the depth of the water at the locality as the inflorescence appear immediately above the water surface) densely bearing several to over 40 spikelets, lightly green. Involucral bract 1, stem like, to 40 cm long, terete, septate. Spikelets cylindrical or oblong, to 14 x 4 mm, pink-fulvous, acuminate. Achenes obovid, triquetrous, with transverse wavy lines and apiculate.

Common in marshy places, shallow waters and also along the margins of ponds. Fl. & Fr.: August – December. Vern.: Tam.: *Poppangorai.*

Kalasamudram RF (ATP), NY 674 (SKU & MH); Devaragattu (KNL), *SS* & *AMR* 22414; Guvvalacheruvu (KDP), *RVR* 8137; Papinenipalli – Perayapenta (PKM), *BR* & *BSS* 33154; Kolleru lake (WG), *KH* & *BR* 10911; Lakkavarapu reservoir (VZN), *MHR* & *MCK* 14888; Balapalle (CDP), *JLE* 15786 (MH); Neelithottikandriga, Satyavedu (CTR), *MCB* 45224 (MH); Gudur (NLR), *KCJ* 18515 (MH); Koradi RF (NLR), *P.Venu* 112126 (BSID); Tirumalayapalem (EG), *SS* 1754 (AU); Gingee RF (SA), *KMS* 12320 (MH); Pelakuppam (SA), *VNS* 40540 (MH); Sathanur dam (NA), *EV* 55652 (MH); Majhipada RF, Satkosia Tiger Reserve, *KCM* 6163 (BSID); Kalasanapur (GJM), *VNS* 4662 (MH);

INDIA: Throughout India.

WORLD: Tropical Asia from India and Sri Lanka through Indo-China to Malesia.

Schoneoplectiella juncoides (Roxb.) Lye, Lidia 6: 25. 2003. *Schoenoplectus juncoides* (Roxb.) Palla, Bot. Jahrb. Syst. 10: 229. 1889; Matthew, Fl. Tamilnadu Carnatic 2: 1784. 1983. *Scirpus juncoides* Roxb., Fl. Ind. 1: 228. 1820; Saxena & Brahmam, Fl. Orissa 4: 2205. 1996. *S. erectus* C.B.Clarke in Hook. f., Fl. Brit. India. 6: 656. 1893; Fischer 3: 1666. 1931.

Annual, erect, caespitose herb without conspicuous rhizome, to 30 cm high; culms slender, terete, obtusely several angled, light green, clothed at base with few sheaths only. Sheaths 3, the lower ones scale like, brownish, the upper one to 5 cm long, pale green, obliquely truncate. Inflorescence a pseudolateral head with 3-6 spikelets; involucral bracts to 13 cm long, rather suddenly subacute at apex, dilated at base. Spikelets oblong to ovoid-oblong, terete, 3 x 8 mm, straw-coloured, acute, densely many-flowered. Achenes broadly obovoid, unequally biconvex, 2 mm long, compressed, faintly transversely undulate, apiculalte.

Common in the margins of rice fields, river banks, swampy and marshy localities. Fl. & Fr.: September – December.

Chintalapalli Road (ATP), *KH* 7469; Batrepalli (ATP), *KRKS* 37949; Kuppam (CTR), *S.India Flora* 10316 (MH); Near Padmapuram gardens (VSKP), *MHR* & *MCK* 14838; Seethampeta (SKLM), *MCK* 25297; Forest near Anantagiri (VSKP), *NPBK* 11006 (MH & CAL); Puliyur (NA), *KS* 6524 (MH); Konetteri (NA), *KS* 6042 (MH); Similipahar (MBJ), *SX* & *MB* 5315(RRL-B); Badrama (SBP), *SX* & *MB* 2370 (RRL-B); Lanjigarh (KHD), *MB* & *Dhal* 7383 (RRL-B); Raigoda, Satkosia Tiger Reserve, *KCM* 6543 (BSID).

INDIA: Throughout India.

WORLD: Indo-China, China, Japan, Malesia.

Note: C.B. Clarke misapplied the same *Scirpus erectus* Poir. to a quite different species occurring in Madagascar, S.E. Asia and Tropical Australia and he was followed by several authors. The correct name of the species is *Scirpus juncoides* Roxb. and the name *S. erectus* Poir. does not belong to its synonymy.

S. erectus is related to *S. supinus* L. than to *S. juncoides* Roxb. It differs from *S. supinus* by the larger spikelets, large, more distinctly mucronate glumes, the bristly appendage of the connective, the bifid style, and the larger, convex, faintly wavy ridged elliptic or suborbicular achenes.

Schoenoplectiella lateriflora (J.F.Gmel.) Lye, Lidia 6: 25. 2003. *Scirpus lateriflorus* J.F.Gmel., Syst. Nat. 2: 127. 1791; Saxena & Brahmam, Fl. Orissa 4: 2206. 1996. *Schoenoplectus lateriflorus* (J.F.Gmel.) Lye, Bot. Not. 124: 290. 1971; Matthew, Fl. Tamilnadu Carnatic 2: 1784. 1983.

Annual, tufted herb, to 30 cm high, culms terete; sheaths often enclosing a long-stigmaed ovary or enlarged aches at base. One leaf blade often developed. Involucral bract to 10 cm long, slender, caniculated, obtuse. Spikelets 4-6, sessile, ovate-oblong, pale-brown, acute with glume in 5 ranks; glumes elliptic, papery, strongly keeled above, apiculate, tinged dark red brown; hypogynous bristle lacking. Achenes broadly obovate, unequally triquetrous, to 2 mm long, sharply transversely ridged, black.

Occasional in plains, in rice fields and other wet areas. Fl. & Fr.: September – January.

Mallapalli (ATP), *BR* & *ANS* 36569; Devaragattu (KNL), *SS* & *AMR* 22444; Madhavaram (WG), *MCK* 22907; Raigoda, Satkosia Tiger Reserve, *KCM* 6534 (BSID).

INDIA: Throughout India except N.W.India.

WORLD: SE Asia, Australia.

Schoenoplectiella roylei (Nees) Lye, Lidia 6: 26. 2003. *Isolepis roylei* Nees in Wight, Contrib. Bot. Ind. 107. 1834. *Schoenoplectus roylei* (Nees) Lye, Bot. Not. 124: 290. 1971. *Scirpus quinquefarius* Buch.-Ham. ex Boeck., Linnaea 36: 701. 1870; FBI 6: 657. 1893. *S. roylei* (Nees) Duthie, Fl. Upper Gangetic Pl. 3: 361. 1921; Saxena & Brahmam, Fl. Orissa 4: 2209. 1996.

Annual, erect, slender, tufted herb, to 30 cm high; culms terete, slightly compressed, conspicuously transversely septate. Sheaths obliquely truncate, leaves absent. Spikelets in a single lateral cluster near the top of the stem, ovoid-oblong, pale-brown, shining; involucral bract transversely septate, longer than the stem. Glumes inflated in fruit, elliptic-lanceolate, apex shortly recurved, membranous, keeled; stamens3, styles 3-fid, stigmas 3. Nut triquetrous, obovoid, faintly transversely wavy, black.

Occasional in drying paddy fields, on muddy or wet soil and in marshes near canals and streams. Fl. & Fr.: August – November.

INDIA: N.W., C. & N.E. India.

WORLD: Africa, Turkey, Afghanistan.

Schoenoplectiella senegalensis (Steud.) Lye, Lidia 6: 27. 2003. *Schoenoplectus senegalensis* (Hochst. ex Steud.) Palla, Bot. Jahrb. 10: 299. 1888; Matthew, Fl. Tamilnadu Carnatic 2: 1786. 1983. *Isolepis senegalensis* Hochst. ex Steud., Syn.

Pl. Glum. 2: 96. 1855. *Scirpus jacobii* Fischer, Kew Bull. 1931: 103. 1931 & Fl. Madras 3: 1666. 1931.

Perennial, erect, caespitose herb, to 30 cm high; culms slender, subterete, transversely septate. Sheaths close, obliquely truncate. Inflorescence pseudolateral, capitate. Involucral bracts greater than 20 cm long, stem like. Spikelets *c.* 20 in a cluster, ovoid-subglobose, 8 mm long, fulvous. Achenes obovoid, triquetrous, to 2 mm long, undulate, sharply angled, apiculate.

Occasional in marshy places. Fl. & Fr.: August – December.

Chintalapally Road (ATP), *KH* 7470; Mallapalli (ATP), *BR* & *ANS* 36570; Devaragattu (KNL), *SS* & *AMR* 22435; Jonnavalasa tank (VZN), *MCK* 18830; Near Komativaricheruvu (KDP), *GVS* 46826 (CAL); Gobanapalem (WG), *KS* 5163 (MH & CAL); Gandigam lake, Pennagaram (SLM), *EV* 22449 (MH); Tindivanam to Villupuram (CPT), *KRM* 60364 (MH); Gingee RF (SA), *KMS* 12344 (MH);

INDIA: Peninsular India.

WORLD: South Africa.

Schoenoplectiella supina (L.) Lye, Lidia 6: 27. 2003. *Schoenoplectus supinus* (L.) Palla, Bot. Jahrb. Syst. 10. 299. 1888. *Scirpus supinus* L., Sp. Pl. 49. 1753; FBI 6: 655. 1893; Fischer 3: 1666. 1931.

Annual, tufted herb to 30 cm high; culms trigonous, greenish, clothed at base with 2 sheaths only. Sheaths 5 cm long, membranous, pale green, the orifice obliquely truncate with a 2 mm long mucro, frequently the upper most sheaths born 3 cm above the base of the culm. Inflorescence pseudolateral, head like. Involucral bract 1, to 10 cm long, ventrally sulcate. Spikelets to *c.* 15 to an inflorescence, wholly congested, ovate-oblong, to 6 mm long, pale green and becoming lightly straw-coloured, subdensely many-flowered. Achenes broadly elliptic, 1.8 mm long, truly triangular, suddenly contracted at both ends, mucronate at apex, the sides transversely wrinkled, maturing black.

Common in wet grassy places, margins of ponds, rice fields. Fl. & Fr.: August – December.

Chintalapally Road (ATP), *KH* 7471; Batrepalli (ATP), *KRKS* 37859; Jaladanki (NLR), *PMR* & *ANR* 22360; Darsi (PKM), *MCK* 22824; Balapalle (KDP), *JLE* 15757 (MH & CAL); Near Komativaricheruvu (CTR), *GVS* 46829 (MH & AL); Gobanapalem (WG), *KS* 5161 (MH & CAL); Nayakkanur, Javadi hills (NA), /993 (MH); Komaratchi to Kattumannargudi (SA), *KRM* 60183 (MH); Bhuvangiri to Parangipettai (SA), *KRM* 53574 (MH); Gingee RF (SA), *KMS* 12319 (MH); Kuthukottai RF (DMP), *EV* 57927 (MH); Gandigam lake, Pennagaram (SLM), *EV* 22447 (MH);

INDIA: Throughout India.

WORLD: Iberian Peninsula, N.Africa, Madagascar, China, Russia, Vietnam, Brazil, Argentina and Paraguay.

SCHOENOPLECTUS (Reich.) Palla

1. Spikelets clustered on lateral or terminal heads .... **S. mucronatus**
1. Spikelets many, stalked, umbellate .... **S. litoralis**

Schoenoplectus litoralis (Schrad.) Palla, Sitzber. Zool. Bot. Ges. Wien 38: 49. 1888; Matthew, Fl. Tamilnadu Carnatic 3: 1785. 1983. *Scirpus litoralis* Schrad., Fl. Germ. 1: 42. t. 5, f. 7. 1806; FBI 6: 659. 1893; Fischer 3: 1667. 1931; Saxena & Brahmam, Fl. Orissa 4: 2207. 1996

Perennial, rhizomatous, stoloniferous herb, the stolon terminated by a tuber; culms to 70 cm high, terete, but obtusely triangular below inflorescence, closed at base with blade-less sheaths. Basal sheaths membranous, cylindrical, to 15 cm long, greyish-brown, mouth oblique with hyaline margin. Inflorescence a psudolateral simple, umbel like corymb with 4 unequal ascending rays, to 3 cm long, terminated by a cluster of spikelets. Involucral bract 1, foliaceous, as long as corymb. Spikelets oblong-obovate, to 8 mm long, many-flowered, rust coloured. Achenes broadly obovate, biconvex, 2 mm long, faces a little shining, chestnut brown at maturity.

Rare in shallow brackish waters near the coast, but also in swampy areas in the interior. Fl. & Fr.: September – March.

Peravali (ATP), *TP* & *NY* 506; Gangalakunta (ATP), *KH* 7479; Tungabhadra River (KNL), *KH* 10943; Balasamudram (KDP), *JSG* 20871 (DD); Bughanka (KDP), *KS* 6449 (CAL); Between Podalakur and Sangam (NLR), *PV* 111118 (BSID); Puliyur tank, on Thirupattur – Dharmapuri Road (NA), *KS* 6519 (MH).

INDIA: Throughout India.

WORLD: From Mediterranean through S. Asia to Australia, Africa, Malesia.

Schoenoplectus mucronatus (L.) Palla, Bot. Jahrb. Syst. 10: 229. 1889; Matthew, Fl. Tamilnadu Carnatic 3: 1785. 1983. *Scirpus mucronatus* L., Sp. Pl. 50. 1753; FBI 6: 657. 1893; Fischer 3: 1666. 1931; Saxena & Brahmam, Fl.Orissa 4: 2209. 1996.

Perennial, rhizomatous herb, to 120 cm high; culms robust, sharply 3-angled, sides deeply green, striate when dry, surrounded at base with 2 to 3 sheaths tightly, the lowest sheath scale like, brown to chestnut-brown, the upper often septate-nodose below, mouth obliquely truncate, brown scarious. Inflorescence of hemispherical pseudolateral head with 10 spikelets. Involucral bract 1, culm like, sharply 3-angled, to 5 cm long. Spikelets sessile, stramineous-brown, ovate, to 13 mm long, densely many-flowered, terete. Achenes broadly obovate, obcompressed, triangular, 2 mm long, glossy, transversely wrinkled with dark coloured patches.

A rare species usually collected from forest areas, shallow waters, marshy places. Fl. & Fr.: September – December.

Araku (VSKP), *DDSR* 15870 (MH); Lake, Yercaud (SLM), *SKK* 23095 (MH); Kaveri peak – Yercaud (SLM), *DBD* 31307 (MH);

INDIA: Throughout India.

WORLD: Eurasia from C. & S. Europe to Japan.

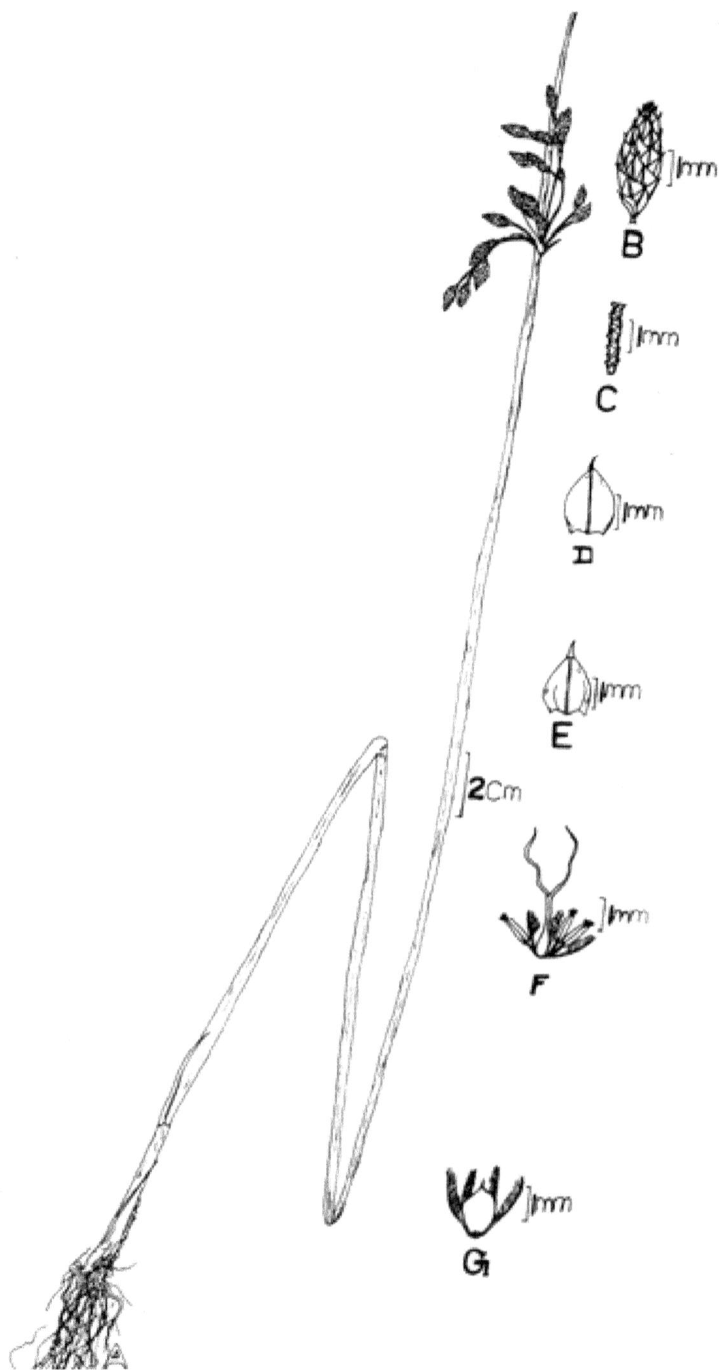

Figure 63. **Schoenoplectus litoralis** (Schrad.) Palla
A. Habit, B. Spikelet, C. Rhachilla, D & E. Glumes,
F. Pistil with stamens and hypogynous bristles, G. Achene with hypogynous bristles.

SCLERIA Bergius

1. Culms from woody rhizome, solitary or close together but not densely tufted; plants perennial:
    2. Achenes globose, as long as wide; hypogynium well developed:
        3. Lobes of hypogynium lanceolate, acute, half as long as the achene .... **S. levis**
        3. Lobes of hypogynium broadly oval, obtuse-tipped, hardly reaching 1/3 the length of achene........................ **S. terrestris**
    2. Achenes elongated, conspicuously ovate, 1.5 times longer than the width: hypogynium much reduced, only strip like:
        4. Robust plant, culms 90-150 cms high; inflorescence corymbose-paniculate, dense .... **S. corymbosa**
        4. Slender plant, culms to 80 cm high; inflorescence very loose with spiciform branches........................ **S. lithosperma**
1. Culms tufted with fibrous roots only, without conspicuous rhizome; plants annual:
    5. Male spikelets with stamen 1; glumes unequal, 7-nerved .... **S. poklei**
    5. Male spikelets with stamens 3; glumes equal, 3-nerved .... **S. parvula**

Scleria annularis Nees ex Steud., Syn. Pl. Glumac. 2: 176. 1855; FBI 6: 687. 1894; Dey & Prasanna in Fasc. Fl. India 27: 92. 2015.

Annual, stems up to 1.2 m tall, slender, solitary or tufted, erect, triquetrous, glabrous, scabrid. Leaves basal and cauline; leafblades 15-40 x 0.3-0.6 cm, linear, herbaceous, glabrous, smooth, scabrid along margins and midvein especially near apex, gradually narrowed to obtuse apex, sheath triquetrous, sparsely pubescent towards middle, glabrous on angles; contraligulate ovoid or deltoid, glabrous or sparsely ciliate. Inflorescence consisting of a terminal panicle and 2 to 3 lateral panicles; terminal panicles 2-4 cm long, lateral panicles 2-3 cm long, eerect, solitary, peduncles either enclosed within the sheath or exserted, smooth or scabrid. Primary bracts 6-20 x 0.3-0.5cm, secondary bracts 3-4 mm long. Spikelets solitary or 2-3 together, bisexual and male. Male spikelets 3-4 mm long, pedicelled, lanceolate; glumes 3-3.2 mm long, broadly ovate, with scabrid keel; sterile glumes aristate, fertile glumes mucronate; stamens 3, filamensts 3.5-4 mm long. Female spikelets 3-4 mm long, sessile, broadly ovate; glumes 2.5-3 mm long, broadly ovate, with smooth keel, sterile glumes acuminate, fertile glumes mucronate. Disc triangular with rounded angles, reddish puncticulate. Nuts 2.8-3 x 1.8-2 mm, ovoid, slightly laterally compressed, truncate at base, obtuse, glabrous, very smooth and shining, milky white-blackish.

Occasional in grasslands, moist places in shallow sandy soil during monsoon months and in paddy fields as a weed in Eastern Ghats of Odisha. Fl. & Fr.: July – November.

Sambalpur, Sonabera plateau, Khariar estate, *HFM* 3690 (DD).

INDIA: Bihar, Jammu & Kashmir, Karnataka, Kerala, Madhya Pradesh, Maharashtra, Manipur, Odisha and Uttar Pradesh.

WORLD: New Guinea and South Central Chhina.

Scleria biflora Roxb., Fl. India 3: 573. 1832; FBI 6: 687. 1894; Dey & Prasanna in Fasc. Fl. India 27: 95. 2015.

Annual, stems up to 55 cm high, slender, tufted, triquetrous, glabrous and smooth. Leaves cauline, leaf blades 4-21 x 0.3-0.5 cm, linear-lanceolate, herbaceous, glabrous, smooth, scabrid along margins and midvein, obtuse; sheath triquetrous, glabrous to puberulous, narrowly winged to wingless; contraligule depressed semi-orbicular, margin membranous, glabrous to ciliate. Inflorescence paniculate; terminal panicles 2-3 cm long; lateral panicles 1-1.5 cm long, erect, solitary or binate, lower ones exserted on compressed-triangular, scabrid peduncles. Primary bracts 1.5-18 x 0.1-0.5 cm, secondary bracts 3-4 mm long. Spikelets solitary or paired, unisexual. Male spikelets 2.5-3 mm long, lanceolate, pedicelled, pedicel 1-1.3 mm long; glumes ovate, acute, sterile glumes 2.5-3 mm long, fertile glumes 2-2.5 mm long; stamens 2-3 per flower, anthers 1.2-1.5 mm long. Female spikelets 2.5-3 mm long, sessile, broadly obovoid, with a sterile male flower. Sterile glumes 2.3-2.5 mm long, fertile glumes 2.8-3 mm long, ovate, acute. Disc deeply 3-lobed, lobes 1-1.3 mm long, appressed, coriaceous, lanceolate-ovate, gradually narrowed upwards to acuminate apex, ferrugineous. Nuts 2-3.5 x 1.8-2.5 mm, globose or slightly depressed globose, beaked with black or purplish style base, cancellate, ferrugienous pubescent on margins between lacunae, dull white to brown.

Occasional in wet grasslands, on the banks of hill streams, along the margins of the forest and as a common weed in rice fields. Fl. & Fr.: June – November.

Cuddapah forests (CDP), *RHB s.n.* (MH, Collected in 1880); Saptasajya RF (DKL), *HFM* 2042 (DD); Purunakote section, Satkosia Tiger Reserve, *KCM* 5171 (BSID); Katrang, Satkosia Tiger Reserve, *KCM* 6791 (BSID).

INDIA: Andaman & Nicobar Islands, Assam, Bihar, Goa, Karnataka, Kerala, Madhya Pradesh, Maharashtra, Meghalaya, Odisha, Sikkim, Uttarakhand, Uttar Pradesh and West Bengal.

WORLD: Indonesia, Malaya, Myanmar, Nepal, Philippines, Sri Lanka, Thailand, Vietnam.

Scleria corymbosa Roxb., Fl. Ind. 2, 3: 574. 1832; Fischer 3: 1677. 1931. *S. ridleyi* C.B.Clarke in Hook. f. Fl. Brit. India 6: 686. 1893.

Perennial, rhizomatous robust herb, to 150 cm high; rhizome woody, horizontal, covered with reddish-brown scales; culms solitary, acutely triquetrous, many-noded. Leaves evenly distributed on the total length of culms; leaf blades broadly linar, to 60 x 20 mm, sub-coriaceous, light green, 3-costate, glabrous, abruptly narrowed to an obtuse apex, the margins smoothish for most length, scabrid towards leaf apices, sheaths elongated, to 20 cm long, somewhat loose, 3-sided, not wigned, the lower ones coloured with brown, the basal sheaths bladeless, reddish-brown; contraligule depressed-deltoid, to 6 mm long. Inflorescence to 80 cm long, cylindrical, leafy; lateral panicles at 3 to 5 nodes, decompound, corymbiform, dense, preduncles long exserted, unequal in length, acutely triquetrous. Bisexual and staminate spikelets intermingled; bisexual spikelets sessile, ovoid, 4-5 mm long, bearing a sessile staminate spikelet at axil of the uppermost glumes. Achenes ovoid, exceeding the subtending glumes.

Sporadically occurring in forest margins. Fl. & Fr.: September – December.

Near Kailaskona (CTR), *GVS* 46053 (MH); Vathangi (EG), *S.India Flora* 12685 (MH); Rampa hill (EG), *VNS* 629 (CAL); Slope of Sesharayi hill (EG), *VNS* 269 (CAL); Maredumilli towards Chinturu (EG), *GVS* 68512 (CAL); Tadepalli, near Adamanishigondi (EG), *NRR* & *TRS* 86!25 (MH);

INDIA: Throughout India except N.W.India.

WORLD: Sri Lanka, Myanmar, China, Laos, Cambodia, Indonesia, Malaysia, Papua New Guinea, Philippines, Thaialand.

Scleria levis Retz., Observ. 4, 13. 1786; FBI 6: 691. 1894. *S. hebecarpa* Nees in Wight, Contr. Bot. Ind. 117. 1834; Fischer 3: 1678. 1931.

Perennial, rhizomatous herb; rhizome woody, horizontally creeping, covered with purple brown scales, culms solitary and approximate, to 80 cm high, 3-sided with slightly concave sides, scabrid on angles. Leaves about evenly scattered on the culm, leaf blades to 30 cm long, herbaceous, 3-costate, scabrous on margins and along costas, sheaths to 8 cm long, 3-sided, with scabrous wings, the lower sheaths tinged with reddish brown; contra ligule depressed, rounded, to 2 mm long, pubescent and hirsute on upper part. Inflorescence scanty and narrow, consisting of 2, loose panicles, usually spaced, the lateral ones solitary at node, the terminal one to 12 cm long; axes scabrous, acutely angular, peduncles exserted from bract sheath. Involucral bracts foliaceous, slightly surpassing the inflorescence. Pistillate spikelets and staminate spikelets in intermingled groups; pistillate spikelets sessile, borne toward the base of branchlets, obovoid; staminate spikelets peduncled, lance-oblong. Achenes globose, 2.5 mm long, smoothish, apiculate at apex; disc lobes acute, lanceolate, half as long as the achene.

Found in wet sands, along ravine slopes of the forest grasslands. Fl. & Fr.: March – October.

Maredumilli (EG), *KH* & *BR* 10901; Papavinasanam (CTR), *MHR* 13956; Gundlabrahmeswaram (KNL), *JLE* 16948 (MH); Chinamantrala (KNL), *JLE* 42289 (MH); Way to Diguvametta (PKM), *JLE* 22185 (MH); Tiger camp to Valamuru, Maredumilli (EG), *MM* 105038 (BSID); Near Tiger camp (EG), *MM* 102587 (BSID); Rampa hills (EG), MSR 1534 (CAL); Purunakote section, Satkosia Tiger Reserve, *KCM* 5266 (BSID).

INDIA: Throughout India.

WORLD: Sri Lanka, Bangladesh, Myanmar, Nepal, China, Cambodia, Indonesia, Philippines, Thailand, Vietnam, Papua New Guinea, NE Australia, Pacific Islands.

Note: This species is often confused with *S. lithosperma* due to its small and slender nature. *S. levis* can be easily distinguished by the presence of a well developed disk, whereas the disk is reduced to a narrow, brown ring in *S. lithosperma.*

Scleria lithosperma (L.) Sw., Prod. Fl. Ind. Occid. 18: 1788; FBI 6: 685. 1893; Fischer 3: 1677. 1931; Matthew, Fl. Tamilnadu Carnatic 3: 1786. 1983. *Scirpus lithospermus* L., Sp. Pl. 1: 51. 1753.

Key to Subspecies

1. Nutlets smooth .... subsp. **lithosperma**
1. Nutlets transversely rugose or irregularly somewhat reeticulaate with wavy ridges .... subsp. **linearis**

Scleria lithosperma (L.) Sw. subsp. **lithosperma**

Perennial, erect, tufted herb, to 80 cm high; culms loosely tufted from short knotty rhizome, rigid, slender, 3-sided. Leaves dense or clustered at the middle of the stem; leaf blades narrowly linear, to 25 cm long, rigid, canaliculated, revolute on scabrous margins, deeply green and glaucescent when dried; sheaths 3-sided, to 7 cm long, brownish-reddish, not winged, pubescent, the basal ones bladeless, reddish-purplish; contraligule convex, hairy. Panicles narrow, very loose, spicate, terminal ones to 15 cm long, axillary ones to 5 cm long, dark brown; involucral bracts leaf like, much longer than the inflorescence. Spikelets bisexual, solitary and clustered (in clustered condition 1 female, 1 or 2 male flowers together, others empty), ovate-turbinate. Achenes obovoid, to 3 mm long.

Common along hill slopes associated with other grasses, rocky surfaces and as the undergrowth of dry forests. Fl. & Fr.: Throughout the year.

S.Konda (ATP), *BR* & *ANS* 44842; Peccheruvu (KNL), *RVR* & *PVP* 2163; Sri Kalahasti (CTR), *BR* & *MVS* 32106; Simhachalam hill (VSKP), *KH* & *BR* 5589A; Balepalle (KDP), *JLE* 14984 (CAL, MH); Kailasakona (CTR), *GVS* 46995 (CAL); Top of Bison hill, Sirivaka (EG), *Bourne* 3399 (CAL); Jaddangi (EG), *S.India Flora* 12639 (MH); Ettukkuru, Maredumilli (EG), *MM* 102617 (BSID); Madgole to Paderu (VSKP), *GVS* 28075 (MH); Sithanur (CPT), *S.India Flora* 11225 (MH); Tulka RF, Satkosia Tiger Reserve, *KCM* 6777 (BSID).

INDIA: Throughout India.

WORLD: South America, Central America, Tropical Africa, Sri Lanka, China, Myanmar, Philippines, Thailand, Vietnam, Australia.

Scleria lithosperma (L.) Sw., subsp. **linearis** (Benth.) T. Koyama in Dassan. & Fosb., Rev. Handb. Fl. Ceylon. 5: 353. 1985. *Scleria lithosperma* Sw. var. *linearis* Benth., Fl. Austral. 7: 430. 1878. *S. lithosperma* Sw. var. *roxburghii* C.B. Clarke in Hook. f., Fl. Brit. India 6: 686. 1893; Fischer 3: 1677. 1931.

Achenes transversely rugulose with wavy ridges that are slightly viscid on top.

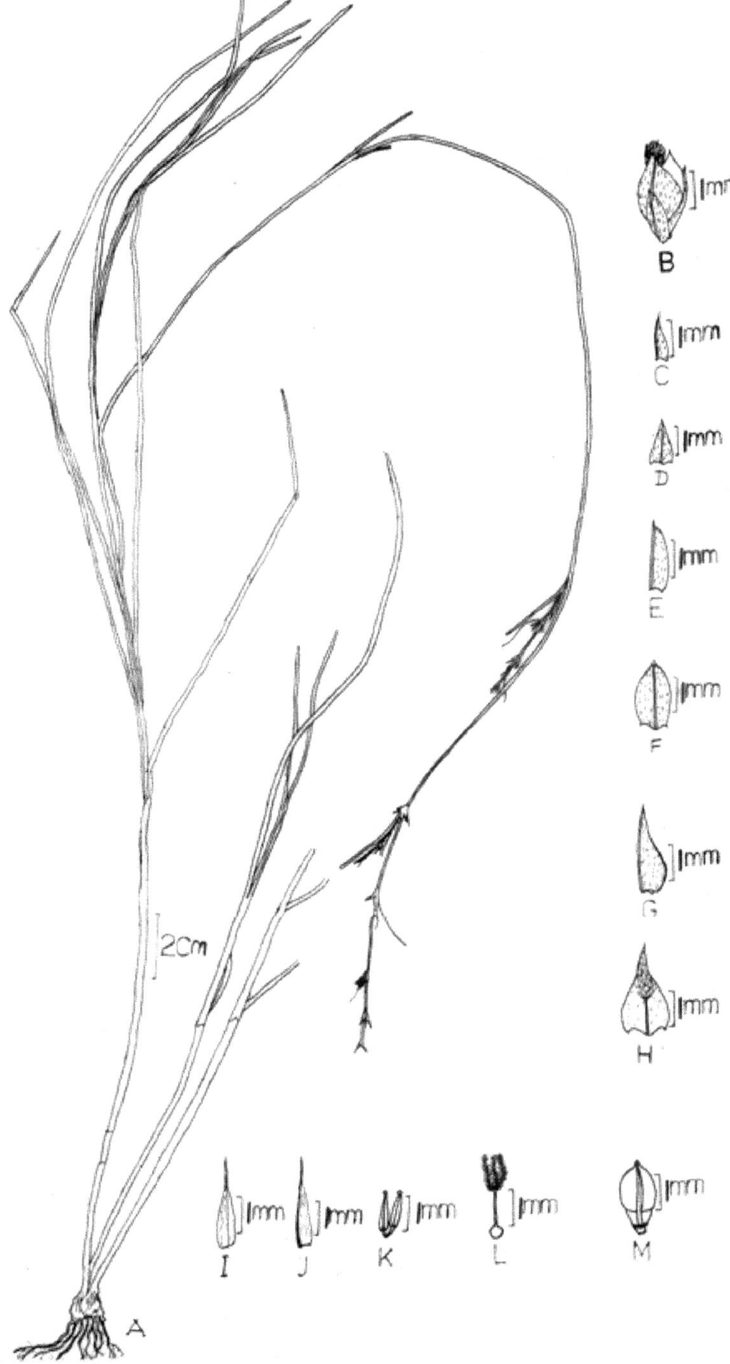

Figure 64. **Scleria lithosperma** (L.) Sw. subsp. **Lithosperma**
A. Habit, B. Spikelet, C-J. Glumes (closed and opened), K. Stamens, L. Pistil, N. Achene.

Not as common as subsp. *lithosperma*. Occasionally seen in dry forest areas. Fl. & Fr.: September – January.

Mahanandi (KNL), *BR* & *BSS* 29365 (SKU & BSID); Simhachalam hill (VSKP), *KH* & *BR* 5589 B; Ettukkuru (EG), *MM* 102617 (BSID); Madgole to Paderu (VSKP), *GVS* 19784 (MH).

INDIA: Rajasthan, Madhya Pradesh, Maharashtra, Andhra Pradesh.

WORLD: Sri Lanka, Indonesia, Malaysia, Papua New Guinea, Philippines, Thailand, Vietnam, N. Australia, Pacific Islands.

Scleria multilacunosa T.Koyama, Bull. Natl. Sci. Mus. Tokyo n.s. 17: 71. 1974; Dey & Prasanna, Fasc. Fl. India 27.100. 2015.

Annual, caespitose herb with purplish fibrous roots, culms to 90 cm high, sharply triquetrous, 3 or 4 nodose, 1 or 2 leaved on the midway portion, clothed at base with 1 or 2 bladeless sheaths. Basal sheaths to 4 cm long, sharply triquetrous, without wings, pale-green and stained with purple, obliquely truncate at orifice. Leaves shorter than the culm; blade 15-30 cm x 3-8 mm, tricostate, fresh green, scabrous on margins and abaxial costa, gradually tapering to acute apex; sheaths to 6 cm long, greenish, sharply triquetrous, the angles scabrous; contraligule rounded-truncate, ciliate with short rusty brown hairs. Inflorescence with 2-4 partial panicles the lowest one distant from the remainder and on a long exserted peduncle, the remainder continguous at the culm apex; partial panicles subspiciform, to 2 cm long, branches bearing 3 to 5 pistillate spikelets and 1-3 staminate spikelets. Lowest leafy bract sheathing for 2 to 3 cm, the blade to 12 cm long, not surpassing the culm; blade of the second bract to 5 cm long, longer than its subtending partial panicle, ciliate at base, the third and fourth bracts much shorter. Staminate spikelets to 3.5 mm long, lanceolate, short peduncled. Pistiliate spikelets to 4 mm log, obdeltoid. Achenes globose, to 3 mm long, rounded at mucronate apex, the whitish ceramic surface irregularly lacunose with many shallow depressions of varying shape and size.

Occasional in wet grassy localities in Araku valley. Fl. & Fr.: September – December.

Araku valley(VSKP), *NPBK* 584 (CAL).

INDIA: Andhra Pradesh, Karnataka, Kerala, Maharashtra and Rajasthan.

WORLD: Northern Sri Lanka.

Scleria parvula Steud., Syn. Pl. Glumac. 2: 174. 1855; Saxena & Brahmam, Fl. Orissa 4: 2218. 1996.

Annual, stem slender, smooth, 30-60 cm high, hairy or glabrate. Leaves linear, 10-25 x 0.2-0.5 cm, suddenly narrowed to the obtusish tip, glabrous, scabrid on the margins in the upper part, sheaths rather loose, winged, glabrous or sparsely pubescent, edges of wings retrorsely scabrid. Inflorescence narrow, elongate, the terminal panicle oblong, 2-4 cm, somewhat longeer than the 1-3 distant lateral fascicles; primary bracts erect, much longer than the panicles in the axils. Spikelets unisexual. Male spikelts shortly peduncled, lanceolate, 4-5 mm long; stamens 3.

Female spikelets 4.5-5 mm long, without a barren or male flower besides the female one. Glumes ovate, acute, stramineous with purplish margins to wholly purplish. Cupula hardly lobed, 1 mm wide. Disc 3-lobed, lobes thickish, appressed, ovate, acuminate. Nut shorter than the glumes, ellipsoid or subglobose, 2-2.5 mm long, slightly trigonous, deeply cancellate, glabrous or ferrugineous.

Saptdadhar, Keonjhar (Panigr. *et al.*, Bull. Bot Surv. India 6: 261. 1964).

INDIA: Widely distributed.

WORLD: Tropical Africa, Sri Lanka, Nepal, Thailand, Indo-China, S. China, Japan, Korea, Malesia.

Scleria pergracilis (Nees) Kunth, Enum. Pl. 2: 354. 1837; FBI 6: 685. 1894; Haines, Bot. Bihar & Orissa 3: 931. 1924; Saxena & Brahmam, Fl. Orissa 4: 2219. 1996. *Hypoporum pergracile* Nees, Edinburgh New Philos. J. 1834: 267. 1834.

Annual, stem 25-45 cm high, very slender, glabrous. Leaves lemon-scented, narrowly linear, 10-25 cm x 0.5-3 mm, glabrous, scabrid towards the top; sheaths narrow, not winged, smooth, glabrous or sparsely pilose, truncate at the mouth or with a very short membranous appendage. Inflorescence linear, unbranched, spiciform, 5-15 cm long, consisting of clusters of spikelets, the lower clusters 1-1.5 cm, distant, upper ones subcontiguous, clusters almost sessile, with 2-5 spikelets; bracts inconspicuous, not or hardly longer than the clusters of spikelets in their axils. Spikelets bisexual, small, obovate, 2.5-3 mm long. Glumess ovate-lanceolate, acute, muticous, glabrous, densely beset with reddish glandular streaks; stamens 2; capsula *c.* 0.5 mm, triangular. Disc obsolete, concrete with the nut. Nut subglobose-ovoid, 1-1.3 mm long, obtusely trigonous, glabrous, shining white.

In wet grassy places near swamps, rice fields *etc.* in Eastern Ghats of Odisha. Fl. & Fr.: September – October.

Borara, Sonabeda Plateau (KPT), *HFM* 3683 (DD).

INDIA: Widely scattered from Garhwal to West Bengal, Bihar, Odisha, Deccaan Peninsula.

WORLD: Tropical Africa, Sri Lanka, Thailand, Indo-China, Malesia.

Scleria poklei Wad. Khan., J. Econ. Taxon. Bot. 22: 559.1998, publ. 1999.

Apparently similar to *Scleria caricina*, sometimes rather less slender and much branched. Inflorescence of axillary clusters, 3-4 mm long and broad with peduncles often cellularly spongious papillose, white. Spikelets unisexual, male ones few, oblong, sessile, 1-2-fid, glumes thinly mmbranous, hyaline; stamens 1; female ones several, turbinate of ovoid, 0.8-1 x 0.5-0.8 mm, distinctly several-nerved, down to base; pedicels 1-1.5 mm long, glumes 2, unequal, ovate, 0.8-1mm long, boat-shaped, 7-nerved (nerves often brown), rounded on the back, not keeld, almost equalling and tightly investing the nut, falling together with it; lobes 3, subulate; side lobes hyaline, slightly shorter than central one, often clasping the nuts, the central lobes herbaceous, *c.* 0.2-0.3 mm long, erect. Nuts obtusely trigonous, subglobose, 0.7-0.8 mm long and broad, strongly reticulate in between the 3 main ribs, sparsely hispidulous at apex, mucronatee, greyish. Disk obsolete, represented by narrow, white ring at base of main body. Style 3-fid, much shorter than nuts.

Rare in Satkosia Tiger Reserve, Odisha. Fl. & Fr.: October – November.

Katrang, Satkosia Tiger Reserve, *KCM* 6797 (BSID).

Scleria rugosa R.Br., Prodr. Fl. Nov. Holl. 240. 1810; Naithani & Raizada, Indian For. 103: 420. 1977; Saxena & Brahmam, Fl.Orissa 4: 2220.1996. *S. zeylanica* C.B.Clarke in Hook.f. Fl. Brit. India 6: 687. 1894 excl. syn. *S. thwaitesiana*; Mooney, Suppl. Bot. Bihar & Orissa 153. 1950. *S. flaccida* C.B.Clarke in Hook.f., Fl. Brit. India 6: 688. 1894 non Steud. 1855.

Annual, stem 10-13 cm high, slender, obliquely erect or decumbent, smooth. Leaves 6-10 x 0.2-0.4 cm, abruptly narrowed towards obtusish tip, glabrous to densely pubescent; sheaths loose, triquetrous to distinctly winged. Inflorescence narrow, elongate, consisting of a terminal panicle and 1-2 lateral, remote fascicles of panicles; primary bracts erect, overtopping the inflorescence. Spikelets unisexual. Male spikelets shortly peduncled, lanceolate; stamen 1. Female spikelts 3-4 mm long; glumes ovate, acute or mucronate, usually long ciliate on the keel, rarely glabrous. Cupula hardly lobed, 0.7 mm wide. Disc thick, appressed, shallowly 3-lobed, densely cellular-glandular, lobes obtuse, semiorbicular. Nut globose, 1-1.5 mm across, white or finally greyish.

In Eastern Ghats of Odisha. Motijharan forest, Sambalpur (Raizada, J. Bombay Nat. Hist. Soc. 48: 679. 1948). Fl. & Fr.: September.

INDIA: Madhya Pradesh, Odisha.

WORLD: Sri Lanka, China, Japan, Australia, Malesia.

Scleria terrestris (L.) Fassett., Rhodora 26: 159. 1924; Matthew, Fl. Tamilnadu Carnatic 3: 1788. 1983. *Zizinia terrestris* L., Sp. Pl. 2: 991. 1753. *Scleria radula* Hance, Ann. Sci. Nat. Bot 18. 232. 1862. *S. chinensis* Kunth. var. *biauriculata* C.B. Clarke in Hook. f., Fl. Brit. India 6: 690. 1893. *S. cochinchinensis* (Lour.) Druce, Bot. Exch. Club. Brit. Isles Rep. 4: 646. 1917; Fischer 3: 1678. 1931.

Perennial herb with short, creeping woody rhizome, to 150 cm high; culms elongated, triquetrous, densely leaved for the most length; leaf blades linear, to 40 cm long, to 20 mm wide, subcoriaceous, 3-costate, gradually tapering to very long acute apex, recurved on scabrous margins; sheaths to 13 cm long, triquetrous, coloured with reddish brown, scabrid on angles; contraligule to 4 mm long, lunate or depressed-rounded, brown. Inflorescence consisting of 4 compound panicles, generally spaced, to 40 cm long, 10 cm wide; partial panicles solitary, ovate pyramidal or broadly ovoid, dense, the branches ascending, scabrid, the peduncle long-exserted up to 20 cm long, compressed, triangular, scabrous. Involucral bracts leaf like, slightly overtopping the inflorescence. Pistillate spikelets and staminate spikelets in groups of 2 or 3; pistillate spikelets sessile, broadly ovoid, to 6 mm long, staminate spikelets lanceolate, to 4 mm long, deeply red-brown. Achenes longer than the two subtending glumes, broadly obovoid-globose, to 3 mm long, shining white or finally dark purplish; hypogynium appressed to achene, 3-lobed or nearly

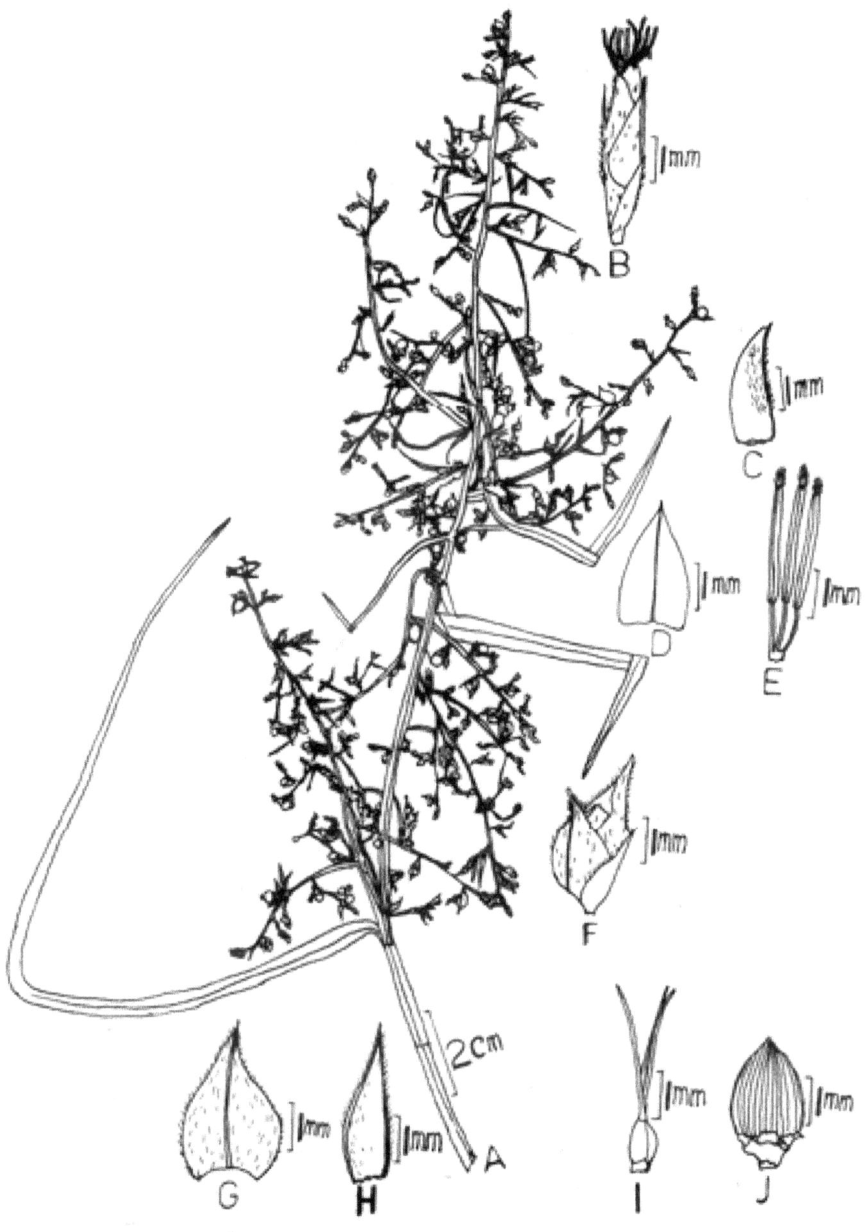

Figure 65. **Scleria terrestris** (L.) Fasset.
A. Twig, B. Spikelet, C. Closed glume, D. Opened glume, E. Stamens, F. Spikelet, G. Opened glume, H. Closed glume, I. Pistil, J. Achene.

obtusely triangular, the lobes ovate, finely toothed at apex, hardly reaching 1/3 the length of achene.

A rare species usually found in wet places in forest areas and scrub jungles as an under growth. Fl. & Fr.: February – December.

Kuppagal RF (KNL), *KH* 10968; Talakona top (KDP), *BR, BSS* & *MVS* 33143; Cuddapah forests (CDP), *RHB s.n.* (MH); Japalathirtham (CTR), *KS* 7869 (MH); Way to Dharawada (WG), *DN* 85517 B (BSID); Nulakamaddi (EG), *GVS* 24531 (MH & CAL); Forest near Sunkarimetta (VSKP), *NPBK* 10921 (MH); Anantagiri (VSKP), *GVS* 21755 (CAL); Kiliyur water fals – Yercaud (SLM), *AVNR* 26901 (MH); Chamundya (GJM), *VNS* 58700 (MH).

INDIA: Uttar Pradesh, Madhya Pradesh, Odisha, Andhra Pradesh, Maharashtra, Karnataka, Tamil Nadu, Kerala.

WORLD: Sri Lanka, Bangladesh, Nepal, Japan, Malaysia, Laos, Indonesia, Philippines, Thailand, Vietnam, Australia.

References

Adsul, A.A. 2015. Taxonomic revision of genus *Chlorophytum* Ker-Gawl. for India. Ph.D. thesis Shivaji University, Kolhapur. pp. 83.

Chandramohan, K., P.R. Sushma and Y.Mahesh. 2018. Note on distribution of *Fimbristylis fimbristyloides* (F.Muell.) Druce (Cyperaceae) in India. Indian J. Forestry 41(3): 227-229.

Chorghe, A.L., L. Rasingam, P.V. Prasanna and M. Sankara Rao. 2017. Three new additions to the flora of Eastern Ghats. Nelumbo 59: 66-70.

Dey, S. and P.V.Prasanna. 2010. The tribe Rhynchosporeae (Cyperaceae) in India. Rheedea 20(1): 1-19.

Dey, S. and P.V.Prasanna. 2015. Cyperaceae: Mapinoideae; Cyperaceae: Cyperoideae, Tribe Schoeneae & Sclerieae. In Sing, P. & Dey, S. (ed.) Fasc. Fl. India 27. Bot. Surv. India, Kolkata.

Kaliamoorthy, S. and T.S.Sarvanan. 2018. *Ledebouria hyderabadensis* M.V.Ramana, Prasanna & Venu (Hyacinthaceae): A new record for Tamil Nadu. Indian J. Forestry 41(1): 57-60.

Kumar, B. 2013. *Cyperus pulchellus*. The IUCN Red List of Threatened Species 2013: e.T177301A7409344. http://dx.doi.org/10.2305/IUCN.UK.2011-1.RLTS.T177301A7409344.en. Downloaded on 09 June 2016.

Mahendra Nath, M., V. Prasad and K. Madhava Chetty 2017. Two endemic species of *Fimbristylis* (Cyperaceae) new to the Flora of Eastern Ghats, southern India. Rheedea 27: 103-105.

Matthew, K.M. 1983. Flora of Tamil Nadu Carnatic. Rapinat Herbarium, Tiruchirapalli.

Panigrahi, G., S. Chowdhury, D.C.S.Raju and G.K. Deka G. 1964. A contribution to the Botany of Orissa Bull. Bot. Surv. India. 6(2-4): 237-266.

Prameela, R., J. Swamy and M. Venkaiah. 2018a. *Typhonium inopinatum* Prain (Araceae): An addition to the flora of Andhra Pradesh, India. Annals of Plant Sci. 7(4): 2147-2149. http://dx.doi.org/10.21746/aps.2018.7.4.12

Prameela, R., J. Swamy and M. Venkaiah. 2018b. *Typhonium roxburghii* Schott (Araceae): A new distributional record from Andhra Pradesh. Biosci. Discovery 9(1): 104-106.

Rasingam, L. and J. Swamy. 2017. *Typhonium inopinatum* Prain (Araceae): An addition to the flora of South India. Indian J. Forestry 40(4); 401-402.

Rehel, S. 2013. *Carex filicina*. The IUCN Red List of Threatened Species 2013: e.T177244A7398983. http://dx.doi.org/10.2305/IUCN.UK.2011-1.RLTS.T177244A7398983.en. Downloaded on 07 June 2016.

Sabu, M. 2006. Zingiberaceae and Costaceae of South India. Indian Association for Angiosperm Taxonomy, Calicut University.

Stearn, W.T. 1943. Royle's Illustrations of the Botany of the Himalayan mountains. J. Arnold Arbor. 24: 484-487.

Suksathan, P., M.H. Gustafusson and F. Borchsenius. 2009. Phylogeny and generic delimitation of Asian Marantaceae. Bot. J. Linn. Soc. 159: 381-395.

Swamy, J. and M. Ahmedullah. 2016. New distributional record of the endemic plant *Dipcadi montanum* var. *madrasicum* (E.Barnes & E.E.C.Fisch.) Deb & S.Dasgupta (Asparagaceae) from Andhra Pradeesh.Plant Sci. Res. 38: 109-111.

Swamy, J., S.Nagaraju, M.Sankara Rao and L.Rasingam. 2016. New distributional record of the endemic plant *Dipcadi montanum* (Asparagaceae) from Seshachalam Biosphere Reserve, Andhra Pradesh, India. J. Econ. Taxon. Bot. 41: 61-63.

Yadav, S.R. and S.P.Gaikwad. 2003. A revision of the Indian Aponogetonaceae. Bull. Bot Surv. India 45: 39-76.

Blyxa aubertii

Burmannia coelestis

Cheilocostus speciosus

Curcuma neilgherrensis

Globba marantina

Hedychium coronarium

Hedychium flavescens

Indianthus virgatus ***Crinum triflorum***

Curculigo orchioides ***Hypoxis aurea***

Dioscorea pentaphylla ***Dioscorea tomentosa*** ***Asparagus racemosus***

Chlorophytum laxum **Chlorophytum tuberosum** **Gloriosa superba**

Iphigenia indica **Lilium wallichianum**

Ophiopogon intermedius **Commelina clavata** **Commelina paludosa**

Cyanotis arachnoidea

Cyanotis axillaris

Cyanotis fasciculata

Cyanotis tuberosa

Dictyospermum montanum

Murdannia dimorpha

Murdannia pauciflora **Murdannia semiteres**

Calamus rotang **Phoenix lourieroi**

Arisaema leschenaultii **Arisaema tortuosum** **Rhemusatia vivipara**

Pothos scandens ***Apanogeton natans***

Eriocaulon cinereum ***Eriocaulon thwaitesii***

Eriocaulon truncatum ***Eriocaulon xeranthemum***

Index to Families, Genera and Species
(Italicized names indicate Synonyms)

Index to Vernacular Names

TELUGU

TAMIL

ORIYA

ENGLISH

www.ingramcontent.com/pod-product-compliance
Ingram Content Group UK Ltd.
Pitfield, Milton Keynes, MK11 3LW, UK
UKHW021447280726
14060UKWH00001BA/278